Addiction and Devotion
in Early Modern England

Addiction and Devotion in Early Modern England

Rebecca Lemon

PENN

UNIVERSITY OF PENNSYLVANIA PRESS

PHILADELPHIA

HANEY FOUNDATION SERIES

A volume in the Haney Foundation Series, established in 1961
with the generous support of Dr. John Louis Haney.

Published by
University of Pennsylvania Press
Philadelphia, Pennsylvania 19104-4112
www.upenn.edu/pennpress

Printed in the United States of America on acid-free paper
1 3 5 7 9 10 8 6 4 2

A Cataloging-in-Publication record is available from the
Library of Congress
ISBN 978-0-8122-4996-5

To Marc and Jasper

Contents

Preface

Addiction is, at its root, about pronouncing a sentence. This sentence might be, as its etymology suggests, an expression of an idea: *ad* + *dīcere*, "to speak, say."[1] Or it might be, as in its legal definition, an assignment, such as sentencing someone to prison; following the term's origin in Roman contract law, an addict was an individual, usually a debtor, who had been sentenced or condemned. *Addīctus* is thus one assigned by decree, made over, bound, or—in one mode of such commitment—devoted.[2]

What, then, does William Prynne mean when he warns against "those who addict themselves to Playes" or cautions readers to avoid those men who strive "earnestly to addict themselves to their trade of acting"?[3] For modern readers he seems to view the theater as a drug, lulling its audiences into narcotic passivity. And indeed, the theater does at times stand as a site of addiction, which, Circe-like, has the power to entrap playgoers: plays are drugs, actors are drug peddlers, and audiences are unwitting victims or eager consumers.[4] Yet this pejorative (even demonic) reading of the word "addict," while arguably at stake in Prynne's description, ignores the word's broader semantic and conceptual history. Eighteenth-century writers deploy the word in its modern signification—"the compulsion and need to continue taking a drug," a usage appearing in 1779 in the work of Samuel Johnson—but sixteenth-century writers instead drew largely on the concept of addiction from its Latin origins to designate service, debt, and dedication.[5]

Unearthing this hidden history behind early modern invocations of addiction, this book offers two primary insights. First, and most important, it illuminates a previously buried conception of addiction as a form of devotion at once laudable, difficult, extraordinary, and even heroic. This view has been concealed by the persistent link of addiction to pathology and modernity: current understandings of, and scholarship on, addiction connect it to globalization, medicalization, and capitalism. Surveying sixteenth-century invocations reveals instead that one might be addicted to study, friendship, love, or

God. Prynne cautions that one might addict oneself to stage plays, but his warning rings differently if addiction in the sixteenth century signals a form of pledged dedication. Within Prynne's caution lies the potential for sincere praise for the act of addiction itself. Rather than rebuking a mode of potentially excessive attachment (addiction), he instead cautions audiences against the wrong kind of addiction: to the false idol of the theater, where actors lure spectators into a form of devotion that should belong to God.

Second, this book uncovers an early modern understanding of addiction as a form of compulsion that resonates with modern scientific definitions. Specifically, the project traces how early modern medical tracts, legal rulings, and religious polemics stress the dangers of addiction to alcohol in terms of disease, compulsion, and enslavement. Early modern debates about tobacco, gambling, and sex also deploy, at times, the language of compulsion and vulnerability that comprises early modern addiction. But this book concentrates on alcohol for two reasons: first, the historical evidence on excessive, habitual drinking is more abundant than for other substances; and second, the scholarship on early modern drinking is well established, providing a critical framework for my own contribution. Certainly, the scholarship on good fellowship and the conviviality of sixteenth-century tavern culture contrasts with an emphasis on the compulsive nature of addicted drinking. Yet a host of early modern writers deploy a language of addiction to describe how the choice and inclination of good fellowship in drinking shifts, through habit and custom, into the necessity of habitual, excessive drunkenness.

The relationship between these two understandings of addiction is not solely oppositional nor can it be so easily mapped onto historical narratives, such as a shift from sixteenth-century devotion to eighteenth-century compulsion. Both meanings of addiction appear in the early modern period. What unites these apparently opposed discourses is a shared emphasis, both rhetorical and experiential, on addiction as an overthrow of the will. Being open to a form of strong inspiration, often described as ravishment, the addict is indeed breathed into by the spirit. This spirit might be God, it might be love, or it might be alcohol. But in an experience of ravishment, the addict is inhabited by another, be it a person, object, or idea.

Addiction is, in its spirituous potential, a form of devotion. Early modern lexicographers helped illuminate this relation by using the terms as synonyms. Glossing "addiction," dictionaries turn to the words "devotion" and "dedication," just as in defining "devotion" they deploy the terms "addict" and "addiction." Even as the word "devotion" is most immediately associated with religious worship, it also functioned—as its connection to addiction reveals—independently of a Christian framework. This is because devotion, like addiction, accounts for a position of loyalty to something or someone: one gives oneself up, as a devotee or addict, zealously and exclusively.[6] Nonreligious use of the word draws on its Latin root: *dēvovēre* (to devote), designated an "earnest addiction or application" and a form of "enthusiastic attachment or loyalty."[7] To be devoted is to be "zealously attached or addicted to a person or cause."[8] One exhibits devotion to a king, to a beloved, to an action, or to a pastime. Both addiction and devotion are forms of service: to be devoted is to exhibit "attached service," to be at someone's command or disposal. Finally, devotion, like addiction, concerns speech: vowing in the case of devotion, and pledging in the case of addiction.

For if early modern addiction concerns an individual subsumed in relation to another, it also involves a dependence on declarative speech.[9] Addiction not only designates a committed relationship of the addict to the substance, spirit, or person to whom he or she is devoted, but also hinges—as noted above—on a verbal contract or pledge. While modern definitions of addiction seem to bear little trace of the term's etymology and early definition, this project uncovers these historical origins, participating in what Jeffrey Masten has called a "renewed historical philology."[10] In his appeal to consider words and their histories, he writes, "We have not sufficiently attended to *etymology*—the history of words (the history *in* words)," urging scholars "to be more carefully attuned to the ways that etymologies, shorn of their associations with 'origin,' persist in a word and its surrounding discourse."[11] In the case of the word "addict," its etymological connection to speaking and pledging, as well as its expression of devotion, might appear entirely buried in modern uses of the term. But this range of meanings persists in early modern usage. Drawing attention to addiction as an utterance uncovers how speaking forth is fundamental to the addictive process. It also reveals such pledging as a challenge to self-sovereignty, as the addict commits to another person or object. Forms of addictive speech—be they pledges, vows, or contracts—track this challenge in their divide between imperative and reflexive articulations: one is attached or compelled by an authority or, alternately and relatedly, one devotes oneself, as with Prynne's caution to those who "addict themselves" to plays or to acting. If Roman and modern invocations of addiction draw largely on the imperative form, in the sixteenth century the reflexive construction proves dominant: addiction represents an exercise of will even in the relinquishing of it, a form of speaking commitment and devotion out loud or in writing. Definitions of "addict" from the period chronicle this interplay. The addict is defined both as the person conscripted by an external authority into service to someone or something, and as the person who devotes and assigns himself or herself to such service.

The result—the layering of Roman, early modern, and modern uses of the term "addict"—is what Roland Greene deems "a semantic palimpsest," in which different meanings of a word appear "in different degrees of availability. Palimpsests suggest one fashion of meanings coexisting with one another, with older ones showing through what comes later."[12] With its origin in contract law overwritten by its devotional invocations, which are then also overwritten by medical uses, the word "addict" offers one such semantic palimpsest, what Masten deems the history *of* and *in* a word. My emphasis on the semantic meanings of addiction—its definition as offered, for example, in a range of early modern dictionaries, and in Latin, French, and English—is coupled in this project with attention to the word's conceptual reach. I read, that is, "both the semantic integers that one finds in a dictionary" and "the concepts that shadow them," as Greene puts it in his study of key words.[13]

Uncovering addiction both as devotional ravishment and as a form of speech helps account for the question that began this project: why is early modern drama so often preoccupied with addictive states?[14] The answer comes, in part, in the parallel between the addict and the early modern actor. Transforming himself in gesture, speech, and dress and adopting the words of another, the actor is bound to his character, to other actors, to the playwright, and to the audience.[15] The actor is, in precise accord with the definition of addiction, assigned and obligated. The apparently oppositional definitions of addiction—as devotion versus compulsion, an exercise of the will versus a relinquishment of it—come together onstage in the figure of the dramatic actor speaking to an audience. The actor at once commands his audience, while also being vulnerable before it. In being "abnormally exposed, abnormally dependent upon us," as Michael Goldman puts it, the actor enters a form of voluntary service that compels him to transform, erase, or shatter himself in relation to another.[16] Dramatic performance is, in these terms, addicted relation: "The drama shapes and is shaped by its expressive instrument: the body, mind, and person of the actor," W. B. Worthen writes of this process.[17]

This link between the actor and the addict has been anticipated by those scholars theorizing acting's relationship to inspiration. The actor, breathed into by the author's script, balances technique with inspiration; she at once releases herself to express passion and trains in her craft. This view of acting was made famous by Konstantin Stanislavsky, who counseled the actor to uncover "inspiration" and "creativity" in order to inhabit the role most fully.[18] But the role of the passions and inspiration in acting predates this modern method. Early modern actors were imagined to release and transform themselves, not only through affect and gesture, but also through bodily comportment. In the process, they also transform the audience and the theater space, the scene of connection between the one and the many. The link between the body and spirit, the actor and audience—both inspired by the playwright and each other—results at times in the "unsettling resemblance between inspiration and disease," as Joseph Roach notes, citing seventeenth-century medical views.[19] Transformation as intersubjective connection instead appears, particularly to a viewer like Prynne, as troubling infection. From this vantage point of acting as both inspiration and disease, the links between acting and addiction seem less unexpected than inevitable. Acting presents a "dramatic paradox" for the Renaissance audience, caught between the actor's creation and his potentially blasphemous deception, or infectious power. The actor, in the creative act, is "both divine and demonic," Worthen argues, "as a magical extension of human potentiality and as a monstrous deformity of it."[20] The doubleness of the actor, like the doubleness of the addict, moves between devoted and compelled, inspired and diseased.

Ultimately addiction, like acting, offers a challenge to models of self-sovereignty, a through-line in this project's argument. If self-sovereignty is often posited as requisite for a life of health and well being, such self-possession eludes the addict. Free will, agency, self-care, and autonomy are given over, often by the addict's active choosing, much as the actor embraces a role or an audience is overtaken by it. Outside of the boundaries of the imagination, such a position of willed compulsion has been largely pathologized by medical experts. It has also been politicized by social theorists: at its extreme, such relinquishment of personal freedom can be taken to justify, as Mary Nyquist illuminates, enslavement on the grounds of the natural servility of some individuals or communities.[21] Yet the valorization of individual autonomy and self-possession can also risk upholding isolation at the expense of community or connection. There is, legal theorist Jennifer Nedelsky writes, "something profoundly and I think irreducibly mysterious about the combination of individuality and 'enmeshedness,' integrity and integration that constitutes the human being."[22] Early modern models of addiction offer one way of rethinking subjectivity through what has arguably proved the ideological and ethical impasse of self-sovereignty and individuality. Lauren Berlant describes the impasse in these terms: the "sovereignty described as the foundation of individual autonomy" overidentifies self-control with the "fantasy of sovereign performativity and state control over geographical boundaries. It thereby affords a militaristic and melodramatic view of individual agency by casting the human as most fully itself when assuming the spectacular posture of performative action."[23] If, as Berlant suggests, we conceive of human agency in concert with militarized action, celebrating productivity and the exercise of control, then it is no wonder that scenes of being that challenge individual sovereignty might invite condemnation and medicalization. Deep attachment or devotion holds the potential to gesture beyond isolated and isolating modes of life. Addiction offers one such model. Drawing attention to addiction as utterance and ravishment, this project illuminates the fundamental dispersal of agency at the heart of addiction itself. In doing so, this project explores how the early modern mode of addictive release might be admired and imitated for offering a form of related living based on connection rather than isolation and on community rather than individuality.

This book begins to tease out such philosophical and ethical resonances of addiction by turning, in the introduction, to the first uses of a word: "addict" and its derivations. The word's use clusters in three arenas: faith, love, and drinking.[24] Analyzing addictions to faith and love, the first half of this project reveals how such addictions require dedication and an exceptional vulnerability that eludes many seekers. To be an addict demands the simultaneous exercise and relinquishment of the will, a paradoxical and challenging combination. One must consent to give up consent, and banish the will, to addict oneself fully. This form of addiction is at once laudable and dangerous, for the addict undergoes a transformation, a ravishment, in pursuit of the addictive object. Examining this process of self-shattering, the project's first chapters expose how addictive release overtakes individuals, bringing them into deep relation with another.[25]

As sixteenth-century audiences actively sought and embraced such addiction to God and love, however, they were also warned of addiction's danger for physical, spiritual, and communal integrity: exceptional attachment or commitment to improper forms exposed the threat of addiction. This book examines, in its second half, such allegedly dangerous addictions, turning to Berlant's theory of "cruel optimism" to understand how an object initially attracting attachment might impede an individual's flourishing. In its study of such cruel attachments, "those binding kinds of optimistic relation we call 'cruel,'" this portion of the project pays particular attention to alcohol as a secondary addiction.[26] The turn from hopeful attachment in friendship, partnership, and community to a compulsive mode of addiction exposes alcohol as an available elixir, one that seems to offer the promise of community and the devotional attachment charted in this book's first half. Yet this study of drinking also anticipates modern notions of addiction. Early modern theological, medical, imaginative, and legal writing directly references habitual drunkenness as addiction, insisting on its link to disease and tyranny and resonating with the work of later medical researchers. Even, then, as my study of alcohol is yoked to this book's primary argument—uncovering early modern addiction's association with devotion and pledging—my work also contributes to the voluminous scholarship on modern addictions, demonstrating the relevance of the early modern period for more familiar notions of addiction as compulsive drug taking. My hope is that this book might help encourage future projects on other addictive relations from this period since, as suggested above, tobacco-taking, gambling, and sex, as well as witchcraft and swearing, appear, at times, as compulsive and ravishing activities. Beyond the necessary limits of this book, I am eager to see what studies my foray into the topic might help encourage.

Addiction and Devotion in Early Modern England

Introduction

Addiction in (Early) Modernity

The scholarship on addiction is vast and capacious. So, too, are the critical bibliographies on early modern faith, love, and drinking. This book, which is indebted to these large fields, charts a path directly between them, clearing the way to a previously obscured area: early modern addiction.[1] This area has remained largely invisible for two reasons. First, critical discourses on addiction tend to emphasize the concept's modernity, as this introduction's opening section reveals. Second, the scholarship on early modern devotion, love, friendship, and drinking—the addictions charted in this project—attends to a wealth of historical evidence beyond what might appear the philological curiosity of addiction's appearance. The study of early modern addiction thus brings together what are otherwise distinct scholarly approaches to the study of modern addiction on the one hand and to early modern practices of faith, love, and good fellowship on the other.

Addiction and Modernity

In her essay "Epidemics of the Will," Eve Sedgwick explores addiction precisely as a feature of modernity. Just as Michel Foucault theorizes how same-sex acts preceded the formation, in the nineteenth century, of the identity of the homosexual, so too with the addict. First came the acts—the drinking, the smoking, and the gambling—then came the character designation of the addict. As Sedgwick writes, "In the taxonomic reframing of a drug user as an addict, what changes are the most basic terms about her. From a situation of relative homeostatic stability and control, she is propelled into a narrative of inexorable decline and fatality," being given "a newly pathologized addict identity."[2] Sedgwick's distinction between acts and identity hinges on the opposition of what she calls the "stability and control" evident in willful choice—the individual who chooses to drink—and the tyranny of compulsion—the "pathologized" addict, who is compelled to consume.

Current addiction research asserts, and at times attempts to theorize, this pathologized identity of the addict. Debates on addiction as choice, predisposition, dependency, and disease move between the poles of free will and undermined agency.[3] Jeffrey Poland and George Graham write, for example, of a "toxic first-person self-pathologizing" that may in fact "undermine a person's efforts to overcome her problems."[4] Their study of addiction and responsibility emphasizes instead the degree of agentive selfhood exercised by addicts, and more broadly, their edited collection features a range of essays on addiction, free will, and choice. Such agency appears compromised, however, to many other addiction researchers. Lubomira Radoilska argues, for example, that "addiction-centered agency is paradoxical by its very nature. For it is eccentric in a self-defeating way: agential control is surrendered in search of a greater, though impossible, control. As a result, a form of passivity or dependence is placed at the heart of an addict's activities."[5] Such passivity or dependence appears, in Radoilska's formulation, as a form of "defeat in action."[6] With diametrically opposed approaches to addiction and agency, theorists struggle to formulate policy in dealing with a perceived health crisis.

"The field of addiction is not short on theories," the authors of *The Theory of Addiction* write: "There are psychological theories, biological theories, sociological theories, economic theories, biopsychosocial theories and more."[7] But the field is arguably short on history. In fact, much of the effort to understand addiction in a modern setting overlooks or radically shortens its history, approaching addiction as if it were a universal or modern phenomenon. This project, while influenced by the range of recent studies, particularly within the philosophy of addiction and the history of science, nevertheless takes a different approach. It uncovers both a longer history of views on addiction and an alternate understanding of addiction as an achievement.

Conventional medical history on addiction dates the concept to the turn of the nineteenth century, when physicians in both Britain and America diagnosed alcoholism as a nervous disorder; no concept of addiction, it is claimed, existed in England or America before this period. Advances in medical science and psychology led to its definition in both countries. First, the British navy physician Thomas Trotter, who has been called "the first scientific investigator of drunkenness," produced a 1788 Edinburgh doctoral thesis arguing that habitual drunkenness is itself a disease.[8] His dissertation was published in 1804 as *An Essay, Medical, Philosophical, and Chemical, on Drunkenness, and Its Effects on the Human Body*, and in it Trotter notes, "In medical language, I consider drunkenness, strictly speaking, to be a disease."[9] This disease manifests in illnesses attendant on overdrinking, including "universal debility, emaciation, loss of intellect, palsy, dropsy, dyspepsia, hepatic diseases, and all others which flow from the indulgence of spirituous liquors."[10] Nearly simultaneously, Benjamin Rush in America (one of the original signatories of the Declaration of Independence and a man deemed the founder of American psychiatry) published *An Inquiry into the Effects of Ardent Spirits upon the Human Body and Mind* (1785), in which he also defines drunkenness as a disease: "Drunkenness resembles certain hereditary, family and contagious diseases."[11] Rush's work theorizes the trajectory from choice to compulsion: "The use of strong drink is at first the effect of free agency. From habit it takes place from necessity. That this is the case, I can infer from persons who are inordinately devoted to the use of ardent spirits being irreclaimable, by all the considerations which domestic obligations, friendship, reputation, property, and sometimes even by those which religion and love of life, can suggest to them."[12] As with Trotter, he names the diseases stemming from drunkenness, including jaundice, dropsy, epilepsy, gout, and madness.

The work of Trotter and Rush ushered in a "new paradigm," as the medical sociologist Harry G. Levine writes. This new paradigm "constituted a radical break with traditional ideas about the problems involved in drinking and alcohol."[13] Specifically, opinion shifted on habitual drunkenness (and in turn on opium use and other addictive behaviors) to a disease model, the key feature of modern definitions of addiction. As the historian of science Roy MacLeod notes: "It was too easy to view alcoholism simply as immoral excess, its cure, simple moral restraint, and its expense, a personal responsibility."[14] As a result, he writes, "the transformation of public attitudes from the conception of alcoholism as a moral sin to its recognition as a nervous disease required concerted effort."[15] In understanding the shift in viewpoint on excessive drinking, scholars not only stress the moralizing of earlier periods, as MacLeod does here, but they also point to earlier conceptions of drinking as a matter of choice. Levine, for example, discusses how "during the 17th century, and for the most part of the 18th, the assumption was that people drank and got drunk because they wanted to, and not because they 'had' to."[16] He elaborates: "In the modern definition of alcoholism, the problem is not that alcoholics love to get drunk, but that they cannot help it—they cannot control themselves."[17]

This paradigm shift in the study of addiction is of a piece with other scientific discoveries of the period, Roy Porter argues: "Building to some degree on the work of precursors such as Erasmus Darwin, nineteenth century doctors set about investigating the pathology of excessive drinking, exploring its associations with conditions such as dropsy, heart disease, cirrhosis of the liver, . . . nervous disorders, paralyses."[18] MacLeod also recognizes this new nineteenth-century paradigm and charts the general impact of this breakthrough over the course of the century:

> Not until the last half of the 19th century did the scientific appreciation of alcoholism become general. Only then, under the guidance of a few doctors and reformers, was the image of the drunkard as a disorderly, ill-disposed social unit gradually transformed into one of a neglected patient suffering from a mental disease with well-marked clinical features. Reformers, who sought to remove the moral stigma from alcoholism and to treat the alcoholic by medical means, led the advance guard of a movement to promote prevention and cure on a public basis.[19]

As part of these reform movements, the first temperance societies appeared in England in the 1830s, and Parliament passed the landmark Habitual Drunkards Act in 1879. That legislation is, in terms of this addiction narrative, the culmination of efforts by physicians and reformers who shifted the notion of inebriation from social condemnation to scientific understanding. In doing so, they redefined a habitual drunkard from a sinner to someone with a disease akin to lunacy.[20]

More-recent historians have put pressure on the pioneering nature of Trotter and Rush's conclusions. The research of both Porter and Jessica Warner on the eighteenth-century gin craze exposes a notion of diseased drinking in the century before Trotter and Rush.[21] The work of Phil Withington and others on intoxication tracks the "modern obsession" with substance abuse, even as it illuminates how contemporary concerns about intoxication have "enduring roots in the past."[22] Yet even though the dating of addiction might vary, and even as historians illuminate the long history of intoxication, a broad consensus remains that addiction constitutes a modern discovery, one connected intimately to familiar features of modernity: the rise of Enlightenment individualism, medicalization, global trade, nation states, and capitalism.[23] Current advances in neurobiological research further reinforce the link of addiction to modernity by suggesting how addiction's discovery is ongoing and dependent upon modern technologies: using newly available scanning devices, such as PETs and fMRIs, to trace precisely how the addicted brain operates, neuroscientists have exposed the long-lasting changes in brain function caused by addiction, including "the pathological usurpation" of the brain's reward-circuit learning.[24] As a result of such usurpation, the rewired, addicted brain releases dopamine in response to the *anticipation* of drug taking, rather than merely as a result of drug ingestion.[25]

Finally, literary histories have underscored addiction's modernity by studying the emergence, in the late eighteenth century, of the inspired writer-addict. Samuel Taylor Coleridge and Thomas De Quincey stand as early examples of addict-writers, with Coleridge linking literary inspiration and drug consumption in his famous preface to "Kubla Khan." From the Romantic's tincture of opium through Eugene O'Neill's and Tennessee Williams's alcoholism to Jim Morrison's acid trips and William Burroughs's heroin addiction, writers offer autobiographical chronicles of how drug addiction might fuel or fell creativity. The addict-writer holds a clear place in the imaginative landscape of the twentieth century, articulating what seems to be a particularly modern, or postmodern, condition of stasis and excess.[26] "Addiction," as Janet Ferrell Brodie and Marc Redfield write, "belongs as a concept to the social and technical regimes of the modern era."[27] Their cultural history draws attention to the ideological ramifications of addiction, a concept that is "little more than a century old."[28] Chronicling drug abuse, Stacey Margolis calls addiction "a particularly modern form of desire."[29] Anna Alexander and Mark S. Roberts argue that "addiction emerges directly alongside modernity," and Jacques Derrida speaks of our "narcotic modernity."[30] These accounts draw on the perception of the modern bodies as uniquely pathologized and incapacitated, precisely as Sedgwick illuminates. Specifically, the modern subject, imbricated in a global economy, finds addiction at once an expression of powerlessness and pleasure.

Yet even as modern medical and psychological research illuminates the workings of addiction in entirely new ways, and even as writers from Coleridge onward experience addiction more acutely than in the past, addiction is not a singular feature of modernity. As this introduction's final section reveals, a model of addiction as compulsion and disease existed earlier than the nineteenth century. Overturning the notion that addiction was "discovered" only a century ago, or even two or three centuries ago, this project demonstrates an early modern awareness of alcohol addiction as a disease along the lines charted by medical pioneers and modern-day neurobiologists: addiction alters the brain and results in a familiar, and oft-repeated, set of related diseases. To imagine that the premodern period remains entirely distinct from modernity when it comes to addiction is to overlook the rich evidence from the sixteenth and seventeenth centuries that suggests the awareness of addiction as a disease.

Further, and perhaps more important, the insistent yoking of addiction, disease, and modernity has allowed us to ignore what is arguably the more compelling half of the addiction story, which becomes evident through study of early modern writings: addiction represents a singular form of commitment and devotion, worthy of admiration as much as censure.

Addiction as Devotion

One of the early examples of the term "addiction" comes, according to the *Oxford English Dictionary*, from a line in Shakespeare's *Othello*: in celebrating a military victory, the play's Herald tells the soldiers "each man to what sport and revels his addiction leads him" (2.2.5–6).[31] In other words, each man can choose to follow whatever activities he pleases. Yet the term "addiction" is deployed widely before this *Othello* reference, and the play's engagement with theories of addiction—as Chapter 4 will discuss at length—is more complex than the lexicographical gloss credits. Addiction is not, it turns out, mere inclination.

Invocations of addiction begin to cluster in printed texts from the 1530s, as in the work of George Joye, who produced the first printed translation of several books of the Old Testament. *The prophete Isaye, translated into englysshe* (1531) offers one of the earliest usages of the term, in a context entirely familiar to modern readers. Joye warns, "Wo be to the haunters of dronkenes which ryse erly to drinke, continuinge in it tyl nighte being hot with wyne: in whose bankets there are harpes and futes taberet & pype washed with wyne."[32] These "haunters of dronkenes" will suffer divine retribution: "The helles haue opened their unsaciable throtes and their mouthes gape beyende mesure that thither mought descende pryde, pompe, riches and al that are addicte to these vices."[33] Joye's warnings at once recall the familiar medieval and early modern schema of the seven deadly sins and predict the century's broader legislative and conceptual interest in pathological addiction.

While the invocation of drinkers, "addicte to these vices," anticipates both the modern definitions of "addiction" in relation to substances and the railings of puritans who attack drunkenness, Joye uses the term more expansively as well. In *The Psalter of Dauid in Englyshe* (1534), he warns of mortal men "addict to this worlde" and against the ungodly who are "addycte unto wyckedness," and "addicte and all giuen to wickedness."[34] He also praises the faithful follower of God as an addict, asking God to "make faste thy promyses to thy servant which is addicte unto thy worshyppe."[35] Further, in *The Unitie and Scisme of the Olde Chirche*, Joye insists on the unity Jesus preached, with the faithful "addict unto none but to christ." He writes Jesus hoped that his apostles "thorow love might consent and godly agree being all one thinge in christe, and that there be no dissencions nor sectis in his chirch unto no creatures being addict unto none but to christe hir spouse dedicatinge hirself."[36]

This range of the term's appearance—to signify excessive drunkenness, inclination to wickedness, overattachment to worldly pleasure, as well as devotion to God and Scripture—suggests its broad association with forms of attachment. Furthermore, the term's appearance in early translations of the Bible and polemics surrounding the Reformed faith indicates its link to religious controversy. Specifically, in the context of post-Reformation England, the term appears most frequently to describe one kind of dedication: to God and the church. In the wake of theological debates following Henry VIII's break from Rome, addiction becomes a sign of study, commitment, and piety, as well as a signal of false attachment to, and dangerous tyranny of, the Pope or Antichrist. Thus, in the 1540s the term appears repeatedly in church histories by writers such as John Bale, Polydore Vergil, and Thomas Becon. Bale, for example, writes of those "addict to their supersticyons," and specifically those "Antichristes addict to the supersticiouse rytes of the heythens in their sacrifices, their ceremonies, their observations, their holy dayes, theyr vygils, fastinges, prayinges, knelinges & all other usages contrary to the admonyshement of Christ."[37] Here addiction signals an attachment to material aids to worship, which were associated with the Roman church. Vergil, too, condemns those "wholy addict to the honoryng of their false goddess," while praising those "men of the laye sort geven and addicted to praiers."[38] The answer, as Philip Nicolls counsels his readers, is to "addict youre selves to the meaneynge of the scripture."[39]

Reformed writings overtly celebrate addiction as an intense mode of devotion and commitment, even as they express concern for misguided addictions to the improper faith. Following the etymology of "addiction" as *ad* + *dīcere* (to speak, to declare), these writings trumpet a model of addictive living that is at once an invitation and a prescription. The Elizabethan "An Homilee of good workes" (1571), for example, encourages addicting oneself to prayer as a means of pleasing God. Those who "did eyther earnestlye lament and bewayle their sinfull lyues, or did addict them selves to more fervent prayer" learn that "it might please God to turne his wrath from them."[40] Such positive invocations of addiction also fill those guides encouraging modes of pious living, as Humphrey Gifford writes in *A posie of gilloflowers* (1580). In his "Farewell Court" of poems, he counsels his readers to "cast away the vile and vaine vanities that the wicked world accounteth as precious, and addict all their doings towards the attainement of lyfe euerlasting."[41] Barnabe Googe encourages, in his translation of Marcello Palingenio Stellato's *Zodiac* (1565), the addict to dedicate himself specifically to prayer: "The mind wel purgde of naughty thoughtes, / in fervent sprite to praye: / And wholly to addict himselfe / the heavenly state to finde / And all the cares that fleshe doth give, / to banishe from his minde."[42]

Even as these guides to pious living encourage addiction, at the same time writers acknowledge its difficulty. Not just anyone can achieve it. The popular text *Of the Imitation of Christ* (1580) encourages parishioners toward addiction, for example: "Learne to contemne outwarde things and to addict thy selfe to spiritual; so shalt thou perceave the kingdome of God to come into thee." Nevertheless, the author concedes the challenge of this charge, writing how "fewe there be which addict themselves to the studie of celestial things, because fewe can withdrawe themselves, wholie from the love of this world."[43] In *The glory of their times* (1640), Donald Lupton acknowledges the struggle by pondering Christ's fortitude in pursuing addiction: "How did hee addict himselfe to watching, fasting, prayer, and Meditation?"[44]

The historian Raphael Holinshed, in his *Chronicles* (1577), draws upon the language of these pious writers, a migration from religious to secular texts that speaks to the scope of addiction's invocations. Holinshed praises those devout, spiritual leaders who are capable of addiction, writing of "the reverende Fathers of the spiritualtie, and other godly men addict to vertue, unto whome the setting forth of Gods worde hath beene committed."[45] More specifically, he praises King Edward the Elder as "in his latter dayes beeying greatly addicted to devotion and religious priests."[46] In *A direction for the health of Magistrates and Studentes*, the reader similarly learns the value of addicted leaders, for the virtuous ruler must "ernestly addicte himself to the studie of Morall Philosophie and of the sacred Scriptures."[47] These secular texts adopt a language of devotional addiction as a means of educating readers on the attributes of good rulership.

Addiction requires a natural disposition and ability; it is not purely a matter of hard work or instruction. As John Huarte writes in *The examination of mens wits* (1594), if a "child have not the disposition and abilitie, which is requisit for that science whervnto he wil addict himselfe, it is a superfluous labour to be instructed therein by good schoolemaisters."[48] Therefore, even as religious prescriptions follow the etymological invitation of addiction as a mode of speech or a form of command that they offer to their pious readers, these writers also understand addiction as an inclination that the individual both does and does not control. Their readers might attempt the form of addictive devotion counseled in the texts, but as theologians expound—most prominently Jean Calvin, as Chapter 1 discusses in detail—addiction is also perceived as a form of grace. Lancelot Andrews states how only by "being so visited, redeemed and saved, we might wholy addict, and give over our selves, to the Service of Him who was Author of them all."[49] Roger Edgeworth, too, invokes election and addiction, writing how only certain men are chosen for priests: "Election and imposition of a prelates hande," a future priest "is piked out & chosen among the moe to be addict and appoynted to God, and to be a minister of God in the Churche or congregation."[50] The ability to addict is both a gift and an effort. For William Baldwin it requires following in Christ's footsteps:

> Wherwith although I be afflict,
> In wurth I take all lovyngly:
> Beyng for Christes sake addict
> To suffre al paynes wyllyngly,
>   Continually.[51]

Baldwin's metered rendition of the *Psalms*, like Sternhold and Hopkins' *Booke of Psalms*, engages a broadly pious audience. As hymns that might commonly appear in church, these psalms offer one way in which the language of addiction-as-devotion appeared in everyday life, independent of study of Reformation theology. Addiction was at once intoned in the church and echoed in the streets.

Finally, addiction is a form of service, as emphasized in the English College of Douai's translation of the book of Psalms, which offers "a general and very fitte prayer, when we addict ourselves by a firme resolution to serve God." Further, the speaker protests that "even by the mortal hate of the wicked I saw, that Gods law is most excellent, and therefore addicted myself so much the more to lone [love] it and to hate al wicked ways."[52] Thomas Taylor's parable of the sower and the seed similarly counsels that the faithful say "in their hearts, Thus much wealth I will attaine unto, and when I have done that, I will addict my selfe to the service of God."[53] Paul, in his letter to Titus, impresses upon his audience the value of addicted service, at least in Erasmus's version of the text. The epistle begins with lines in which Paul casts himself as an addict: "I Paule my selfe the addict servant & obeyer, not of Moses lawe as I was once, but of God the father, and ambassador of his sonne Jesus Christ."[54]

For most of the sixteenth century, addiction, in its link to God and service, was not a problem; it was an achievement. To be an addict indicated commitment, vulnerability, hard work, and courage. To be an addict meant to devote oneself entirely to a calling—to be addicted to scripture, to scholarship more generally, or to Christ. Nevertheless, in its derided invocations, such addiction might signal enthrallment, the relinquishment of good sense and true faith: one might be addicted to the pope or superstitious practices, or one might, as Joye suggests, be addicted to alcohol. Thus addiction appeared both laudable and dangerous, a commitment to salvation or degeneration. Furthermore, the term "addict" contains at once a sense of obligation (as in its Latin origin, in contract law), as well as a sense of choice (to bequeath or give). These alternate, competing, but connected senses of addiction—as compulsion and choice, as the right path or the reprobate one—resonate with the philosophical and theological questioning familiar to readers of Reformation literatures: what is the role of free will in faith? If the godly seek to will away the will, hoping to receive grace, addiction encapsulates this struggle in the desire to give oneself over to a higher power. The struggle remains, ultimately, an active, unresolved one, because the concept of addiction leaves profoundly unsettled this question of devotional agency: the addict might will himself or herself toward God; equally, however, as the voluminous literature on pious living suggests, an earthly authority might attempt to command or dictate (*dicare*) such dedication, or God might offer it through grace.

Addiction as Abuse

The profound uncertainty surrounding both the agent propelling addictive devotion (be it the believer or an external authority) and the object of devotion itself (the godly or heretical path) invites wary understandings of addiction's power. Thus, in addition to the view of addiction as an extraordinary form of commitment, sixteenth-century religious polemics warn against the dangers of fervent attachment to the wrong object. Such warnings take the form of cautions against idolatry and of a more general fear of material forms of worship associated with Catholicism. Thomas Bilson, writing in support of the English church, claims that Catholic "writers were all addicted to images," while William Charke criticizes the "willful addiction to the olde translation" of the Bible.[55]

The investigation of errant addiction is particularly evident in the works of John Foxe, who derides those who "addict themselves so devoutly to ye popes learning," singling out individual stories of those "worshipping of Idoles" to which they are "addict."[56] Depicting the adoration of icons as a kind of addiction, Foxe writes, as in the case of the Catholic Lord Cobham: "If any man do otherwise abuse this representation, and geve the reverence unto those Images, which is due unto the holy men whom they represent . . . or if they be so affected toward the domb Images, that they do in any behalfe addict unto them, eyther be more addicted unto one Saint then another, in my minde they doe little differ from Idolatrye, grievously offending agaynst God the author of all honor."[57] Foxe links addiction and abuse here, deriding those believers who mistakenly "abuse" representations and "addict" themselves to "domb images" or "one Saint then another." Such a form of addiction is a grievous offense, for it establishes a deep but improper commitment to idols over God.

The Reformers' concern with addiction to physical forms of worship connects to their suspicion of addiction to physical pleasures more broadly. The Elizabethan "An Homilee agaynst gluttony and drunkenness," for example, directly links improper forms of worship and gluttony: "Neyther woulde we at this day be so addict to superstition, were it not that we so much esteemed the fillyng of our bellies."[58] Such a concern for idolatry and appetite appears from the inception of reformist movements in England. Henry VIII writes in *A glasse of the truthe* against those worshipers exhibiting "a great lacke of grace, and an overmoche addiction to pryuate appetites."[59] Attachment to physical pleasures produces, these authorities speculate, misguided religious faith, or vice versa.

The suspicion of material devotion expands from the attacks on papal dictates, iconography, and other earthly aids to worship into the preoccupation with addictive worldly lures. And in the process, puritan railers condemn lust, gaming, tobacco taking, stage plays, and any number of other material pleasures. Early modern historians have called this phenomenon the Puritan

Reformation of Manners, because "pious pleadings" in the 1580s led, by the seventeenth century, to what Keith Wrightson has deemed "a programme of national significance," regulating behavior from tavern haunting to May games.[60] "Scores of pamphlets and printed sermons," Martin Ingram concurs, gesture toward "a 'national' movement," one that is innovative in its reach: "Save for preaching from the pulpit and the circulation of statutes, proclamations and town ordinances," he writes, "there was no late fifteenth-century equivalent."[61] This reformation of manners had a dramatic legislative and administrative impact on the early modern landscape, an impact achieved largely by the vehemence of the reformers and their "narrowness of concern."[62] The legislative effect was particularly felt in the arena of drunkenness, which was increasingly perceived as a national problem. Beyond tavern regulation, early modern legislators developed laws against drunkenness itself, following the decade of puritan attacks on drinking and tavern haunting. As A. Lynn Martin puts it, "If the moralists are to be believed, drunkenness reached plague proportions in the sixteenth and seventeenth centuries, especially in England."[63] Indeed, the early modern period saw the passage of England's first national law against drunkenness itself, after a forty-year parliamentary battle. The "modern" discovery of addiction and the resulting legal regulation with the 1879 Habitual Drunkards Act (discussed above) was in fact mirrored centuries earlier with the 1606 Statute Against Drunkenness.

Concerns about drunkenness have, of course, a long history, extending from classical literature forward.[64] From Beowulf to Langland, medieval writers chronicle drinking rituals and abuse. The English abbot and homilist Aelfric of Eynsham, anticipating later puritan detractors, cautions that "drunkenness is a vice of such magnitude that . . . drunkards are not able to obtain the kingdom of God."[65] And *The Trinity Homilies* compare the gluttonous man to a swine in language resonant with early modern descriptions: "Some men pass their lives in eating and drinking, as swine, which foul themselves, and root up and sniff ever foully."[66] Yet even as medieval examinations of drinking accord with early modern discussions on drunkenness as beastly, ungodly, and dangerous, their framework differs notably, frequently hinging on the language of the seven deadly sins and the eight sins before them.[67] The deadly sin of gluttony—with attendant drunkenness—attracted special attention as a gateway sin, a point Chaucer's Parson makes clear in *The Canterbury Tales*: those guilty of gluttony, as he puts it, "may no synne withstonde."[68] From the character of Gluttony in medieval dramas such as *The Castle of Perseverance* to Langland's extended tavern portrait in *Piers Plowman* to the

Parson's and Pardoner's tales about the seven deadly sins in Chaucer, the warnings against drunkenness are often predictable: the drinker fails to attend church, saps family's finances, endangers health, and commits other sins as a result of being drunk.[69]

As powerful as this medieval framework of the deadly sins might be, by the sixteenth century writings on drunkenness shift away from the schema to a more pointed view of drunkenness as disease and reprobation.[70] From Shephard's attack on Catholic curates in "Doctor Double Ale" to Skelton's infamous misogyny in "The Tunning of Elynour Rumming," allegations against drinkers offer satirical portraits of corruption and hypocrisy.[71] What had been deemed errancy and sin, in need of salvation and forgiveness, is increasingly figured as compulsion, the inability to shift away from drinking. To John Downame, as for other puritan writers, those who "addict themselves to much drinking" prompt a spiritual, economic, and legal crisis.[72] William Perkins thus warns his allies "not to addict ourselves to drinking," while William Prynne chidingly writes, "The people given to idlenesse and vaine discourse doe in these dayes addict themselves more to drunkennesse, surfetting, Playes and wantonnesse, than to divine things."[73] Writers bolster their arguments with medical diagnoses, anticipating the modern notion of addiction as compulsive, pathological attachment. Texts such as *The Drunkards Cup*, *Diet for a Drunkard*, *The Drunkard's Character*, and *The Condition of a Drunkard* speak of drunkenness as a physical disease and as a defining identity much earlier than current narratives on the rise of "modern" addiction suggest.[74]

Downame elaborates on the two ways that drinkers abuse themselves: "First by drunkenness, when by immoderate swilling and tipling they are deprived of the use of their reason, understanding, and memory; so as for the time, they become like unto beasts. Second, by excesse, when as they addict themselves to much drinking, and make it their usual practice to sit at the wine or strong drink."[75] While "drunkenness" and "excess" might seem synonymous to modern audiences, for Downame they are distinct. "Drunkenness" describes the phenomenon of overdrinking, regardless of how often—drunkenness produces, he says, substance-related problems (deprivation of "reason, understanding, and memory"). "Excess" denotes *habitual* overdrinking or compulsive use (a "usual practice" of excess, as men "sit at the wine or strong drink"). Thus, both singular drunkenness and habitual drinking are part of the dangers of alcohol. Addiction appears, Downame claims, when the drinker can no longer abstain: "They who addict themselves to this vice, doe finde it so sweet and pleasing to the flesh, that they are loath to part with it."[76]

Downame's concern for those addicts of alcohol stands in contrast to the widespread Galenic prescription of alcohol in promoting good health.[77] Alcohol's role in that regard was especially crucial since daily beverages such as beer and ale provided both clean water and calories to their consumers. Indeed, as Louise Hill Curth and Tanya M. Cassidy write, "Alcohol, consumed in moderation, was thought to be an important ally in the fight against disease. Ale, beer, and wine were all touted for their preservative properties."[78] Excessive and habitual drunkenness challenged the Galenic prescriptions for self care and provoked increasing concern from medical, legal, and religious authorities, prompting examination, as Jennifer Richards illuminates, of precisely what constituted "enough."[79] This concern is expressed not only in sermonizing writings but also in historical chronicles such as Holinshed's, where he notes the role of beer in the English diet and the resulting "ale knightes so much addicted thereunto, that they will not cease from morow untyll even."[80]

Addicted drinkers suffer, many writers argue, from disease. The daily bouts affect the drinkers' brains, leading to greater toleration for alcohol. John Hoskins, in concert with his contemporaries, describes how wine gives "the braine a blow, that like a subtil wrastler, it may supplant the feet afterwards."[81] While these authors had none of the modern tools, such as brain-scan technology, that are available to twenty-first-century neurobiologists, they nevertheless anticipate modern research in their preoccupation with the drunken brain, as well as the drunken body. Further, these writers catalogue a set of diseases familiar to modern researchers and chronicled in Trotter and Rush's work, cited above. The drinker, Downame and other critics of drunkenness argue, "is brought unto grievous diseases, as dropsies, gouts, palsies, apoplexie, and such like."[82] This catalogue of diseases appears in nearly all discussions of drunkenness in this genre of religious polemic.[83] Drunkenness leads to such "diseases in the body of man, as apoplexies, falling sicknesses, palsies, dropsies, consumptions, giddinesse of the head, inflammation of the blood and liver, distemper of the brain, deprivation of the sense, and whatnot," as the anonymous author of *A looking glasse for drunkards* (1627) writes.[84]

Of course, this language of disease does not indicate that early modern notions of diseased alcoholism map easily onto modern ones. To early modern writers, drunkenness is also a sign of errancy, not least because, as Roy Porter argues, "sickness was largely seen as personal, internal, and brought on by a faulty lifestyle. . . . Careful attention to all aspects of 'regimen' or lifestyle, would prevent 'disease' (literally 'dis-ease') in the first place."[85] Yet ultimately these authors find the language of vice and condemnation insufficient, an important point considering that later addiction studies label the early modern period as a strictly moralizing one in its descriptions of drunkenness. Wrestling with the agent behind the lure of drinking, reformers alternate between blaming the drinker and the power of alcohol. Ostensibly the "drunkard" brings this infirmity "upon himself." Yet equally, in trying to account for the radical changes in a drinker's condition, these writers turn to language on the overthrow of the subject: what had previously signaled dedicated commitment when directed at God indicates instead a form of slavery when linked to alcohol. As we learn in *The odious, despicable, and dreadfull condition of a drunkard*, "drunkards" suffer from a "slavish condition," tied to the "taphouse."[86] The language of tyranny and enslavement illuminates the strange condition of drunkenness, in which a subject is both himself and not himself. Arguably "the outcome of weakness or self-indulgence on the part of the paradigmatically 'free' agent," drunkenness resonates in these writings with a condition problematically deemed to be voluntary slavery, the socially and politically stigmatized failure of mastery on the part of an individual. "When a higher faculty of the free self falls subject to a lower faculty, or when the free self as a whole becomes hopelessly enamored of inferior, mundane pursuits," Mary Nyquist writes, "ethico-spiritual 'slavery' is the inevitable result."[87] This peculiar notion of slavery's voluntarism allows the drinker some agency: the addict is not a slave in a political or legal sense, but rather is reflexively attached and ravished by an object or activity of choice, becoming diseased and abused in the process.

In their complex invocations of drunkenness as a disease of body and spirit, these early modern theologians are at the heart of a historical irony: it is the largely religious preachers who explore the empirical connection between habitual drunkenness and a set of disorders linked, today, with alcoholism. Yet in the context of twentieth-century addiction discourse, these writers will be dismissed as ignorant moralizers, as proto-temperance fanatics, and as biased evangelicals, even as their writings anticipate modern medical definitions of addiction far more precisely than the work of their contemporary physicians. Physicians will eventually speak of drunkenness as disease, but not until fifty years after these religious writings. Notably, Dr. Everard Maynwaringe takes up the concern with drunkenness. In his *Vita sana & longa the preservation of health and prolongation of life proposed and proved* (1669), he writes "that drunkenness is a disease or sickness, will appear in that it hath all the requisites to constitute a disease, and is far distant from a state of health . . . the eyes do not see well, nor the ears hear well, nor the palate relish, etc. The speech faulters and is imperfect; the stomach perhaps vomites or nauseates; his legs fail . . . an unwholesome corpulency and . . . plentitude of body does follow: or a degenerate . . . and a decayed consumptive constitution . . . as well as imbecility of the nerves."[88] Maynwaringe, like the godly polemicists before him and modern researchers after, links excessive drunkenness to precisely those diseases that continue to be associated today with alcoholism. Indeed, as Jessica Warner has illuminated, these pamphlets directly anticipate the work by the addiction pioneers Trotter and Rush centuries later: "We ultimately owe our own habit of identifying heavy drinkers as addicts and alcoholism as a disease not to physicians but to the clergymen of preindustrial England."[89] This is because, she argues, "it is in the religious oratory of Stuart England that we find the key components of the idea that habitual drunkenness constitutes a progressive disease, the chief symptom of which is a loss of control over drinking behavior."[90] Yet, as Christopher C. H. Cook has argued, "under the influence of the Enlightenment, the vast interdisciplinary literature that surrounds addiction and alcohol studies has come to exclude theology."[91]

Drinking and Good Fellowship

The embrace of one form of addiction, to God, and the censure of another, to alcohol, creates the appearance of an oppositional logic structuring the conceptions of addictive attachment. But the story of early modern addiction is more complex than mere opposition. For against the puritan concern about addictive drinking as a form of diseased compulsion lies a contemporaneous discourse on drinking as laudable commitment to community and nation. When examined through the ubiquitous early modern conversations on good fellowship, certain aspects of drinking culture—namely, the community ties, friendships, and national allegiances—parallel the devotional addiction to God or love, a point taken up in Chapter 3 of this project. Mark Hailwood's recent study provides a succinct definition of this capacious category of good fellowship and, in doing so, highlights its links to drinking and to loyalty to community: "It was an activity structured by a number of rituals—toasting, drinking contests, games and gambling, songs—and by a series of behavioural conventions that encouraged liberal spending, heavy but controlled drinking, and the maintenance of a jovial—or 'merry'—disposition and atmosphere. These rituals and conventions expressed a number of values: courage, self-control, loyalty, financial prosperity and independence, a pride in hard work, a bold defiance of dominant gender norms."[92]

In sharp contrast to the godly condemnations of diseased drinking, the rituals of good fellowship and its attendant values, including "courage" and "loyalty," attest to the cultural benefits of drinking culture. Exclusive friendships sustain drinking communities in times of strife. In his poem "Good Fellowship," for example, Hugh Crompton's speaker trumpets his dedication to communal drinking:

> Fill, fill the glass to the brim,
> 'Tis a health unto him
> That refuses
> To be curb'd, or disturb'd
> At the power of the State,
> Or the frowns of his fate;
> Or that scorneth to bark or to bite at our Muses:
> And that never will vary
> From the juyce of the Vine, and the cups of Canary.[93]

The emphasis on exclusive loyalty—one who "never will vary" in his drinking—appears in a range of writings on good fellowship that are structured around those who "refuse" to be daunted in their commitments to each other. Thomas D'Urfey's "The Good Fellow" offers a similar rallying cry:

> A pox on the times,
> Let 'em go as they will,
> Tho' the taxes are grown so heavy;
> Our hearts are our own,
> And shall be so still,
> Drink about, my boys, and be merry.[94]

The speaker upholds his unity with his drinking "boys," who still claim ownership of their loyal "hearts" even in times of political isolation. "To quaffe is fellowship right and good," writes William Hornby; such drinking fellowship helps "maintain friendship and nourish blood."[95] Even the critics of good fellowship recognize its connection to forms of loyalty and devotion. Thus drunkenness goes, William Prynne writes, "under the popular and lovely titles of hospitality, good-fellowship, courtesie, entertainement, joviality, mirth, generosity, liberality, open house keeping, the liberall use of Gods good creatures, friendship, love, kindnesse, good neighbour hood, company-keeping, and the like."[96]

The language of good fellowship resonates with a model of addiction in urging one's release into a spirit—the spirit of alcohol—as a sign of loyalty, with the alehouse functioning as an alternate family.[97] "To consider seventeenth-century drinking," Adam Smyth writes, is "to consider friendship, community, conviviality."[98] Tavern drinking helped establish affiliation and loyalty, often to a community structured around shared gender, class, regional, or political affiliations made evident in drinking habits.[99] Since one's choice of alcohol helped to signify one's class status, drink functioned as a mode of social recognition.[100] Drunkenness, or claims of drunkenness, might serve as a way for the gentry to distance themselves from those impoverished visitors who haunted the alehouse; or drinking could help designate political affiliation, either through the types of beverage consumed or the spaces of consumption.[101] Smyth's collection, *A Pleasing Sinne: Drink and Conviviality*, elucidates the range of such sociable drinking, from the English versions of Anacreontic verse studied by Stella Achilleos, to the ballads analyzed by Angela McShane, to the drinking communities illuminated by Michelle O'Callaghan, Marika Keblusek, and Charles C. Ludington. A range of studies of admirable drunkenness also expose the flourishing of carnivalesque ritual and the politics of mirth, both dependent on drinking culture for political and social union.[102]

Despite the seemingly opposed (and equally voluminous) cultural responses to drinking from puritan critics and inspired tavern haunters, both groups share an embrace of the spirit as connection to community and fellowship. Ravishment, be it through transforming God on the one hand or inspired drinking on the other, reshapes the devoted addict.[103] Indeed, these drinkers of divine and alcoholic spirits wrestle over the claim to good fellowship itself, as the godly attempted to assert their form of pious "good fellowship" in their communities of the faithful. The link of good fellowship to drinking thus provoked particular ire, with puritan critics calling out the deception of secular calls to "good fellowship" that served merely as a synonym for drunkenness. As Henry Crosse claims in *Vertues common-wealth* (1603): "If we looke into the monstrousnesse of sinne in this age, we may see every abhomination sport it selfe, as though there were no God. Drunkennesse is good fellowship." Indeed, he warns, one might "carrie the verie badge of good fellowship upon his nose."[104] As William Perkins asks, in dismay, "Is not drunkennes counted good fellowship"?[105] George Benson also derides the "drunkennesse of good fellowshippe," while Thomas Cooper condemns how "drunkenesse is counted good fellowshippe."[106]

The oppositions and countercurrents in discourses of drinking help reveal such ideological clashes between the godly and the good fellow as less directly oppositional than fraternal or sororal, two intimately connected, if fractious, impulses. Further, it is precisely in the vexed responses to drunkenness that we see addiction's range and pliability. Both devotional and compulsive at once, drunkenness provoked the variety of cultural responses that are at stake in early modern addiction and then are buried in later centuries. For even as conversations on excessive habitual drinking increasingly insist on drunkenness as a disease and pathology, they nevertheless retain the notion of addiction as a laudable pastime, a form of good fellowship that proves constant amid cultural changes.

Project Outline

Each of this project's chapters wrestles with addiction in relation to devotion, compulsion, agency, and authority. To do so each chapter offers a discussion of addiction in a different arena, moving from theology and lexicography to medical writings and puritan polemic to legal tracts and national politics. In the process, rather than concentrate solely on those texts that repeatedly deploy the word "addiction," I instead select texts that pose the broader philosophical issues at stake in invocations of addiction, drawing particularly on the imaginative richness afforded in the study of literary texts.[107] "While other discourses may be compromised by ambiguity, literature," Roland Greene writes, "is drawn to it—and can fashion it into something new, granting the premium of fresh perspective to old problems."[108] Sermons attempt to convert readers; literature, by contrast, serves to "entertain," both in the sense of offering a pleasurable pastime and in the sense of considering new ways of thinking about familiar issues. Through concentration on a literary text in concert with cultural and political writings, each chapter illuminates an aspect of addiction's rich possibilities.

When *The English Faust Book* describes Faustus as "addicted" to study, and when Marlowe's *Doctor Faustus* depicts necromantic study as "ravishing," these texts draw on classical and Renaissance notions of addiction as a beneficial and laudable form of commitment. Tracing the invocations of addiction in classical and theological writings ranging from Cicero and Seneca to Calvin and Perkins, Chapter 1 overturns modern, pathological conceptions of addiction by exposing the concept's classical and Renaissance meanings. More specifically, the chapter establishes how the influence of Calvin and Calvinist-minded Cambridge divines appears in *Doctor Faustus*, not just in the drama of election—as has long been argued—but also in the play's preoccupation with the challenge of commitment. Dedication, the play reveals, paradoxically requires both effort and surrender. If early modern theologians encourage such release, Marlowe illuminates addiction as a process of both wonder and terror.

This project's second chapter moves from theology to lexicography, and specifically to Shakespeare's *Twelfth Night*, examining addiction through the figure of the devoted lover. In relinquishing self-sovereignty in favor of another, the lover transforms into an addict, an achievement unavailable to other, more self-serving characters. For an embrace of devotional addiction is an embrace of the magic and serendipity of love, a process uncontrolled by human will. Furthermore, this foregoing of control requires dedicated commitment; it is a sustained process of giving oneself repeatedly through time. This mode of loving contrasts with what might appear, to modern readers, to be the more obviously addictive practice in the play: drunkenness. Yet this chapter delineates the difference between addiction and drunkenness precisely through a comparative study of Olivia, Viola, and Toby: Toby is too much himself to allow addictive transformation.

Henry IV stages the complex and contradictory invocation of addiction as both the laudable attachment chronicled in Chapters 1 and 2, and a compulsion, anticipating the modern era. Chapter 3 studies Falstaff as an admirable addict, dedicating himself to Hal. The play's markedly self-possessed rulers throw his mode of attachment into sharp relief: Hal, like his father before him, rises through the addictive energies of dedicated men only to abandon them. With his addictive relation forestalled and his devotional pursuit failing, Falstaff turns from Hal to the material conditions of their friendship, namely drunken good fellowship. As a result of this shift, the lauded ability to release oneself as an addict appears less dedicated than compelled, resonant with contemporary attacks on drunkenness as disease, examined in this chapter. The very language that designated loyal commitment comes to signify a form of bondage and is used to chronicle the compromised will of the drinker.

The project's fourth chapter turns from reflexive addictions—those actively chosen and embraced—to imperative addiction, studying *Othello*'s staging of incapacity through legal debates on responsibility. While Othello's love for Desdemona represents a form of laudable addiction as he dedicates himself to another, Iago's polluting attachment results in a transformation of addiction, from Othello's primary devotion to his new wife to his secondary addiction to his villainous ensign. Prompted to murder by love and loyalty, Othello proves both compelled and free to act. His criminal action, mitigated by incapacity, is anticipated in a much more minor key by Cassio earlier in the play. Read through early modern legal debates on drunken incapacity, Cassio's actions—like Othello's—should receive the full force of the law. Shakespeare, however, challenges such strict legal responsibility by staging addiction's double bind: how can one be both strictly responsible and *non compos mentis,* or incapacitated? The legal insistence on responsibility even at moments of madness contrasts with Shakespeare's more nuanced interrogation of addictive possession, in which the addictive propensities of both Cassio and Othello stand as a form of heroism: they open themselves to others and allow themselves to become possessed, in stark contrast to the excessive exercise of the will showcased in Iago.

This project's fifth and final chapter, rather than analyzing addiction through one exemplary play, instead turns to a single addictive practice: health drinking. This binge-drinking ritual helped to bolster beleaguered communities, as drinkers pledged themselves through expressions of loyalty and faith. Studying this addictive practice through a generic and historical range—surveying drama and poetry over an eighty-year period, from the 1580s through 1660—reveals both the longevity of addiction as devotion and the variability of attitudes toward one addictive practice. Health drinking was initially condemned in the 1580s and 1590s as a deplorable and foreign practice, but by the 1630s it was celebrated for its loyalist potential in uniting politically isolated royalists. Health drinking exposes, then, not simply the range of attitudes toward addiction as a mode of attachment, but divergent responses to one addictive practice that appears at once compulsive and dedicated. The book thus ends with a chapter that, despite its methodological distinction from the rest of the book, condenses many of the paradoxes evident throughout: early modern addiction represents choice and tyranny, devotion and disease.

Each of the project's chapters takes up, as suggested above, a different arena of addiction discourse. Showing the range of addiction's reach, each chapter save the last is also rooted in a popular play that deploys addiction discourse in an especially rich and nuanced staging. Most specifically, the plays under discussion dramatize the addict-actor relationship, which is explored in this book's preface, by simultaneously reinforcing and challenging their connection: Mephastophilis's power over Faustus, Viola's over Olivia, Iago's over Othello, and to a complex and different degree, Hal's over Falstaff's. Each of these relations expose the intimate connection—and opposition—of acting to addiction. Through a counterfeiting character who uses theater to his or her own ends—through a character who can claim, with Viola and Iago, "I am not what I am"—the play's hero transforms. Willing away his or her will, the hero shapes him or herself into an addict, one bound, tied, and obligated to the play's counterfeiting actor. In devotional relation, this heroic addict proves both sincere and dependent, in contrast to the potentially duplicitous freedom exercised by the counterfeiting actor. Yet this counterfeiter, who deploys deceit over sincerity, is of course played by an actor who is himself—in his own relation to the play's script—an addict, a figure bound to enact his own role onstage, just as the play's addict, in his sincerity, is an actor who counterfeits.

Refracting the relation of actor and addict, the plays under discussion defend theater through the resulting dramatic effects: pitting the character of the counterfeit actor against the devoted addict, these plays uphold the addict who, in his or her sincerity and vulnerability, is overcome. The addict, whether in the form of Faustus, Olivia, Falstaff, or Othello, asserts the power of theater to move and transform the audience against the theatrical rival, a condensation of anti-theatrical concerns. In pitting—or even reconciling—the actor and the addict, the counterfeiter and the devotee, these plays draw on the model of addiction to recuperate the theater and produce devotion from the audience, shaping an actor-addict so sincere and dependent upon us that we allow ourselves to be moved.

Chapter 1

Scholarly Addiction in *Doctor Faustus*

> Faustus, being of a naughty mind and otherwise addicted, applied not his studies, but took himself to other exercises.
>
> —*The English Faust Book*

What does it mean to say, as *The English Faust Book* does in 1592, that Faustus is "addicted"? Faustus, it seems, should apply himself to the study of divinity but is otherwise inclined, embracing alternate fields as the infamous version of the legend by Christopher Marlowe depicts in detail. Marlowe's *Doctor Faustus* opens with Faustus weighing the merits of divinity, a field in which he "profits," "the fruitful plot of scholarism grac'd."[1] But his very talents snare him, for, "excelling all" his peers, he becomes "glutted" with "learning's golden gifts" and begins to seek another form of scholarly sustenance, ultimately "surfeit[ing] upon cursed necromancy" (1.1.18, 24, 25).

If Faustus's appetite for scholastic heights differs from narcotic addictions, his surfeit nevertheless resonates with modern notions of addiction as pathology. As Deborah Willis writes in her study of the play, "It is not hard to draw an analogy between Faustus's evolving relationship to magic and modern narratives of addiction."[2] Marlowe's play, she argues, anticipates modern, medical definitions of the addict in staging the diminishing will of the individual in the face of compulsive behavior. Yet early modern addiction, as this chapter will explore, also appears in *Faustus* and in a series of sixteenth-century tracts to be beneficial and even laudable. As a result of what could be called compulsive addiction, but which one might equally deem devotion or dedication, Faustus proves an able and talented scholar, adopting a profession for his "wit" and excelling in it (1.1.1–2, 11). He thus fulfills the Latin root of the word *addīcere*, which was discussed in the preface: in Roman law to addict was to bind someone to service or to affix or attach oneself to a person, party, or cause. In Latin writings more broadly the term "addict" came to connote giving oneself over, or dedicating oneself, to a master, lord, or a vocation. Following these Latin origins, sixteenth-century writers use "addict" to designate service, debt, dedication, and devotion. In chronicling scholarly pursuits, for example, early modern translations of Cicero and Seneca invoke addiction to help account for the devotion necessary to follow an academic path, as this chapter reveals. So, too, with Reformation theological texts from Jean Calvin through English reformers such as John Foxe and William Perkins, in which addiction signals the state of deep dedication and surrender through which the believer receives grace.

Marlowe attended Cambridge at the height of the controversy over Calvinist theology, and his reaction to his education deeply marks his play.[3] In exploring the influence of Calvin and Calvinist-minded Cambridge divines on Marlowe, scholars have debated the play's staging of the doctrine of predestination and election, asking whether Faustus's damnation serves as a warning for spectators or a critique of Calvinist determinism.[4] Reading the play as a drama about election (whether or not it endorses Calvinist theology) proves challenging, however, because as Alan Sinfield has noted, "The predestinarian and free will readings of *Faustus* . . . obstruct, entangle, and choke each other."[5] The play does not make its representation of reprobation or election clear, but instead elusively hints at and refuses to resolve this question.

Given the play's provocative but at times contradictory presentation of theological doctrine, it is worth considering the question of free will and determinism from a different vantage point. As the following pages will explore, the Calvinism of Marlowe's education proves useful because it illuminates not only the theological influences on his play but also, more broadly, how he might have understood the nature of scholastic and theological commitment itself. Calvin, like his English followers Perkins and Foxe, outlines the doctrine of predestination and election through reference to—and celebration of—a single-minded devotion deemed addiction.[6] Faustus is, in line with this form of devotion, addicted to study, giving himself entirely to his chosen field: he signs a legal contract, professes his dedication, and exclusively commits himself to his studies. Marlowe stages scholastic devotion as a laudable addiction, drawing on classical and Christian evocations of the term, even as Faustus's choice of necromancy illuminates one of the dangers of such devotion: attachment to the wrong faith or field. Marlowe's Calvinist contemporaries acknowledge precisely this danger, suggesting how the surrender and release associated with addiction, while potentially saving, can lead to damnation when directed to the wrong spirits or forces. Thus, one might be addicted to sin or carnal pleasures, or more frequently, suffer from addiction to idolatry and popery, a condition Calvin writes of enduring before his conversion by God.

Tracing the invocations of addiction in the theological writings influential to Marlowe, this chapter thus approaches *Doctor Faustus* not as a drama of election but as one about the challenge of commitment. In drawing on and questioning contemporary invocations of addiction, Marlowe stages Faustus's perilous attachment to bad religion while never condemning his title character for his devotional aptitude in the first place. The tension of the play lies precisely in how Faustus's devotion and surrender to necromancy might have signaled his predisposition to what his contemporaries deemed a positive addiction, namely to God. To condemn Faustus's constancy to Mephastophilis, or to necromancy more generally, disregards the very predisposition for addiction that might have led him to God, for it is Faustus's paradoxical willingness to forego the exercise of free will, and his resolve to release into the supernatural, that marks him as open to receiving grace. His dedicated resolve might have flourished in the proper direction, as the play's epilogue notes—Faustus might have "grown full straight" (epilogue). Instead, he follows magic—and as a result, the moralizing voice of the chorus attempts to frame Faustus's path as pathological or sinful, deeming him a glutton who surfeits on necromancy.

Yet the play vigorously depicts Faustus's relation to magic as a sign not of his compulsive appetite, but of his scholarly drive.[7] Even as the Chorus warns that Faustus serves as an emblem, an Icarus burned by magic or a fierce God, the play itself stages a different (albeit related) drama, one not preoccupied with magic—after all, Faustus's magic tricks have proved disappointing to generations of audiences—but with the struggle inherent to devotion. Overpowering dedication, and the individual release of oneself to an external force, is at once necessary, dangerous, and potentially pathological. If to early modern writers such surrender is often laudable and desirable, Marlowe, through Faustus, pushes early modern conceptions by staging both the wonder and terror of addictive release. That grace might enter in the form of the devil proves the play's most haunting challenge to Calvinist invocations of addiction. The drama of addiction thus hinges on the longing for, yet also the regret surrounding, true faith, as Faustus finds himself—through the very process that might have offered salvation—contractually bound to hellish companions instead.

Addicted to Study

Tracking the first appearances of the term "addiction" in English reveals its use in two contexts: classical study and Reformed theology. Early modern translations of Cicero and Seneca both evoke addiction to study as a positive pursuit. In Cicero's *A panoplie of epistles* (1576), as translated by Abraham Flemming, he recounts fondly the "knowledge, learning, and exercises, whereunto from my childehoode I haue béen addicted."[8] The Latin original ("iis studiis eaque doctrina, cui me a pueritia dedi") deploys the term "dedication," signaling that the early modern translator found "addiction" an adequate cognate. Further epistles underscore Cicero's attachment to study as a form of addiction. Writing of the "study, to which I was addicted," Cicero calls scholarship the "letters to which I have ever been addicted."[9] Addiction here signals sustained attachment and devotion, as Cicero expresses his commitment to his course of study and his singular application of his talents. Cicero's son seems to have inherited, or reproduced, this addiction to study, at least according to a letter to Cicero from Trebonius, who, on seeing Cicero's son in Athens, reported him to be "a yong man addicted to the best kinde of studie . . . , and of a passing good reporte of modesty: which thing, what pleasure it ministred unto me, you may wel understand."[10]

Seneca, too, describes study as a form of addiction. In the translation by Thomas Lodge, *The workes of Lucius Annaeus Seneca, both morrall and natural* (1614), the young philosopher pursues his studies, as Faustus himself does, against the wishes of his family: he "addicted himselfe to Philosophie with earnest endeuor, and vertue ravished his most excellent wit, although his father were against it."[11] Just as Faustus challenges the promptings of his professors with his "wit" and finds ravishment in his studies, so too does Seneca (1.1.6, 111). Indeed both descriptions employ the term "ravish" to describe an intense relationship to a field of study. In doing so they suggest the force of scholarship in overwhelming, transporting, or capturing the scholar. Faustus, like Seneca, is carried away, but willingly and pleasurably. For both, the tension between family and worldly concerns, on the one hand, and the dedication to study, on the other, structures their understanding of vocation, further illuminating the exclusivity and captivation of addiction: "I will wholly dedicate my selfe, and . . . I will addict my selfe unto studie. Thou must not expect till thou have leasure to follow Philosophie. Thou must contemne all other things, to be always with her."[12] This exclusivity—condemning other pursuits for one's field—separates addiction from mere instruction. Seneca rejects other intellectual, and presumably familial, lures in favor of a singular relation to philosophy. Faustus, too, models such dedication. "I wonder what's become of Faustus, that was wont to make our schools ring with *sic probo*," his friends demand (1.2.1–2). He retreats into necromancy, dismissing, as the opening soliloquy dramatizes, all other fields. Addiction to study is an extreme form of dedication and requires one to clear away all other obligations.

Addiction, as deployed in these early modern classical translations, is a crucial component of scholarship: only with clarity and dedication can the philosopher find his calling. Furthermore, addiction represents a process of culling away rival pressures, be they worldly or even intellectual. Lodge's translation of Seneca's essay "The Tranquilitie and Peace of the Mind" reads, for example, "A multitude of bookes burtheneth and instructeth him not that learneth, and it is better for thee to addict thy selfe to few Authrs, then to wander amongst many."[13] Addiction as dedication stands in contrast to flighty, unfocused pursuits: "He then that hath all his commidities in their entyre, may stay in the hauen, and addict himselfe readily to good occupations, rather then make saile and to go and cast himselfe athwart the winds and waves."[14] The scholar is not, Seneca argues, an explorer visiting new ports. One cannot "wander," but one must hone, cull, and focus. Committing to one location, one "haven," the scholar studies deeply. Wide-ranging study is a burden and distraction. Better to "addict thy selfe to few Authrs." So, too, with Faustus, who, in narrowing the available fields, announces he will "sound the depth" (1.1.2) to find a pursuit that will envelop or ravish him. He will "profess" his art, proving a "studious artisan" and a "sound magician" (1.1.2, 56, 63). From this vantage point of addiction, Faustus's desire to "sound the depth" of his studies, and his interrogation of fields in search of the proper path, seems not fickle but ultimately focused. Rather than choosing necromancy out of a kind of boredom, as Kristen Poole argues—"his descent into the black arts at first seems to be the product of his intellectual ennui, as he searches for new challenges and intellectual heights"—he instead seeks his Senecan "haven."[15] While Poole's phrase "intellectual ennui" aptly accounts for Faustus's fear of death and stasis, which is evident in his condemnation of divinity as "hard" (1.1.40), nevertheless he dismisses certain forms of scholarship not out of exhaustion but because he seeks to immerse himself in a limitless field. He needs to aim at the unknown, the unseen, and the unachievable.

Addiction, then, is a particular form of scholarship—it involves commitment, focus, depth, and stillness. Of course, these authors also concede the dangers of such single-minded dedication. Seneca announces the dangerous power of addiction when he writes, for example, that one must be cautious in one's pursuits: "For the minde being once mooued and shaken, is addicted to that whereby it is driven. The beginning of some things are in our power, but if they bee increased, they carie us away perforce, and suffer us not to returne backe: even as the bodies that fall head-long downeward, have no power to stay themselves."[16] Seneca teases out the complex relationship of surrender and free will in scholastic addiction. Initially, the addict exercises choice: in the beginning "some things are in our power." One might choose one's path, as Faustus does—he elects to practice necromancy over divinity. But, Seneca writes, once the mind heads in a certain direction, addiction can carry one away. Addicts "have no power to stay" themselves. Momentum threatens but also fuels the addicted mind. Once on a path, the scholar progresses along it, gains speed, and moves forward even against his or her own will. Thus addiction is at once desirable, since it provides the dedicated resolve that propels the scholar forward, and potentially dangerous, since the power of addiction pulls one along the chosen path, for good or ill. The title of a text by the lawyer William Fulbecke betrays this double link of addiction and study: *A direction or preparatiue to the study of the lawe wherein is shewed, what things ought to be observed and used of them that are addicted to the study of the law, and what on the contrary part ought to be eschued and auoyded* (London, 1600). If the pursuit of learning is admirable, then the deeper the devotion, the greater the addiction and the more accomplished the scholar proves. "Driven," "carr [ied] away," "fall[ing] head-long downward," the scholar demonstrates a lack of control admirable and overwhelming at once.

Addicted to God

Why does the scholar choose one path and not another? Seneca suggests that scholarly addiction emerges from one's choices: "The beginning of some things are in our power." But Marlowe's contemporaries would answer differently. Addiction—whether to divinity or necromancy, to scholarship or to sex—comes from predispositions that, at least in a post-Reformation Europe influenced by Calvin, come not from human will but from God's. [17]

As with Seneca, Calvin praises addiction as a form of careful study, in this case not of philosophy but of scripture: "They are then apt to receive the grace of the Gospell, which not regarding any other delightes, do wholy addict themselves and their studies to the obtaining of the same."[18] Like Faustus, the believer dismisses all other fields and devotes himself to his chosen path. In the case of the Christian reader, the fruits of study lead to addiction to Christ: "Therfore no man shal ever go forward constantly in this office, save he, in whose heart the love of Christ shal so reigne, that forgetting himself, and addicting himself wholy unto him, he may overcome al impediments."[19] Followers of Christ "addict themselves unto him, so that they did acknowledge him to be that Messias."[20] Further, "those are truly gathered into Gods sheepefolde . . . addict themselves to Christ alone."[21] The singularity of the commitment is clear: one is addicted to Christ "alone" "wholly." Moreover, addiction to God compels the believer to follow a path, eschewing individual thought or will in favor of discipleship. Calvin writes, "For whosoeuer doe simplye addict themselves to Christe, and doe not strive to adde anye thinge of their owne head to the Gospell, the true lyghte shall never fayle them."[22]

Throughout his Latin writings, including the biblical commentaries and sermons that comprise the vast majority of his published works in England and on the continent, Calvin deploys the verb *addīcere* to designate godly devotion, writing "se totos addicunt" of these dedicated readers. [23] These Latin commentaries appeared in multiple editions and translations in England and dominated university libraries to the degree that, as Philip Benedict notes, "by the last decades of the century, Calvin's works had eclipsed those of all other theologians in the library inventories of Oxford and Cambridge students." [24] Further, the importance of the English translations of Calvin, in addition to the French and Latin editions also published in England, can hardly be overstated. Between 1570 and 1590, forty-three editions appeared: "No author would be as frequently printed in England over the course of the second half of the sixteenth century as Calvin," Benedict continues.[25] *Bibliotheca Calviniana*, the table of editions of Calvin by language, reveals the prominence of English editions within a European frame: they are second only to Latin and French (Calvin's original languages) and far exceed German, Italian, Dutch, Spanish, and other European-language translations.[26]

To study Calvin's invocation of addiction in these publications is to find comfort in addictive surrender: being singularly focused, at the expense of other beliefs and relationships, brings the potential for redemption. Calvin writes, "GOD in his mercie dealt so lovingly with his people, when he redeemed them, that is, that they being redeemed, should addict and vow them selues wholy to worship the aucthour of their salvation."[27] Ministers of the word "may addict & give themselves wholly to the Church, whereto they are appoynted."[28] This reflexive construction—Christ's followers "addict themselves" or "give themselves" (se totos addicant & deuoueant) to him and his church—might seem to indicate will and agency on the part of the believer in choosing the addiction. And Calvin does encourage his readers and audiences to foster the complete devotion encapsulated in addiction: his sermons and biblical commentaries repeatedly admonish listeners and readers to pursue utter devotion. But ultimately, he argues, addiction speaks not to individual will but to God's favor. Only those who are "disciples of God" can "addict themselves"; those who are "unapt to be taught" reject Christ: "It cannot be but that they shal addict themselves unto Christ, whosoeuer are the disciples of God, and that they are vnapt to bee taught of God who do reject Christ."[29] The appearance of a reflexive construction in his Latin and its English translation ("addict himself," "addict themselves") seems, on the one hand, to counsel the believer to prepare him or herself for grace: "he must doe his diligence." On the other hand, to modern readers the reflexive nature of the construction may be misleading in implying the believer's role in his own addiction, since the agency of addiction does not lie with the addicted believer but in God's grace. In this chicken-and-egg construction, those who are unapt to be taught cannot be taught; those who reject Christ have been rejected.

Calvin's language is subtle. His rhetoric suggests, at moments, a form of free will in which the believer might stray from addiction to Christ toward another, less laudable attachment: "They do corrupt the power of Christ, who are addicted to their belly and earthly thinges: hee sheweth what we ought to seeke in hym and for what cause we ought to seeke him."[30] We "are ofte withdrawn" into lusts, being "addicted to [our] belly and earthly things."[31] But, at the same time, he makes it clear that only God can "correct that disease" before one can act: "Because by reason of the grossenes of nature, we are always addicted unto earthly thinges, therefore he doth first correct that disease which is ingendered in us, before he sheweth what we must doe."[32] Once God cures, he "sheweth what we must do" and "sheweth what we aught to seeke in hym." In other words, one cannot even see the right path until God cleanses the natural depravity evident in one's misguided addictions. The dedicated mind, with its resolution, is an illusion. Addictions are signs of grace or reprobation that one does not control.

Calvin establishes, then, a complex relationship between the compulsion to follow Christ and the lure of material life. On one level, these desires are clearly opposed to one another since one's addiction, be it to Christ or to worldly pleasure, indicates elect or reprobate status. But on another, more fundamental, level, Calvin acknowledges that everyone struggles with pure addiction. Even the most faithful wrestle with competing desires. He claims, "It is true that the faythfull them selues are never so wholly addicted to obey God, but that they are ofte withdrawne with sinfull lustes of the flesh."[33] One might aspire to be "wholly addicted" but might err. In other words, addiction to God and addiction to the belly are at once opposed and yet connected as two sites for devotion that might both hold the believer. The tension between these two opposed forms of addiction can be reconciled only by acknowledging the inevitability of one's dependence on God's will. Humans, Calvin implies, struggle with some form of addiction. It is just a question of whether abandonment to addiction leads one to or away from God. Through grace, one might be able to embrace as firmly as possible servitude and obedience to God. This form of service and addiction is uplifting. The strength of one's embrace of this addiction, however, depends on God's grace. Without such a gift, one struggles with the earthly appetites and compulsions shackling the human body to its baser nature. The resulting addictions represent debasing tyranny.

This dangerous aspect of addiction appears in the English translations of Calvin's French sermons. In his Latin writings Calvin deploys the term *addīcere* routinely, and the English translation faithfully tracks the term, rendering it as "addiction," as noted in the citations above. By contrast, sixteenth-century French lacks modern French's *addiction* and *dépendence*. When Calvin attempts to describe the phenomenon of addiction, he turns to potential cognates in terms ranging from *adonner* to *attacher* to *dedier*. Yet his English translators reintroduce the term "addiction," illuminating the absent presence of the concept in the French sermons. Translating *adonner*, a term that designates a willful giving over of oneself, Arthur Golding turns to the term "addiction"—but only in special circumstances. Calvin, for example, uses versions of the word *adonner* 206 times in his *Sermons de M. Jean Calvin sur le livre de Job*. In Golding's sizeable English folio translation (produced in six discrete impressions between 1574 and 1584 and amounting to what Stam calls a "bestseller," which "achieved a popularity beyond that of any other Calvin commentaries"), he translates *adonner* as "addiction" in only three cases, each describing a special instance of attachment.[34] When Calvin writes, "Or nous ayant acquis si cherement, il ne faut pas que nous soyons plus adonnez à nous mesmes, mais que nous soyons du tout dediez à son service," Golding translates it as "Hee hath purchased vs so dearly, we must no more be addicted to our selues, but be wholly dedicated to his service."[35] Further, Calvin writes, "Et pourtant ce n'est pas raison que doresenavant nous soyons plus adonnez à nous mesmes: mais qu'un chacun soit prest de se dedier pleinement au service de Dieu," and the translation reads, "It is not meete that henceforth we shoulde be any more addicted to our selues, but every man should bee readie wholly to dedicate himselfe too the service of God."[36] If active, dedicated devotion to God is the praiseworthy goal, Calvin here also describes an improper form of donation or giving over, a form deemed addiction in English, as in Latin. In this English version, "addiction" and "dedication" appear as synonyms, used interchangeably—one should not be "addicted" to the wrong path but "dedicated" to the proper course—and at the same time indicate addiction as a potentially dangerous form of attachment. *Addīcere*, *adonner*, and "addiction" signal mistaken attachment to oneself and the world, even as this attachment might mirror a more desirable form of addiction to God.

Calvin warns of the dangers of this improper addiction because he experienced them, at least according to his personal account of conversion. In the address to the reader prefacing his commentaries on the Psalms (first translated into English by Golding in 1571, following the Latin original published in Geneva in 1557 and French editions in 1558 and 1561), Calvin offers one of his few explicitly autobiographical statements about his conversion. He claims that he had been groomed for the ministry from a young age, but his father directed him to study law instead. While he endeavored to satisfy his father's wishes, "God with the secret bridle of his providence did at the length turn my race ageine the other way," toward divinity. Calvin relates how initially he "was more strictly addicted to the superstitions of the Papistrie, than I might with ease be drawn out of so deep a puddle." His sudden conversion freed him from such bondage. God's grace literally turned Calvin around, redirecting and reshaping him: God "sodenly turned my mind (which for my yeeres was over muche hardned) and made it easie to be taught."[37] Calvin here views himself as "strictly" or, in an alternate translation, "obstinately" addicted to the pope and superstitious belief.[38] By "superstition" he indicates, as Alexandra Walsham notes, devotion to relics, saints, and other material manifestations of faith: "The ease with which the populace had been deceived by these tricks was itself a just punishment from God for its gullibility and natural addiction to 'this most perverse kinde of superstition,' and to a carnal religion that revolved around visible, physical things."[39] Here, Walsham's terminology draws attention to the link between religious devotion and addiction. Only with God's help to redirect and reshape him can Calvin relinquish his obstinate attachments: God did "turn my race" the right way: he "turned my mind" away from the papacy. God turns him around, softens his heart, and frees him from earthly lures so he can dedicate himself to the divine, a more compelling and liberating addiction.[40] As a result, after his conversion he "burned with so great a desire of profiting: that although I did not quite give over all other studies, yet I followed them more coldly."[41]

Calvin's English followers, including Foxe and Perkins, take up and extend his model of addiction to scripture and superstition, continuing to tease out the depravity inherent in misguided addictions even as they trumpet the joys of addictive devotion to the divine. Foxe's *Actes and monuments* (after the Bible, arguably the second most popular religious book in Elizabethan England following a 1572 government order requiring copies to be placed in all cathedrals in the country) invokes the devotional aspects of addiction when he narrates the lives of Protestant martyrs such as John Frith and William Tyndale, two figures who dedicated themselves to the study of scripture. These Reformed theologians demonstrate the addictive potential celebrated by Calvin himself. Frith, Foxe writes, "began hys study at Cambridge. In whose nature had planted being but a child maruelous instructions & love unto learning, whereunto he was addict. He had also a wonderful promptnes of wit & a ready capacitie to receaue and understand any thing, in so much that he seemed not to be sent unto learning, but also borne for the same purpose."[42] Like Seneca and Faustus, Frith has a "promptness of wit" and proves "addict," "borne for" rather than merely acquiring learning.

Tyndale, too, proves addicted to study, which he pursues at Oxford, "where he by long continuance grewe up, and increased as well in the knowledge of tounges, and other liberall Artes, as especially in the knowledge of the Scriptures: wherunto his mind was singularly addicted."[43] Foxe praises the divine pursuits of Frith and Tyndale in terms that resonate with the scholastic addiction of Seneca and especially the devotional addiction of Calvin: these religious men are "singularly" focused, "borne" with aptitude and "turned" by God. Foxe also, and indeed more frequently, illuminates the dangers of addiction as expressed in mistaken attachments, especially when the believer proves addicted not to God but to Catholic idolatry: "These which addict themselves so devoutly to ye popes learning, were never earnestly afflicted in conscience, never humbled in spirite nor broken in hart, never entred into any serious feeling of Gods judgment, nor ever felt the strength of the law & of death."[44]

Foxe's contemporary, the Calvinist William Perkins—deemed by the end of the sixteenth century to be one of England's most popular religious writers, with seventy-six editions of his work appearing before his death in 1602—also highlights the danger of Catholic attachments over devotion to God.[45] "Perceived as translating Calvin for the masses," as Poole puts it, Perkins praises those who "addict themselves unto Diuinitie," yet cautions against the study of exegesis over scripture.[46] He writes, "Hence come dissentions and errors into the schooles of the Prophets, which cannot be avoided while men leave the text of scripture & addict themselves so much to the writings of men, for thereby hee can more cunningly conuey strange conceits into mens minds: and therfore every one that would maintain the truth in purity and syncerity must labour painfully in the text."[47] The opposition of "purity and sincerity" to "error" and "dissentions" indicates the struggle of addiction. Perkins, even more pointedly than Seneca, explores how addiction to scholarship can go awry when the object of study is inappropriate.

Embracing Reformed theology, Perkins is particularly keen to encounter scripture directly. Study and translation of "the word of God" is the scholar's appropriate calling. "The writings of men" only detract from the truth, and "popish writers" in particular lead audiences astray. Divinity students, he writes, "within this sixe or seven yeeres, divers have addicted themselves to studie Popish writers, and Monkish discourses, despising in the meane time the writing of those famous instruments and cleere lights, whom the Lord raised up for the raising and restoring of true religion, such as Luther, [and] Calvin."[48] Religious dedication, indeed dedication to God, is no longer enough; one must turn away from the Catholic version of God to celebrate that of Martin Luther, Calvin, and their followers. Reformed writings, like scripture, ring with "true religion," "cleere lights" and purity. The *copia* of Erasmus must cede to the crystalline prose of Luther.

If Catholic writings corrupt the reader, only godly conversion cures. As Perkins puts it: "Againe, after conversion it is not an idle power in them: 1. Ioh. 3.9. *He that is borne of God sinneth not*, that is addicteth not himselfe, nor setteth himselfe to the practise of sinne; and the reason is given, because *the seed of God remaineth in him*."[49] Commitment to God and interest in worldly pleasures prove mutually exclusive. Perkins writes, "The love of the trueth, and of the world, the feare of the face of man, and the feare of God can never stand together. As also howe dangerous a thing it is to be addicted to the love of the world: for it hath beene alwaies the cause of revolt."[50] This is the power of addiction—it is a singular devotion that defines us, for good or ill. If Calvin understands abandoned devotion as a source of salvation as well as reprobation, then Foxe and Perkins more explicitly praise addiction to God in contrast to errant addiction to Catholic idolatry.

Finally, Marlowe's contemporaries warn against necromancy itself as a form of addiction. In *A dialogue of witches* (1575), Lambert Daneau writes of addiction to Satan in these terms: "Whosoeuer were seruisable or addicted to Satan, were called by the name which is wel knowne and commune, that is Sorcerers," who forged an "agréement with the diuel . . . & to be short, have wholy addicted them selves to Satan."[51] The active "agréement with the devil" proves, however, a form of ensnarement in which the sorcerer is victimized by the devil: they "fall into the snares of Satan, and become Sorcerers, that is to say, addicted unto Satan."[52] Condemning the sorcerer, Daneau includes a spirited call for another form of addiction, for if "the serpent is more addicted or subject to Satan, then the other beastes," humans at least have the choice to turn away.[53] Here the story of a convert who embraces Christ delivers Daneau's point: "That he was converted to the fayth of Christ, it is read of him how earnestly and diligently he was addicted to that studie [of necromancy], which afterwarde, through the great goodnesse of god, he forsooke and renounced."[54] The parallels to Faustus are evident here. The scholar's dedication, longing, and effort, directed initially to the wrong field, shift to worship of God instead, through whose goodness the convert is saved.

Addicted to Magic

If writers from Calvin to Foxe and Perkins insist on the double-edged quality of addiction as a firm commitment that may or may not lead to grace depending on the form of the devotion, Marlowe stages both the danger of choosing the wrong field and the struggle of committing in the first place. The play's opening acts, from the first scene to the signing of the necromantic contract, chart Faustus's devotional struggle as he seeks the addiction lauded from Seneca to Calvin to Perkins, hoping to lose himself in a vocation by relinquishing reason, soul, and body to a higher power. The play opens with Faustus, sitting in his study, surveying a range of scholastic pursuits and famously dismissing them all as inadequate to his purposes. In doing so he illuminates the challenge before him as he pursues a field of limitless endeavor. He wants to "*level* at the end of every art" (1.1.4, emphasis added), namely, aim at—but never reach—an end. Thus he condemns those fields that limit his striving. Logic, medicine, law, and divinity fail to attract his devotion because they result in a mere "end" rather than an imaginative expanse. While Aristotelian logic might have "ravish'd" him at one point, it now bores him: "Read no more: thou hast attain'd the end" of that field (1.1.6, 10). So, too, with medicine: "Why Faustus, hast thou not attain'd that end?" (1.1.18).[55] These lines suggest the scholar's desire to strive forward rather than to complete his studies. Law and divinity, too, limit his striving. Law "fits a mercenary drudge / Who aims at nothing but external trash," while divinity, also, offers apparent certainty: "we must sin, / And so consequently die. / Ay, we must die, an everlasting death" (1.1.34–35, 45–47). If the audience might recognize divinity as offering unlimited grace (as the scriptural passage he reads goes on to promise), to Faustus its end in "everlasting death" mirrors the finality and nearsighted "aims" of other fields.

Faustus wants "to live eternally"; as Seneca writes, he wants "to be always with" the field of choice, perpetually moving forward so that, as Faustus puts it, "being dead" he might be raised "to life again" (1.1.24, 25). If these lines seem blasphemous—he desires, after all, to raise the dead in the manner of Jesus—they also speak to his desire for scholarship to offer him an unending path for life. Of course, as Genevieve Guenther argues, Faustus seems to crave a resolutely material life, seeking not everlasting salvation in heaven but instead life on earth, thereby making his comment doubly blasphemous.[56] He wants to raise the dead back into their own bodies, she states, not into heavenly union. But what Guenther underplays, and what is notable in this opening soliloquy, is Faustus's striving. His experience of embodiment is not static or fixed but mobile, for lack of curiosity or ambition is a kind of death, a mere "attain'd end" (1.1.18). By "end," as Edward Snow argues, Faustus signals a "termination" rather than "an opening upon immanent horizons." As a result, "having 'attained' [an] end means that he has arrived at the end of it, used it up, finished with it."[57] Magical texts, by contrast, allow him to imagine an unachievable, continually receding goal, a mystical form of knowledge just beyond his reach: it is "necromantic books" that "Faustus most desires," for they are "heavenly" (1.1.51, 53, 51).[58]

Faustus ultimately choses necromancy because it offers not dominion but the ravishment of addiction: "'Tis magic, magic that hath ravish'd me" (1.1.111). The scholar seeks to be overcome and, as Calvin writes, "not regarding any other delightes," to "wholy addict" himself and his "studies to the obtaining of" his goal.[59] Even as Faustus wants to revel in "power," "honor," and "omnipotence," he is fundamentally a "studious artisan" (1.1.55–57). Flourishing in his studies he hopes to be, as Cornelius promises, more consulted than the Delphic oracle. That is to say, he desires to be a source of knowledge that is invisible, empty, and devoid of will, reflecting instead the voice of the divine. This *omphalos*, or navel, of the world delivers its messages from a divine power that Faustus, too, wants to channel, "forgetting himself, and addicting himself wholy" as Calvin writes, so as to "ouercome al impediments."[60] Of course Faustus's attraction to necromancy does not arise solely from his ambitious spiritual goals; as Luke Wilson has argued, he chooses necromancy with an expectation of its returns. He speaks of "gold," "pearl," "pleasant fruits," and "princely delicates" (1.1.83–86). More specifically he seeks to command: necromancy offers servile spirits "to do whatever Faustus shall command" and to be "always obedient to my will." "I'll be a great emperor of the world," he claims (1.3.37, 97, 104).

Yet scholars have tended to overlook how Faustus—perplexingly and contradictorily—seeks such power through the utter surrender of himself, releasing his own mind into a metaphysical, even divine, relationship. However much he might claim to pursue magic for material gain, his more sustained desire centers on metaphysical union. He seeks this merger through study, searching out the field that promises ravishment and then submitting himself to that field's masters: Mephastophilis and Lucifer. Just as Calvin counsels ministers to "addict & give themselves wholly to the Church, whereto they are appointed,"[61] so too does Faustus give himself: he "surrenders up to [Lucifer] his soul" (1.3.90). As with Calvin, Marlowe stages the complex exercise of the human will: Faustus strives and seeks, he labors in his field, but he must also surrender himself to it. Even as he proves eager to see if devils will obey him, and even as he celebrates his own skill in conjuring ("who would not be proficient in this art? / How pliant is this Mephastophilis, / Full of obedience and humility, such is the force of magic and my spells!"), ultimately Faustus "dedicates," "surrenders," and "give[s]" himself (1.3.28–31, 90, 103). On finding that Mephastophilis serves not himself but Lucifer, Faustus dedicates himself to Lucifer too; on finding his conjuration was *per accidens* rather than a sign of necromantic skill, Faustus responds not with disappointment but by pledging himself further: "There is no chief but only Beelzebub, / To whom Faustus doth dedicate himself" (1.3.58–59).

Faustus's embrace of metaphysical merger appears in two ways: first, in his willingness to forego the logic he has mastered at the opening of the play, and second, in his signing of the contract.[62] In choosing necromancy and binding himself to its masters, he exhibits the single-minded, exclusive attachment to his calling typical of the lauded addict: he follows faith (however dubious it might be). Aristotle's logic and Ramus's methods celebrate reasoning and critical thinking, but Faustus, despite his proficiency in logic and rhetoric, ignores such skills. Instead Faustus wants a "miracle," he seeks to "be eterniz'd," and to revel in "heavenly" books (1.1.9,15). This is not the ambition of a logician or lawyer. Imagination, emotion, hope, and faith, not logic, fuel his desires, arguably mirroring the devotion of the Christian faithful whose addiction to God defies earthly reason: "mine owne fantasy, / . . . will receive no object, for my head / But ruminates on necromantic skill" (1.1.104–6). A. N. Okerlund writes of these lines: "Faustus is telling us his mind is made up and not to be confused by critical analysis. . . . Distinguishing the valid from the invalid statement is the problem here—the problem to which Aristotle, Ramus, and their scholarly followers devoted their lives. But Faustus apparently cares not at all about the irreconcilable meanings of the Angels' statements and hears only the words which excite his desires." As Okerlund concludes, "Marlowe intends to call our attention to Faustus's deliberate violation of formal logic."[63] While such a failure of logic might seem foolhardy, and indeed damnable, when viewed from the vantage point of addictive dedication Faustus's illogical willingness to embrace magic appears as a sign of his faith: he refuses to be swayed from his path, in a manner Perkins himself might praise, by the writings of men. "We must no more," Calvin writes, "be addicted to our selves, but be wholly dedicated."[64]

If Faustus's language of dedication, surrender, and ravishment—the language of addiction—expresses his scholarly ambition to lose himself in his studies, in surprising contrast (and throwing into high relief the scholar's addictive devotion) Mephastophilis proves a cautious, reasoned, and even logical partner in magic. One finds reason and logic, for example, both in Mephastophilis's answer to Faustus's queries (he is, as many critics have noted, disarmingly straightforward in his answers) and in his effort to draw up the contract. Mephastophilis twice demands a "deed of gift" from Faustus (2.1.35, 60). The precision of Mephastophilis's "deed of gift" is Marlowe's addition to his source. In the *English Faust Book*, the term is "covenant," which has greater resonance with biblical than English or continental law.[65] Deploying a category of contract in the highly legal phrase "deed of gift" and emphasizing Mephastophilis's logic rather than obfuscation, Marlowe creates a figure more sympathetic than the trickster of medieval mystery plays. At the same time, Marlowe draws heightened attention to Faustus's failure—his inability to deduce or even hear the patently evident error of his choice.

Yet, Marlowe reveals, Faustus's failure is also a triumph, for it exposes further his desire to addict himself to his field of choice precisely as Seneca and Calvin counsel. He embraces the contract as an opportunity to realize his addictive goals, constructing a document baffling in its terms but satisfying in its potential. This contract is another sign of Faustus's longing for integration over autonomy, addiction over willpower. If Foxe, as noted above, derides those members of the early church who have "never entred into any serious feeling of Gods judgement, nor ever felt the strength of the law & of death," Marlowe stages Faustus's willing embrace of such deep feeling, encountering the strength of the law eagerly.[66] For Faustus acknowledges that he will be proficient, indeed "great," only to the extent he gives himself up entirely, donating his soul to another "as his own." It is when Lucifer claims and owns Faustus's soul that the magician merges with the devil he follows: "bind thy soul that at some certain day / Great Lucifer may claim it as his own, / And then be thou as great as Lucifer" (2.1.50–52). Far from shying away from such terms, Faustus designs them: he offers Mephastophilis his soul before the spirit has even requested the gift deed. In the play's first act he tells Mephastophilis, "Go, bear these tidings to Lucifer: . . . Say he [Faustus] surrenders up to him his soul" (1.3.87–90). Then, in drawing up the contract's terms in act 2, Mephastophilis's request of "a certain day" (1.3.91) becomes, under Faustus's design, "four and twenty years" (2.1.108), while the demand that he "bind [his] soul" (2.1.50) becomes Faustus's more elaborate offering of "body and soul" (2.1.106) and further, "John Faustus, body and soul, flesh, blood, or goods" (2.1.110). In not just signing the contract but designing its terms, Faustus paradoxically wills away his will, resolving to surrender himself to the greater force of magic. Mephastophilis proves the beneficiary of Faustus's longing for union and dissolution: "Had I as many souls as there be stars, / I'd give them all for Mephastophilis" (1.3.104).[67]

Contracted Faustus

Faustus's contract is notable for its omissions as much as its guarantees. Indeed, the contract has generated significant critical discussion because its rewards for Faustus are so vague. Faustus appears, critics argue, to be unaware of how bad a bargain he constructs. In exchange for essentially two things—the ability to be a spirit and the service of Mephastophilis, both for twenty-four years—Faustus gives his body and soul to Lucifer. While the terms of the contract seem unfavorable to Faustus, it is nevertheless worth asking, what if the contract actually articulates precisely what Faustus seeks? In posing a version of this question, Guenther suggests that Faustus, in discounting the metaphysical realm, embraces the contract without recognizing its repercussions.[68] But this chapter answers differently, by saying that if Faustus indeed seeks the devoted union he trumpets, he finds the contract a means of articulating this desire, if not securing it. Faustus's ostensible goal—to be "great emperor of the world" (1.3.104)—cedes to his deeper aim, which is stated in the contract itself. Rather than securing his own "command" or empyreal power, he instead signs a contract ensuring that his own form will disappear and be supplemented by the continual presence of another. Indeed, he repeatedly insists that the contract include body and soul, even as Mephastophilis seems unconcerned with Faustus's physical remains. Mephastophilis tells Faustus, "Thou hast given thy soul to Lucifer," to which Faustus responds, "Ay, and the body too" (2.1.132–33). If his body and soul will be Lucifer's after death, before that time Faustus will be physically joined to Mephastophilis, who will come—as the contract states—to Faustus "at all times" (2.1.103–4).

Forging a contract securing constant companionship with his magical mentor, on signing Faustus immediately asks (after first enquiring about the location of hell) to be married. He deflects his desire for a mate by claiming, "I am wanton and lascivious" (2.1.142), but this earthly request arguably tips his hand in betraying longing not for empyreal power but for union, precisely what the contract with Mephastophilis offers. Through marriage, as through magic, he seeks companionship on earth, to be overcome by relationship even as he also constructs a metaphysical union. The necromantic contract thus doubly satisfies Faustus, by offering him earthly company and spiritual merger: he enjoys Mephastophilis's company for twenty-four years and then joins Lucifer, who elevates Faustus's soul in claiming it as his own.[69] For a character so ostensibly preoccupied with his own glory, Faustus proves surprisingly eager to lose himself in his field of study and devotion to the field's masters. He seeks to be ravished, consumed, and overcome by the study of magic and the companionship of its practitioners. The contract's terms thus illuminate the paradox of Faustus's devotion: he is choosing to give up choice; he is exercising his right to surrender himself. Rather than seeking legal protection and securing his own claims, Faustus uses the contract to voice his loyalty, his surrender, and his willingness to give of himself to magic. Through the contract, in other words, Faustus attempts to announce, and secure, his addiction.

It is perhaps not surprising, then, that Faustus takes the contract more seriously than anyone might reasonably expect. The legal scholar Richard Posner puzzles over Faustus's "sanctity of contract," exploring the numerous ways Faustus might have wiggled out of his obligation. First, the contract does not involve an immediate exchange but instead relies on Mephastophilis serving Faustus for twenty-four years before Faustus delivers his soul. "Such a contract," Posner argues, "establishes a long-term relationship; and since not every contingency that might arise over a long period of time can be foreseen, it is understood that the parties will act in good faith to resolve problems as they arise rather than stand on the letter of the contract."[70] Even if Mephastophilis does exercise a "good faith" effort to fulfill every request, the contract remains riven with other weaknesses. As Posner writes, "The law refuses to enforce contracts that are against public policy, and a contract with the devil fits the bill."[71] If challenging the contract at the last moment would seem an unfair gain for Faustus, even here he could have nullified the bargain by offering restitution to the devil in the form of his body, his estate, and his service for the remaining years of his life, as the legal scholar Daniel Yeager argues in his analysis of the play.[72] Faustus's repudiation of the contract would be all the easier given the weakness of Mephastophilis's position. The legal insistence of Mephastophilis that Faustus sign a contract in the first place might alert audiences—if not Faustus himself, who dismisses law as "too servile and illiberal" (1.1.36)—to the illegitimacy of his argument. "Mephostophilis's insistence on formalities," Yeager writes, "reveals his doubt about the validity of the contract."[73] Posner, too, concludes, "The devil could not argue either that he didn't know that contracts with him were illegal or that the primary wrongdoer was not himself but Faustus. . . . So Faustus might have wiggled out of his contract after all."[74]

Faustus does not seek, of course, to wiggle out of the contract. The question then becomes why Faustus upholds what Posner deems the "sanctity of contract" at all. Why believe the contract is, as Yeager writes of Faustus, "inviolable," especially when Faustus has studied law and might recognize the legitimate challenges he could mount against Mephastophilis? He upholds the contract, this chapter answers, because this unmistakably legal exchange demonstrates the eagerness with which Faustus seeks—and perceives himself—to be bound. The issue is not, as Posner puts it, the "irrevocability of Faustus's contract," but rather Faustus's *perception* and *desire* that his choice should be irrevocable. Once committed, Faustus remains convinced of the legitimacy of this commitment and strains to maintain his half of the bargain.[75]

If Faustus's addiction were secure, surely neither he nor Mephastophilis would need a document signed in blood. But Faustus and Mephastophilis turn to these legal measures, one realizes as the play continues, because Faustus's initial efforts to pursue addiction through willpower and resolve failed. At the start he repeatedly tells himself, "Be resolute" (1.3.14), reassuring Cornelius and Valdes of his commitment. When questioned by Valdes, who tells Faustus he can be a magician only "if learnèd Faustus be resolute," Faustus responds, "Valdes, as resolute am I in this / As thou to live. Therefore object it not" (1.1.134, 135–36). Resolution to study and life go hand in hand for Faustus. As he conjures for the first time he again repeats: "Fear not, Faustus, but be resolute" (1.3.14). But resolve is not enough. Willpower alone cannot sustain Faustus in his commitment to magic. The contract represents, therefore, his second-order attempt to bind himself, offering more of himself than Mephastophilis demands. He designs a deed that will keep him dedicated to magic and overcome his hesitations. Logic ravished Faustus, as he admits at the opening of the play, and yet the scholar rejects this field anyway. He was resolved on divinity, until he was not.[76] In embracing magic, in allowing himself to be ravished again, Faustus attempts to ensure his commitment through firmer means than he had exercised with his earlier devotions—hence, the contract's specificity, and its insurance of his merger with Lucifer and Mephastophilis, not twenty-four years in the future but from the very moment of signing. And he must ensure (or at least attempt to secure) this continued obligation contractually because he knows what Seneca, Calvin, Foxe, and Perkins have illuminated before him: devotion is difficult.

If to some viewers Faustus's failure to challenge the contract signals his reprobation (he literally cannot see what the audience is able to recognize—he's making a terrible bargain in selling his soul to the devil), this chapter suggests how the play offers a more complex portrait of the hero than this answer allows. Faustus is not merely an emblem of Icarus, even if the Chorus might frame him this way. What makes Faustus's situation at all sympathetic is his *drive* to devote himself to his studies, and through the contract he attempts to demonstrate—indeed, bloodily performs—precisely this devotion. Despite challenges to logic and reason, despite isolation from friends and distance from the heavens, Faustus binds himself to his field of study. The dilemma he faces—whether to commit himself to his path despite all of this evidence against it—is a compelling and inherently dramatic one not because it involves summoning the devil and being devoured by a hellmouth, but because it mirrors the travails of all aspiring addicts. Faustus wants to be bound, compelled, reshaped, and overcome by a metaphysical force. He seeks, as he repeatedly states, ravishment. While Faustus's commitment to his contract might be, as Yeager calls it, "numbingly self-defeating," Marlowe's play illuminates, in this drama of self-defeat, the nature of attempted devotion.[77] "Self" defeating might, in another context, be the precisely desirable outcome of devotion. The dissolution of the self in the supernatural is what the Christian faithful pray for and what the addict seeks. Indeed, even the bodily inscription warning Faustus away from the contract serves, arguably, to remind him of his desire for merger. When Faustus finds "*Homo fuge*" inscribed on his arm, he responds, "Whither shall I fly?" (2.1.77). This phrase of course refers to the biblical invocation, "man of god, flye," from 1 Timothy 6:11.[78] But one might also read "fuge" in its musical sense, which originated in the sixteenth century. A fugue, or *fuga* (out of *fugere*), is a form of composition weaving together two distinct threads contrapuntally. In this case, "fuge" resonates with Faustus's broader desire to be subsumed or ravished by a greater power. Man, were he "fuge," might turn into the music of the spheres. The word teasingly evokes an ideal, nonviolent form of union: just as the music emerges out of intertwining two strands of sound, producing harmony and depth, so too might Faustus be taken up into a relationship greater than himself.

Yet, tragically, in attempting union through a legal contract, Marlowe exposes Faustus's desired but ultimately failed addiction. Like the Roman slave contractually bound to a master, Faustus becomes an addict through the law. But the addiction celebrated from Seneca to Calvin is not legal but vocational. It involves a calling. A contract upholds Faustus's rights, even if they seem paltry. A contract can be negotiated and annulled, as Posner and Yeager note. One does not, by contrast, "wiggle out of" addiction. Thus, Faustus's attempt to secure his addiction through contract exposes his devotional failure before he even begins. True devotion requires no contract, no promptings, and no threats. In the same way a beloved might erroneously hope a marriage contract could secure a lover's fidelity, Faustus relies on the necromantic contract to fix his own insufficient desires.

Wavering Faustus

Faustus's signing of the contract, ironically, betrays his own failed addiction. The document that secures his damnation fails to—and could never—represent his devotion. Certainly, the play's remaining scenes offer the fulfillment of Mephastophilis's promise: viewers see the rewards of necromancy in Faustus's adventures. But as critics have long noted, the fruits of magic are rather slim. If Faustus hopes to command nations, he finds himself playing parlor tricks, leaving the audience, if not Faustus himself, disappointed.[79] He mocks the pope and the horse-courser, he brings grapes to a duchess and conjures historical figures for the emperor and scholar friends. Why Marlowe, who stages Tamberlaine's march across Europe and Asia, would hesitate to stage more satisfying magical triumphs has rightly preoccupied critics and audiences. The most evident answer, provided by the Chorus and ostensibly in concert with Calvinist theology and Elizabethan authorities, finds Faustus to be an emblem for misguided ambition. The failure of magic supports readings of the play as a cautionary tale (why sell one's soul for mediocre magic?) insofar as one finds the play's middle section to be an extended lesson on Faustus's bad choice.

This chapter offers another answer, one—as suggested above—that finds the drama of the play to lie not in its subject matter of magic but, in properly Aristotelian fashion, in its action. For the drama of the play's middle acts lies in Faustus's wavering: the scholar with heroic resolve, a man who signed a contract he refuses to challenge, nonetheless falters. Indeed, perhaps more surprisingly than critics have noted, having made such a dramatic deal with the devil and having offered up his blood in signing, Faustus must nonetheless continually remind himself of his pledge. Faustus reassures himself, "Fear not, Faustus" (1.3.14). This imperative presages a series of reminders that Faustus offers himself as he wavers. "No go not backward. No, Faustus, be resolute. / Why waverest thou?" (2.1.6–7). This wavering, he claims, is because "something soundeth" in his ears, a voice that counsels, "Abjure this magic, turn to God again!" (2.1.7, 8). He keeps entertaining the possibility of repentance, or rather, the possibility of escape from his chosen commitment. In the first glimpse of the scholar after signing the contract, he cries, "When I behold the heavens then I repent" (2.3.1). Even though he might lament "my heart's so harden'd I cannot repent," he also actively embraces magic again on recalling the "ravishing sound" of the Ampion's harp making "music with my Mephastophilis" (2.3.18, 29–30). He cries, "I am resolved: Faustus shall ne'er repent. / Come Mephastophilis, let us dispute again" (2.3.32–33). Wanting to dedicate himself entirely but pulling away, and wanting to repent but returning to magic, Faustus seems insecure in the very bargain he designed.

Faustus *both* picks the wrong field *and* can't quite commit himself to it. For a man who begins the play wanting to be obliterated through integration into necromancy, he never achieves full surrender or release but instead wavers between professions and masters. He tells Charles V, "I am content to do whatever your Majesty shall command me" (4.1.15–16), and in doing so receives "a bounteous reward" (4.1.92–93); the Duke of Vanholt, too, tells him, "Follow us and receive your reward" (4.3.33). Even as he is bound to Mephastophilis and Lucifer, Faustus relates to earthly authorities as a pandering courtier seeking favor. He obsequiously calls Charles V "my gracious sovereign" while deeming himself "far inferior to the report men have published, and nothing answerable to the honor of your imperial Majesty" (4.1.12–14). Seeking favor and accepting rewards from earthly authorities, Faustus then relishes his power to humiliate his social equals or inferiors. The mocking knight and the horse-courser experience Faustus's high jinx. These comic interludes strain against Faustus's initial desire to be ravished, enveloped, and devoted: he seems preoccupied with his own status and reputation. Rather than dissolving his self, he seeks to protect and amplify it.

Fluctuating between authorities and erratic in his devotion, Faustus then begins to reproach others for his choices. As Poole writes, "Faustus has the unattractive habit of blaming others for his actions, often positioning himself as a passive entity."[80] He blames his own fall on reading: "Oh would / I had never seen Wittenberg, never read book" (5.2.19). Or, alternatively, he blames his fall on Mephastophilis, claiming he was tricked: "go accursed spirit to ugly hell: / 'Tis thou hast damn'd distressed Faustus' soul" (2.3.77–78). Finally, Faustus claims that his relationship to Lucifer and Mephastophilis is incomplete, since he has not experienced magical power but only indulged his appetites. He, like the Chorus, condemns himself as a glutton, surfeiting on his desires: "The god thou serv'st is thine own appetite, / Wherein is fix'd the love of Belzebub" (2.1.11–12).[81] He revels, he claims, in "a surfeit of deadly sin, that hath damned both body and soul" (5.2.10). He doesn't even have the satisfaction of full, spiritual devotion to necromancy—it is his appetite that governed him, he claims, nothing else.

Finally Faustus calls out to God, in direct defiance of his contract: "Ah Christ, my Saviour, / Seek to save distressed Faustus' soul!" (2.3.83–84).

Having questioned faith but yearning for God, Faustus here proves a more complex and sympathetic character than the static scholars who fail him. Here he is not merely wavering—he wavers toward the divine, and in doing so admits the challenge of true faith. His prick of conscience, like the potential intervention of God in the form of the Good Angel or Old Man, teases the audience with hope for Faustus's salvation. Indeed, for a Christian audience Faustus's wavering toward repentance, while unrealized, is admirable, even heroic. The audience's strong desire for Faustus's conversion is modeled both by characters internal to the play and by the Chorus. Scholars cry, "God forbid!" (5.2.35) on learning of the contract, asking, "O what shall we do to save Faustus?" (5.2.46) and lamenting that the doctor had not turned to them earlier: "Why did not Faustus tell us of this before, that divines might have prayed for thee?" (5.2.40–41). The Good Angel, the Old Man, and the Scholars unite in attempting to sway Faustus back to salvation and devotion to God. They counsel Faustus, "Call on God" (5.2.26) even as the scholar understands it is too late.

Such wavering might demonstrate Faustus's residual faith. Indeed, his necromantic addiction will always, one might argue, be compromised by his awareness—from his studies of theology and his emersion in Christian Wittenberg—of God's divinity. Yet, at least to the extent Marlowe engages with Calvin's theology, such mixing and mingling of Faustus's devotions is as much troubling as hopeful. "No man shal ever go forward constantly in this office," Calvin writes, "save he, in whose heart the love of Christ shal so reigne, that forgetting himself, and addicting himself wholy unto him, he may ouercome al impediments."[82] Calvin's emphasis on exclusivity—the believer is constant, overcome, and subjected, while his love of Christ is entire, whole, and unfailing—precludes wavering. The faithful might be tempted, certainly: "The faythfull them selves are never so wholly addicted to obey God, but that they are ofte withdrawn with sinfull lustes of the flesh."[83] But Calvin clarifies that "ofte withdrawn" signifies not recantation but instead mere temptation, as the faithful remain steady in their dedicated service to God. For Calvin, all humans struggle with addiction: "By reason of the grossenes of nature, we are always addicted unto earthly thinges." Nevertheless divine intervention might "correct that disease which is ingendered in us," turning earthly into godly addiction.[84] Marlowe, by contrast, depicts not conversion from one addiction to another but the incompleteness of attachment itself, whether to necromancy or to God.

If Calvin's writings affirm the power of addiction to overcome the believer entirely, Marlowe instead stages—in his gnarled, questioning universe—a believer with an incomplete addiction. The play's central conflict thus concerns Faustus's attempt but ultimate inability to addict himself to supernatural forces. As he claims, "I do repent, and yet I do despair" (5.1.63). For even as Marlowe depicts the potential heroism of striving toward Christian conversion, he equally challenges it, by making repentance on Faustus's part a form of spiritual and legal betrayal. For Faustus to reject the very path he surrenders to, by taking alternate advice and rejecting magic when its outcomes are insecure, would be to signal his infidelity to faith more generally, be it to the magic he embraces or to the God he does not. Tragically, then, even as Faustus's wavering might be read as a sign of his potential for salvation, it nevertheless betrays his failed devotion not just to Mephastophilis and Lucifer but to anything: God, necromancy, friendship, or study of any kind. Staging the gap between the desire for addiction and its realization, the play illuminates how a character allegedly predestined for hell, overcome by desire for magic, and contractually bound to necromantic masters still cannot achieve addiction.

Yet in staging Faustus's failure, Marlowe depicts not the depressing or powerless spectacle of the damned but the monumental difficulties of the addiction Calvin trumpets. Addiction, it turns out, is hard. If, as Rasmussen writes, "the central problem with most orthodox interpretations of *Doctor Faustus* is that they often verge on lack of sympathy, even open hostility," viewing Faustus as a failed addict instead illuminates his wavering not as a sign of weakness but as indicative of the challenge of his task.[85] Calvin sidesteps the effort necessary to achieve total surrender. Is addiction to faith really as simple as he makes it sound? Indeed, is addiction to sin that easy? Even as Calvin notes the ways in which the elect might stray from their addiction to God, he also describes addiction as effortless; it is simply a question of which addiction one might follow. Calvin's theory, which is evident in his conversion story and his theory of election, seems to promise that addiction is everywhere—and more potently, that God is everywhere, as seen in all one's addictive predispositions.[86] But, Marlowe reveals, this theory of God's dominant will falls short. If humans are so passive before this all-powerful God, then where is He? Mephastophilis works throughout the play to secure the soul of a character who is all too eager to give it away; God, by contrast, may or may not speak through the conscience, the Good Angel, or the Old Man.

It is perhaps perverse, then, that despite his inadequacies as a devotee, and despite his wavering, Faustus nevertheless reaches the promised end. He achieves final integration into Lucifer's kingdom, and he does so not because of his own devotion but because of Mephastophilis's extraordinary efforts. Again and again Lucifer and Mephastophilis counsel Faustus toward right belief, toward the kind of behavior expected of their "faithful." Toward the end of the play, as Faustus tries to repent, Marlowe stages a divine figure literally holding the tongue and hands of the devotee, prohibiting him from straying: Faustus cries, "The devil draws in my tears. . . . O, he stays my tongue! I would lift up my hands, but see, they hold 'em, they hold 'em" (5.2.59–63). This staging of Faustus's damnation, even as it shocks viewers, also arguably appeals to them.[87] The fantasy of God accompanying the faithful through every hour of the day, staying their hands, holding their tongues, and distracting them with spectacles when they think of straying—this God exists only in reverse fantasy, in Marlowe's play in the form of Lucifer. If even Faustus fails as an addict, despite receiving both direct encouragement from Mephastophilis and tangible material benefits from magic, imagine the challenges facing the godly. Tormented by popish regimes, ridiculed for restrained living, besieged by existential melancholy, and plagued by mortal questions, the godly must endure worldly troubles without a divine Mephastophilis by their side.

Conclusion

Marlowe's play stages a supernatural universe in which even the unfaithful, weak, and wavering subject might meet his desired end. Understanding how Faustus's addiction falls short illuminates the treacherous illusion of free will in the play. An attempt to exercise free will in defiance of his contract—indeed, the need to bind himself in a contract in the first place—reveals Faustus's failure to lose himself in his devotional pursuit. The resulting opposition between free will (as it might allow him to turn from necromancy) and devotion (as it might demonstrate the fidelity of his commitments) is thus a Catch-22. Even as the evocation of free will might seem to dramatize Faustus's potential to turn from sin, it also—to the degree that he successfully turns—demonstrates his propensity to infidelity and inconstancy, regardless of the devotional field. It is Faustus's problematic inconstancy that signals his fall, as much as his failed exercise of what one might or might not take to be free will. Indeed, one might argue that Faustus should express even more commitment to Mephastophilis than he does, for only through this full exercise of addiction might he reveal his predisposition for true faith.

Rather than viewing the play as hinging on the tension between faith and free will—a tension that casts Faustus as *either* predetermined in his damnation *or* capable of saving himself—the study of addiction in Faustus illuminates instead the drama of his attempted devotion and his failed surrender. His desire to release his will to Mephastophilis indicates a predisposition to precisely the kind of radical faith required of the righteous believer; but his failure to achieve the form of commitment he trumpets indicates his fall. Viewed from this vantage point, the real question in the play is not whether Faustus has free will, but rather *why* Faustus has such a hard time committing. The answer suggested above is that devotion does not come easily. Even as the play illuminates the horrors of following the wrong path, it even more potently stages the challenge of, and fortitude necessary to, surrender to an addiction. Individual desires, combined with the external promptings of community, culture, and law, might still prove inadequate to the task. Faustus's wavering exposes his incapacity for addiction, evident in his all-too-human propensity for wandering. After so many centuries, what remains admirable about Faustus is precisely his repeated attempts to give himself away to his pursuits in the face of his own fear and hesitation. This sort of addiction is clearly dangerous, but it is also extraordinary and compelling.

Chapter 2

Addicted Love in *Twelfth Night*

Love is that of which we are not masters.

—Jean-Luc Nancy

Chapter 1 analyzed addiction to God and to study in Calvin and Marlowe. This chapter continues to explore the challenge of addiction by turning to another form of devotion: secular love. The drama of addiction in *Doctor Faustus*—moving through incantations, willful service, and contractual donation of the self—finds surprising parallels in Shakespeare's *Twelfth Night,* a play in which frustrated love yields to service and devotion.[1] Shakespeare offers a broader range of potential addicts than Marlowe, as multiple characters strive to give themselves over to the spirit, in this case of love or drink. Nevertheless, the challenge of relinquishing control proves as challenging in comic Illyria as in tragic Wittenberg.

Twelfth Night stages, in order to embrace, the loyalty and fidelity of firm addictions: the devoted lover is the play's most powerful force, and it demands relinquishing sovereignty of oneself to risk loving another. Fostering this addiction—this willingness to forego self-rule in favor of a stronger force or attachment—is an achievement. It is an attachment requiring both commitment and devotion expressed not temporarily, but over and through time. Addiction as a mode of loving shifts attention from a momentary experience—I fell in love, I feel love—to an extended connection. As David Schalkwyk argues, "Love is *not* an emotion, even though it does involve emotions. Love is a form of behavior or disposition over time; it involves . . . 'commitment and attachment.' But such dispositions are not given; they are navigated, negotiated, even discovered in the course of what we think of as their 'expression.' "[2] Thus love is not a bodily condition, such as a humor; it is not a complexion but an inclination that has turned to what the early moderns would deem an addiction. *Twelfth Night,* as Schalkwyk goes on to write, "embodies love through dedicated behavior and action, rather than the causal interiority of bodily heat or humor."[3] This form of loving is a sustained inclination that transforms: through addicted loving, characters go through the process of becoming themselves, offering a range of loving expressions that are at once, as N. R. Helms argues, surprising and what he calls "expectable." In *Twelfth Night* a character can, he writes, "change before our eyes, while remaining the same character."[4]

Such addicted loving stands in opposition to another practice more obviously associated with addiction for modern audiences: drunkenness. Sir Toby and Sir Andrew prove incapable of surrendering their own material desires—for money, status, and pleasure—even as they surrender consciousness and physical capability when drinking. Theirs is not addiction, at least not in early modern terms, but instead appetite, a craving at odds with the vulnerability and release of devotion. *Twelfth Night* interrogates and ultimately exposes the limitations of what we might call the humoral or appetitive inclinations, including those practiced by Toby, Andrew, and Malvolio. The humoral predispositions of these characters amplify their preexisting states, so that the more they indulge their inclinations, the more they become the bodies and affects that define them. Indeed, the humoral fixity of these characters is such that, as Jason Scott-Warren argues, they appear animalistic rather than entirely human.[5] Toby and Andrew are most themselves when drunk. Their festive antics reinforce familiar ideological fault lines and stock characterizations. These characters *become* themselves, to follow the analysis of Helms, only in the sense that they perpetually reinforce or agree with their own past actions, in contrast to the play's addicted characters, who experience transformation, *becoming* a new iteration of themselves.[6]

Twelfth Night thus offers, for this project, a productive study of the distinction between addiction and habitual drunkenness: as the play reveals, addiction celebrates—or in Roman terms, requires—the release rather than the exercise of self. Olivia, Orsino, and Viola experience an initial fixity of character, as the addicted melancholic, the Petrarchan lover, and the shipwrecked sister. But they come to release themselves into love and experience what it means to be overcome by devotion at the expense of one's desires, former attachments, and identity itself.

The Melancholy Addict and the Comedy of Humors

The countess Olivia suffers from an addiction to melancholy. Conjuring an image of the ill-suited Malvolio before his mistress, Maria suggests that his smiles will be "unsuitable to her disposition, being addicted to a melancholy as she is" (2.5.195–96).[7] And Olivia is not the play's only melancholic. Orsino most obviously exhibits the features of melancholic love in his preoccupation with Olivia. As Feste tells the duke, "Now the melancholic god protect thee" (2.4.73). Viola, too, begins the play in a state of melancholic grief and describes her condition of "green and yellow melancholy" (2.4.113) to Orsino. Indeed, as Keir Elam puts it in his Arden edition of the play, "the comedy offers a veritable anatomy of the most fashionable of humours, melancholy."[8] In its fascination with the melancholy body, the play joins in the diagnostic interest and philosophical speculation surrounding this humoral state, what Drew Daniel calls the melancholy assemblage. This social network includes, he writes, those "who spectate and speculate upon the interiority of an allegedly melancholic body."[9] *Twelfth Night* participates in and invites such spectatorship and speculation on melancholic states. It does so not only because it offers anatomy of the melancholy humor, but also, more surprisingly, because it comes to celebrate change and alteration away from this state, precisely what a melancholy addiction seems to foreclose. After all, how can one prove both "addicted to melancholy" and open to change?

We might begin to investigate this question by turning to Shakespeare's contemporaries, many of whom yoke—as Maria does—melancholy and addiction. The surgeon John Banister prescribes "a decoction for such as are weake and addicted to melancholie," while Thomas Dekker cautions against those who are "hard fauour'd, dogged, addicted to melancholly, to diseases, to hate mankind."[10] For Banister and Dekker, melancholy addiction signals a disease, a permanent state that might only be overcome with management or medical intervention. Yet in *Amadis de Gaule*, we learn of a character who exercises a degree of choice: she suffers from "the extreame melancholie, whereto (over-much) shee addicteth her selfe." As a result of "being so continually sad," she threatens to "fall into some dangerous disease."[11] While each of these examples yokes melancholy, addiction and disease, the agency behind addiction is tangled, since the melancholic "addicteth herself," and proves willfully "addicted to melancholy, to diseases," even as an addict might also "fall" into such illness and prove "weake."

Part of the challenge in understanding addiction to melancholy comes in determining its condition as a fixed or chosen state. For some writers, both humors and addictions are predetermined. When Gervase Markham speculates on humoral predispositions, for example, he writes of "the predominance or regencie of that Element" in which the body "dooth moste entyrelye participate, so for the moste parte are his humours, addictions, and inclinations; for if he have most of the earth, then is hee melancholie, dull, cowardlye, and subject to much faintnesse."[12] Markham offers a medical narration of sorts, accounting for the fixity of melancholic addiction: a creature has a "complexion"—based in a natural element—that is evident on his body in the form of "colours" that betray his "nature." Such a "humour" or complexion is described as an addiction. Melancholics, he goes onto explain, are "kyteglew'd, blacke, both sortes of dunnes, Iron-gray, or pyed with anie of these colours," for they are connected to "the earth." Thus despite the word "inclination" seeming to indicate preference rather than determination, Markham's description suggests an inevitability to the humor or addiction—it is a condition determined by one's physical state, it influences our tastes and character, and it can be regulated but not overcome.

Michel de Montaigne also views a humoral state such as melancholy as a fixed addiction, and directly cautions readers against its entrapping power. Montaigne writes: "We must not cleave so fast unto our humours and dispositions. . . . It is not to bee the friend (lesse the master) but the slave of ones selfe to follow uncessantly, and bee so addicted to his inclinations, as hee cannot stray from them, nor wrest them."[13] Montaigne, like Markham, links addiction to inclination, humour, and disposition, arguing even more vehemently that such fixity is a form of slavery. If we are addicted to our own inclinations, we "cleave" to our humor, we become a "slave" and "follow uncessantly," we "cannot stray from them, nor wrest them." This humoral addiction is limiting; it designates a particular character, one that is fixed and rigid. It is worth noting Montaigne's language of bondage: to be "addicted to . . . inclinations" is to "be tied," to be "the slave of oneself." With this enslavement comes compulsion, "necessity," and "incessant" bondage, a form of the ethico-spiritual slavery that attests to the "weakness or self-indulgence of the paradigmatically 'free' agent," to invoke Nyquist's formulation rehearsed in the introduction.[14]

In the above examples, melancholy threatens to overwhelm the individual, who becomes defined through a humoral disease. Such is the case with Robert Burton, who famously takes a thousand pages to catalogue melancholy's contours. A predisposition becomes enslaving without proper management. This view of humoral addiction as a form of tyranny recalls the Roman understanding of addiction, noted in the Preface. The contracted "addict," bound to service, is enslaved and compelled. It might seem, then, that addiction and melancholy function as synonyms, both accounting for a tyrannical, entrapped state: addiction is the state of being enslaved, and melancholy is the enslaving condition or power. Montaigne's French, and Florio's translation of it, reinforces this link. When Montaigne cautions against clinging to humors, what Florio translates as "addicted to his inclinations" appears in the original as "être prisonnier de ses propres inclinations."[15] To be imprisoned is to be addicted. Taken this way, Maria's use of the term "addicted" helps illuminate Olivia's condition: she is in a state of incarceration, overcome by a humor. The yoking of addiction to disease and melancholy in the citations above further bolsters this reading. Addiction, in anticipation of its modern applications, appears as pathological or enslaving compulsion.

Yet Florio's use of "addicted" in his own dictionary suggests a slightly different signification to the term, and challenges the linkage of addiction and humor. In his *A Worlde of Wordes: or, Most copious, and exact Dictionarie in Italian and English*, Florio uses the word "addicted" not as a synonym for imprisoned but instead in connection with words like dedication and affection. He defines *Dédito* as "given, addicted, dedicated, enclined," while *Affettionato* appears as "affected, affectionated, addicted."[16] To be addicted, here, seems to be attached or dedicated. Similarly he designates *Dedicare* as "to dedicate, to consecrate, to addict" and *Dicare* as "to vowe, to dedicate, to addict, to promise."[17] Notably, the word "addicted" never designates a state of confinement. It does, however, help define the more dramatic terms *Revólto* and *Volgiuto, volto*, appearing in their definitions as "turned, overturned, tossed, tumbled, transformed . . . addicted, converted."[18]

To be addicted, at least in Florio's lexigraphical glosses, is to experience a converting, transforming devotion. And this resonance of "addict" with devotion and dedication can be further elucidated through other early modern dual-language dictionaries, where addiction suggests less compulsion than devotion, a form of fixity and determination but one with a positive valiance. As in the case of Calvin's writings, explored in Chapter 1, "addiction" often appears as a translation for *adonner*. Randle Cotgrave in his *Dictionary of the French and English Tongues* uses addict to define *s'adonner*, writing "*s'Addonner à*. To give, bend, addict, affect, apply, devote, incline, render, yeeld himselfe unto."[19] Guy Miège, in his *New Dictionary French and English, with another*

English and French (1677) defines *s'adonner* similarly: "to give (addict, or apply) himself to something," as in "*s'addonner à la virtue,* to give himself to virtue."[20] Drawing on these framings of addiction not as compulsion but as devotion helps illuminate the danger, as well as the promise, of melancholy addiction. The individual suffers from the humoral disease of melancholy. But the propensity for addiction, namely devotional attachment, presages the ability to give or apply oneself fully. Addiction at once signals agency and excess, giving oneself over to a condition voluntarily and entirely. Florio's translation of Montaigne's imprisonment as "addicted" arguably reveals this, shifting from a permanent state to one that is chosen through devotion.

Turning to English language dictionaries further exposes the common understanding of addiction as, initially, a form of choice. Indeed, it is English language lexicographers who illuminate precisely the devotional and transformative potential of the addict in terms that resonate with Shakespeare's stagings. In dictionaries, "addiction" is defined largely as a laudable preoccupation; far from signaling a form of slavery or tyranny, addiction signals the deepest form of chosen attachment. In Thomas Cooper's thesaurus, he defines "addiction" (namely "Addico, addîcis, pen. prod. addixi, addictum, addícere") as a form of giving over or bequeathing: it is "to say: to avow: to deliver: to sell" or "to alienate from him selfe or an other, and permit, graunt, and appoint the same to some other person."[21] In his examples of such addiction, Cooper turns to Latin invocations of the term from Cicero, Quintilian, and Caesar: "*Addicere se alicui homini, siue cuipiam rei.* Cicer. To addict or give him selfe: to bequeath. *Addicere se sectæ alicuius.* Quintil. To addict or give himselfe to ones sect or opinion. *Seruituti se addicere.* Cæsar. To bequeath him selfe." The Latin *addīcere* becomes "approve," "allow," "give," or "bequeath" in English. Furthermore, the English and French terms "devote" in turn draw on the English word "addict." Cooper defines "to devote Deuoueo, déuoues, deuôtum. pe. pro. Denouêre)" as "to vowe: addict or give: solemnly to promise: to bequeath."[22]

Each of these definitions grants agency to the addict. Addiction is an active process of giving oneself over, or delivering oneself, precisely as Calvin counsels in the writings examined in Chapter 1. The addict consents to be overtaken. These definitions also frame such consent in terms of promising, bequeathing, allowing, and giving—transactions that are relational and potentially generous. Cooper describes, in accordance with the term's original usage in Roman law, addiction to service: "*Addicere quempiam pro debito dicitur Prætor.* Cic. To deliver a debtour to his creditors to be vsed at their pleasure. *Addicere in seruitutem.* Liu. To judge one to be bonde: to deliver as a bondman."[23] It is only in this last definition that any sense of bondage or compulsion appears, although even such apparently servile enslavement can signify, as in the case of Faustus, a desire for metaphysical merger. More frequently than bondage, addiction resonates with notions of faithful devotion, a kind of bequeathing that evokes friendship and marriage. Thus John Baret defines "addicte" as a form of devoted giving: "to addicte & geue him selfe to ones friend ship for ever."[24] Thomas Thomas, too, defines devotion as a form of addiction or giving: "*Dēvŏvĕo, es, ōvi, ōtum, ére.* To vow, to addict or give, solemnlie to promise."[25] Most specifically, this addiction evokes notions of love. Cotgrave employs the term "addict" to help define giving, attaching, and affecting, both in the example of *s'Addonner à* rehearsed above, and with *Affectionner*, which signifies "to affectionate, beget a liking, breed an affection; excite, incite, or animate, unto. *s'Affectionner à.* To affect, or love; to addict, or devote himselfe; to give his mind, unto."[26]

If Cooper and Cotgrave define addiction in terms of bequeathing and giving, Thomas amplifies it with a sense of delivering over or confiscating; in other words, for Thomas, addiction can designate both voluntary and compelled forms of service:

> *Addīco*: To deliver up unto him that offereth moste: to put to saile: to confiscate: to deliver some worke upon a price: to addict, bequeath or give himselfe to something: to saie: to avow: to alienate from himselfe to another, and permit, graunt, & apponit the same to another person: to condemne: to approoue or alow a thing to be done, to deliver, depute or destinate to; to judge, to constraine, to pronounce and declare.[27]

Thomas's definition at once presents addiction as a kind of constraint and as a gift. This complexity of something that is and is not voluntary, something initially free but ultimately constraining, appears to Montaigne as a condition to be avoided, a form of slavery. But viewed through the vantage point of devotion to God or to a beloved, this language of devotional constraint encapsulates the rights and responsibilities, the volition and compulsion, at stake in deep, extended intimacies.

Recognizing how the term "addicted" might not *amplify* melancholy (Olivia is excessively attached to it) but instead *temper* it (she has chosen to be attached) offers a new perspective on Shakespeare's famously multi-perspectival play. In addition to viewing the play as an anatomy of melancholy, we might also see how it toys with or stages the issue of addiction as willful release. The play's title, in pairing the Epiphany of *Twelfth Night* with the "will" of *Or What You Will*, embeds this paradox, for it insists upon the interplay of choice and release, willfulness and devotion. This interplay structures the love relations of the play to a degree that "what you will" stands both in opposition to *Twelfth Night*'s celebration of Epiphany and in connection to it. "The very word *epiphany*," as Bruce R. Smith writes, "means an appearance or a revelation and suggests that on that special day celebrants could expect something visionary, a miracle, a manifestation of divinity."[28] Such emphasis on the divine and magical might conventionally oppose the will, but Shakespeare announces their intimate connection in his play's very title and, in doing so, anticipates the link of will and release at stake in the addictions he stages.

In the lexicons above, addiction appears as a form of designating, giving, bequeathing, serving, or devoting akin to marriage or religious faith. Thomas indicates the ways in which such devotion might not be entirely voluntary: one might be given over to service. But most frequently addiction appears as a commitment to something; a dedication to an activity, person, or relationship; a devotion "to" or "unto." Thus addiction is and is not an act of will; it represents, as the lexicographical definitions suggest, a radical form of giving oneself, what Tim Dean, following Levinas, calls "unlimited intimacy," and what Leo Bersani calls "self-shattering."[29] As Dean writes of such unlimited intimacy, "Not only the envelope of selfhood but the very distinction between self and other is undone."[30] For even as lexicographers yoke giving and bequeathing within addiction to devotion, their definitions also hint, as Thomas reveals, at a condition of release and donation of the self. One is delivered over to someone or something, one is constrained even as one also consents to this gift giving.[31]

Yet how might one consent when one no longer signifies or exists? How do consensual relations emerge out of forms of vulnerability and servile devotion?[32] How might sharing intimacy with a stranger represent the deepest, most radical, and spiritual form of loving available? These questions resonate strongly with *Twelfth Night*, a play in which Olivia excitedly marries the wrong man, Sebastian weds a woman he does not know, Orsino agrees to marry a woman he's never seen, and Viola finds love through a form of anonymous service. The early modern concept of addiction—which equally insists on the link of devotion, service, gift-giving, and love—helps account for the play's famously serendipitous depiction of loving: love in *Twelfth Night* offers a radical challenge to identity and character, dissolving the boundaries of self in relation to another. Addiction—to love or melancholy—takes characters out of themselves; addiction challenges the self, if not in the specifically physical terms signified by Bersani's "self-shattering," then in equally conversionary terms, as love transforms identity and character. Addiction, the play reveals, is not a governing humor requiring, at best, skillful management.[33] Instead, it is an ability to foster deep attachment, presaging exactly the propensity to love needed in Illyria.

Devoted Attachment and Eager Appetite

Twelfth Night begins with gluttony. Orsino is overcome by love, and he seeks to surfeit on it:

> If music be the food of love, play on,
> Give me excess of it, that, surfeiting
> The appetite may sicken, and so die.
> (1.1.1–3)

The language of binging and purging here evokes the compulsive ingestion of material substances, from food to liquor to drugs. The desire to indulge so heavily that cravings will finally end might even be called an addictive fantasy. But appetite is precisely not addiction. Compulsive ingestion, the play reveals, is merely habitual and customary consumption, rejected even as it is embraced: "Enough, no more, / 'Tis not so sweet now as it was before" (1.1.7–8). Orsino glosses his own rejection of love, his sense of "enough, no more," by turning to the image of the sea. Yet his image of the "spirit of love" as the sea is not, despite his rhetorical attempts, parallel to his own process of loving. For if he hopes to purge himself of his amorous appetite, his watery image attests to love's limitless capacity:

> O spirit of love, how quick and fresh art thou
> That, notwithstanding thy capacity
> Receiveth as the sea, naught enters there
> Of what validity and pitch soe'er
> But falls into abatement and low price
> Even in a minute.
> (1.1.9–14)

Love, he postulates, has no limits. It overpowers any object or being that enters into its domain. Orsino's images are contradictory: he hopes to purge himself of love through binging, yet he also believes that love overpowers all, being a sea of infinite capacity. Perhaps most obviously this image of the devouring sea resonates with Petrarchan rhetoric of the initially idealized and subsequently abject love. As the idealized beloved falls to "low price," so the "quick and fresh" spirit of love survives as a force more powerful than any individuated object, as a poetic expression—in the Petrarchan sequence itself—more lasting than the beloved's body.[34] What simultaneously distinguishes and reconciles these images, even if Orsino himself does not seem to recognize it, are the opposite approaches to the lover's agency. Orsino initially attempts to overpower love, casting himself as an appetitive lover who can control how much he ingests: "give me" "enough." He manages his desire through imperatives. By contrast, the sea of love overpowers the object (both lover and beloved), redefining them entirely.

The challenge for Orsino at this early point in the play comes in sorting through his opposing yet related views of love. Although he celebrates love's "quick and fresh" *spirit*—namely, love's metaphysical capacity to transform all devotees by drowning and reforming them—he does not acknowledge his own potential transformation, his own "fall" into the spirit. Instead he attempts to govern the power of love through forceful wooing. His arguably uncommitted, gluttonous, and fickle feeling rejects rather than embraces love's vulnerability. Furthermore, when Orsino rehearses his attraction to Olivia, he betrays his fantasy of domination, not release:

> O, she that hath a heart of that fine frame
> To pay this debt of love but to a brother,
> How will she love when the rich golden shaft
> Hath killed the flock of all affections else
> That live in her – when liver, brain and heart,
> These sovereign thrones, are all supplied, and filled
> Her sweet perfections with one self king!
> (1.1.32–38)

Orsino's language of tyranny—of usurping "sovereign thrones," of killing "affections," and of triumphing as "one self king"—exposes his desire for conquest more than attachment. He imagines, in other words, love as an even more profound exercise in self-sovereignty, an amplification of the power he already feels.

As a further sign of such sovereign love, Orsino advises Viola in courting Olivia: "Be clamorous and leap all civil bounds / Rather than make unprofited return" (1.4.21–22). The notion of love as "profit" strains the play's more metaphysical or devotional sense of love as surrender. Finally he problematically announces his superiority to his own beloved when it comes to capacity. Despite having praised Olivia's love for her brother, he proceeds to dismiss women as shallow lovers:

> Alas, their love may be called appetite,
> No motion of the liver but the palate,
> That suffer surfeit, cloyment and revolt.
> But mine is all as hungry as the sea,
> And can digest as much.
> (2.4.97–101)

Orsino defines himself as the devourer of love, as if such an image might speak to the strength of his affections. But he misunderstands what the play makes obvious through Olivia: love comes with acceptance and surrender, not conquest. In trumpeting his voracious appetite, and in attempting to maintain self-sovereignty in relation to love, Orsino's form of appetitive love lies in material, not devotional, presence. If the sea transforms ("naught enters there / . . . but falls"), his own love continues to consume ("hungry as the sea, / And can digest as much"). Thus perhaps unsurprisingly Orsino, when spurned, shifts from devotion to accusation, condemning his beloved for leading him on with false faith, even as Olivia reassures Cesario "he might have took his answer long ago" (1.5.255). Orsino proves unable to respond to what his beloved announced—again and again—because he remains convinced of his own sovereignty over the process of love, viewing love as a conquest, not a relation.

In a play that famously twins its characters through anagrammatic names and physical doubling, Orsino's mode of loving, in its insistence on sovereignty and profit, finds its pair in Malvolio's. Both lovers present themselves as tyrants, enjoying the power that such love brings in amplifying their own sovereign reign. Malvolio's imaginative discourse on married life exemplifies this: "To be count Malvolio" (2.5.32), he exclaims. In his fantasy, after three months of marriage, he will find himself "sitting in my state" (2.5.42), "calling my officers about me" (2.5.44); he will "have the humour of state" (2.5.49) and "know [his] place" (2.5.50–51). As critics have long noted, Malvolio's desire for marriage is a desire for place and status. His repetition of "state" betrays this interest in position. Furthermore, his fantasy has a notable proliferation of possessives: "my state," "my place," "my branched velvet gown" (2.5.44–45), "my people" (2.5.55), "my watch" (2.5.57). With marriage amplifying—or indeed securing—his power, he will be able to domineer household dwellers, those who, "with an obedient start" (2.5.55), fetch Toby, to whom Malvolio famously pronounces, in his fantasy, "you must amend your drunkenness" (2.5.71). Like Orsino's appetitive love, promising to devour the beloved as a sign of its power, Malvolio's possessive love draws attention to its own self-interest.

Yet—and this is crucial—unlike Malvolio, Orsino also experiences his own vulnerability and self-emollition, as evident in his recognition of love's overwhelming power. In noting he is "hungry as the sea," Orsino's phrasing suggests his control of, or at least identification with, his appetite. But such voracious inclinations are precisely not in the individual's control. His cravings arguably control him and thus hint at his own addictive propensities. Such agency and vulnerability appears in his image of himself as Actaeon, who was chased by his own hounds. He claims that on seeing Olivia, "That instant was I turned into a hart, / And my desires, like fell and cruel hounds, / E'er since pursue me" (1.1.20–22). With Orsino pursued by his own desires, this image is a closed circuit, at once frustrating the process of love as a giving over of oneself yet leaving open the possibility of his overthrow, his vulnerability as a "hart." This paradox of Orsino's self-absorbed vulnerability is not Shakespeare's invention—beyond its Ovidian source, it clearly resonates with the model of the Petrarchan lover, who longs for a beloved while writing of himself. This Petrarchan mode of loving foregrounds, as Giuseppe Mazzotta has established, "an autonomous, isolated subject who reflects on his memories, impulses, and desires." Through his meditations on his love for Laura, the speaker finds "his pure self."[35] Yet as Ross Knecht has more recently suggested, this emphasis on poetic self-consciousness in the Petrarchan tradition overlooks the inherently passive nature of passion itself: " 'passion' is derived from the Greek verb *paschein,* meaning to suffer or undergo. . . . It characterizes the states of mutable beings, which either act or are acted upon. To be 'passionate,' in this broadest sense of the term, was simply to be subject to some foreign influence."[36] As a passionate subject, Knecht argues, Petrarch's speaker is not a self-absorbed subject insistent on his own agency, but instead a lover at once passive and acted upon. Indeed, as Tim Reiss puts it, the Petrarchan mode of loving depends on a form of anti-personhood, with no recognizable agent. The "subject," he writes, "in any modern or even slightly earlier western sense, one does not find in Petrarch."[37] Furthermore, the English Petrarchan tradition, developed by a poet like Sir Philip Sidney, offers "a fine summary of the self-abnegation already inherent in the Petrarchan lyric."[38] Thus Sidney, in "Astrophil and Stella," develops a loving speaker who proves most himself when selfless: "that wholly hers, all selfness he forbears; / Thence his desires he learns, his life's course thence."[39]

If Sidney's speaker offers an extreme model of such selflessness, Orsino retains a preoccupation with his own feelings and desires. Yet Orsino's expression of Petrarchism, in its gesture toward selflessness and the melancholy attendant on unrequited love, exposes his aptitude for addictive transformation. His mode of loving at least entertains a giving of the self. While the textual form of loving in Petrarchism—replete with the hackneyed blazons and struggles for sovereignty—frustrates requited love, it is precisely through the deployment of Petrarchan tropes, whether sincerely in the case of Orsino or mockingly with Olivia, that the self-sacrificing model of loving is first articulated in Illyria. Petrarchan stasis proves surprisingly less opposed than connected to the model of addicted loving offered by Viola.

Thus, in telling Viola of his own true love for Olivia, Orsino notes the constancy of his devotion to a beloved, in contrast to his images of consumption:

> If ever thou shalt love,
> In the sweet pangs of it remember me;
> For such as I am all true lovers are,
> Unstaid and skittish in all motions else
> Save in the constant image of the creature
> That is beloved. (2.4.15–20)

Orsino here describes precisely the addict in love. The addict is distinguished by constancy in loving: in contrast to the flux and fickleness of other attachments ("unsaid and skittish in all motions else"), the attachment to the beloved remains. The irony of this speech lies in the gap between Orsino's articulation of this theory of addicted love and the practice he pursues more vigorously: his musing about love betrays less the "constant image of the creature / That is beloved" than his own fantasies in loving. His experiences of appetite and drowning, and his desire for conquest and sovereignty, have been described in detail; Olivia has not.

A paradox distinguishes Orsino's mode of loving, then. He simultaneously gestures toward his capacity as a "true lover" and repeatedly fails to achieve such a love. He understands love's overwhelming power—he just mistakenly imagines his ability to control it, even as he admits at other moments his vulnerability to such desires and appetites. We might call his paradoxical challenge typically Petrarchan, in the rich tradition outlined above: one rife with the possibilities of devotion yet riven by the tension of tyranny. Orsino's lovesickness is filled with Petrarchan language—being turned into a "hart" pursued by his own desires (1.1.20), overcome with the "spirit of love" (1.1.9)—and he experiences the classic dilemma of the Petrarchan lover in its paradox of enslavement (of the lover) and liberty (poetic inspiration resulting from his willing addiction to the beloved).[40] Orsino's contradictory viewpoints on love—as a process that he controls, but one that overwhelms—expose the challenge of loving, his inclination toward it, and his failure to release himself just yet. He must experience this challenge directly, not only in the form of rejection, but also more potently in the form of acceptance that models for him the love he might learn to embrace: Orsino's experience of, or capacity for, love shifts as a result of his relationship with Cesario to such a degree that he can recognize his own loving potential. "If the play shows us anything," as Schalkwyk writes, "it is the qualitative difference between Orsino's desire for Olivia and his love for Cesario."[41]

Call upon My Soul

If Orsino imagines *receiving* devoted love, the play offers, in Viola, a character who practices *giving* this love. Viola is the play's most energetic example of love as devoted service. Rather than insisting on self-sovereignty in the face of adversity—as Orsino does—she surrenders herself entirely. Transforming from a woman to man, noble to servant, she proceeds even to woo another on behalf of her own beloved. Yet this willingness to serve, indeed to undo herself, only makes her more compelling, more capable of transforming the stagnant Illyria she enters. The decision to relinquish consent, the willingness to give up free will, is what it means to be a devoted addict.[42] The play gives us this form of love in Viola, who enters into service with Orsino, performing a job that speaks entirely against her own interests. Yet she plays this role so authentically that she manages not only slowly to transform Orsino, but Olivia as well: the addict of melancholy becomes an addict of love. Viola's wooing of Olivia succeeds precisely because she frames it as devotion, which Olivia herself already understands: devotion for another, the willingness to be overtaken. This is not the ego-driven compulsion of the insistent wooer, aware of his own appetite. If Viola promises a message that is "to [Olivia's] ears, / divinity," Olivia, enticed, claims, "Give us the place alone, we will hear this divinity" (1.5.211–12).

Viola's message proves to be one of addicted love, perfectly expressed in her image of the willow cabin, which is an airy, nearly immaterial testament to devotion:

> Make me a willow cabin at your gate
> And call upon my soul within the house;
> Write loyal cantons of contemned love
> And sing them loud even in the dead of night;
> Hallow your name to the reverberate hills
> And make the babbling gossip of the air
> Cry out 'Olivia!' O, you should not rest
> Between the elements of air and earth,
> But you should pity me.
> (1.5.260–68)

In contrast to Orsino's view of love as the hungry, devouring sea, Viola sketches an alternative theory, one based on love's spirit. The speech is marked by airiness and ephemerality—love is the "soul," it is "loyal cantons," "reverberate hills," and "babbling gossip." It is singing and air, it exists between heaven and earth. This nonmaterial love gently but firmly compels: "You should not rest," "you should pity me." The landscape is transformed by an unseen but heard and felt force of sound, penetrating without devouring, transforming without destroying.

True service is a form of generosity that can potentially shift even rigid, tyrannical power. Yet such behavior also unnerves, as William Gouge reveals in his critique of service that nonetheless speaks to addiction's laudable potential: "To be a servant in that place is not simply to be in subjection under another, and to doe service unto him, but to be obsequious to a man, so addicted to please him, and so subject to his will, as to doe whatsoever he will have done."[43] This alignment of slavery and love uncomfortably yokes debasement with erotic ties, as Schalkwyk illuminates: "Subservience . . . encompasses much more than simple social powerlessness: it empties out the very subjectivity of the subordinate, transforming him into the hollow, thoughtless instrument of the addressee's desire."[44] For Lorna Hutson, this willow-cabin speech—and the response to it by Olivia—prompts precisely such concerns about service, in staging a fantasy for the auditors on the advancement possible for the rhetorically gifted servant: "The transgressive 'glimpse' being offered [is] . . . that of the opportunity for social advancement and erotic gratification afforded by education for any servant of ability entrusted with missions of intimate familiarity."[45] Viola's rhetorical success gains credit with Olivia, independent of material proofs. "*Twelfth Night*," she continues, "endorses the notion of rhetorical opportunism, or individual enterprise insofar as it expresses the mastery of fortune and of the occasions of civil life as the metaphorical equivalent of heroic enterprise on the high seas."[46] The fantasy of rewards for good service essentially traps the subject in a convenient fiction: selfless service to tyrants and ruling classes will yield impossible rewards.[47]

Yet it is worth considering the content of Viola's speech in concert with its rhetorical mastery. If the play entwines concerns about social advancement through service in the Viola and Malvolio plots, as Huston argues, in doing so it draws attention to the quality of devotional and transformative love offered by Viola, and in turn by Olivia. Where Malvolio seeks elevation in station, and indeed imagines his love in material terms, Viola and Olivia do not. Malvolio's ambitious attempts at what Hutson calls "individual enterprise" are greeted with ridicule, perhaps not solely for his rhetorical ham-handedness but also for the nature of his self-interested desire. After all, Orsino's more skillful bids, rhetorically poised but equally self-aggrandizing, are dismissed as well. What distinguishes Viola's form of loving—and Olivia's response—is its devotional quality. Viola evacuates herself of any identity, save that of devoted lover. Furthermore, her image of love remains metaphysical and immaterial, even in its insistence. This "divinity" makes no claim on Olivia but patiently waits for the transforming force of pity.[48] Viola offers a form of love parallel to Olivia's addiction to melancholy—a dedicated, committed, unwavering love for a metaphysical being, offered without expectation of reciprocity and effectively emptying the lover of traces of identity itself. When Viola claims she will "call upon my soul," she references Olivia "in the house," devoting herself to another.

What the willow cabin speech stages is less successful rhetorical manipulation than a theory of loving hitherto absent from Illyria. It transforms the landscape by exposing a mode of loving that is other-focused. Viola communicates this form of love not only to Olivia, but also to Orsino, effectively coaching, and indeed transforming, the play's two addicts of melancholy into addicts of love. When she educates Orsino on the capacity of women for dedicated love, she invokes her "sister's" history, saying it is

> A blank, my lord. She never told her love,
> But let concealment like a worm i'th' bud
> Feed on her damask cheek. She pined in thought,
> And with a green and yellow melancholy
> She sat like Patience on a monument,
> Smiling at grief. Was not this love indeed?
> We men may say more, swear more, but indeed
> Our shows are more than will, for still we prove
> Much in our vows, but little in our love.
> (2.4.110–18)

If Olivia's soul becomes Viola's in the willow cabin speech, here a lover's history is erased, it is "a blank," an empty slate about another. The lover surrenders to a force of love itself. This physical description speaks precisely to Orsino's own preoccupations, as he both seeks a beloved on whom he might "feed" in the way this passive sister allows, and insists on his own melancholy patience in waiting for Olivia. Viola's image thus proves doubly appealing: it appeals to aspects of Orsino's compulsive desire, while also resonating with the willow cabin speech in its articulation of selfless love, which is dedicated to another without insistence on return. It is this later aspect of the speech to Orsino that so effectively educates him. As in Chapter 1, in which Faustus's contract signals not his devotion but his propensity for wavering, so here Viola dismisses those lovers' "vows" that are quickly mouthed but prove only "shows." Deeper love, devoted love, is not in the mouth—saying or swearing or showing; instead, it comes with silence, stillness, and surrender. As Maurice Hunt writes, "Selflessly serving her beloved Orsino's passion for Olivia, the disguised Viola makes possible the liberating love of Olivia for Sebastian and eventually of Orsino for herself. A love offering no self-serving advantage becomes the providential instrument for breaking the chains of self-love."[49]

Not Too Fast

Viola's addicted love inspires transformations in Illyria. Orsino responds slowly, almost glacially, to Viola's love. Olivia is moved immediately.[50] Early in her addiction to melancholy, Olivia reveals her own capacity for—and propensity to embrace—love. Yet the phenomenon of her devoted mourning does not prepare the audience, or her, for the dramatic transformation that occurs on meeting Viola. Olivia responds to the willow cabin speech in this way:

> Not too fast, soft, soft—
> Unless the master were the man. How now?
> Even so quickly may one catch the plague?
> Methinks I feel this youth's perfections
> With an invisible and subtle stealth
> To creep in at mine eyes. Well, let it be.
> (1.5.285–90)

Her language betrays her sense of being overcome by an outside force: infected by "plague," she has been violated, diseased, and transformed, descriptions that speak to her own lack of will and agency in choosing this path. Concerned that she might move from a position of mastery to service, from self-sovereignty to bondage, she acknowledges how love threatens to overpower and, indeed, topple her.

Such a challenge to her own sovereignty upends Olivia's initially witty exchange with Viola. On first encountering Orsino's servant, Olivia invokes—even if to mock—Petrarchan modes of loving by offering a blazon of her own features. Presenting herself as a portrait to be admired ("we will draw the curtain and show you the picture," 1.5.226), she then describes herself as an itemized list, in "diverse schedules," "inventoried" (1.5.237). This blazon demonstrates, as she insists, her own will: "every particle and utensil labeled to my will, as, item, two lips, indifferent red; item, two grey eyes, with lids to them; item, one neck, one chin and so forth" (1.5.238–40). Olivia mocks Viola and the Petrarchan tradition through which she is at once praised and wooed. As Viola claims, in typically Petrarchan language, her "beauty" is "truly blent, whose red and white / Nature's own sweet and cunning hand laid on" (1.5.231–32).

In Olivia's highly textual exchange with Viola, her insistence on her own objectification as a portrait or inventoried poem resonates with the mannered position she has adopted as a melancholic mourner. She "hath abjured the company / And sight of men" (1.2.37–38), "she will admit no kind of suit" (1.2.42). As Elam notes of Olivia, her "behavior is a catalogue of the symptoms of female bereavement," revealed in Robert Burton's analysis of the "Symptoms of Maids, Nunnes, and Widows Melancholy" in *Anatomy of Melancholy.*[51] Yet her timely precision—walking once a day veiled and crying for seven years—speaks more to a willed performance than to a natural process. For "seven years' heat" she will "like a cloistress" walk, "veiled" and watering "once a day her chamber round / With eye-offending brine" (1.1.25–29). If melancholy is "caused by an excess of black bile secreted by the liver," nevertheless, as Elam writes, in the play this humor "more often than not turns out to be a behavioural pose."[52] The predetermined quality of Olivia's mourning need not undercut it or compromise its authenticity: rather the adopted pose of her mourning appears less a humoral condition than a willful state of giving herself over to grief. Being "so abandoned to her sorrow" (1.4.19), she has chosen to offer herself entirely. This affect is, as Drew Daniel argues, a mode of connection: "How we understand the term *melancholy*" moves from "connotations of solitude and interior essence in favor of a model based on social extension."[53]

In mocking Petrarchism and embracing melancholy, Olivia signals a kind of poetic self-consciousness, her choice of a devoted pose. But by the end of her encounter with Viola, Olivia has fallen into another form of devotion, ironically the very form of romantic service she jeeringly invoked through Petrarchan rhetoric. She experiences a Petrarchan understanding of love at first sight: her senses are overwhelmed, she is powerless in the face of love. As Knecht writes of *Rime* 94, "Here the cliché of 'love at first sight' is something more: it is an overwhelming visual experience that usurps the offices of the soul."[54] Experiencing such usurpation, Olivia's own reaction to this shocking event is mild and accepting: "Well, let it be." Indeed, despite her description of love as a pestilence, Olivia's affection proves at once secure and transformative, shaking loose the stalemate with which the play begins. By the end of her first encounter with Viola, Olivia proves free of her grief. Of course Olivia's love might be taken as a sign of her fickle or insincere affection for her brother, as she casts aside her former addiction for devotional love instead, a point Elam makes in his analysis of Olivia: "The authenticity of her melancholic disposition is immediately placed in doubt by the rapidity with which she abandons it on meeting Cesario, who in turn alludes to the 'green and yellow melancholy' of his 'sister'."[55]

Viewed from the vantage point of addiction, however, Olivia's initial emotional state anticipates rather than contradicts her embrace of devotional love. She is overcome, overtaken, and addicted as she allows herself to fall entirely into relationship, redefining herself as a result. As she claims,

> Cesario, by the roses of the spring,
> By maidhood, honour, truth and everything,
> I love thee so that maugre all thy pride,
> Nor wit nor reason can my passion hide.
> (3.1.147–50)

Olivia's devotional love is especially notable in that she at once gives herself entirely to Viola, against all reason; and yet she remains chaste, offering up her spirit and heart. As she tells Viola, embarrassed at having revealed her own strong feelings:

> I have said too much unto a heart of stone
> And laid mine honour too unchary on't.
> There's something in me that reproves my fault,
> But such a headstrong potent fault it is
> That it but mocks reproof.
> (3.4.196–200)

While Olivia has been overcome and proves willing to lay out her "honour too unchary," her transgression is notably a verbal one: "I have said too much." The form of her devotion is spiritual, as she strives to soften the "heart of stone," an image familiar from Reformed writings. The unrepentant, unsoftened heart has yet to receive grace.

Olivia seeks to open Cesario's heart through speech, service, and devotion, but notably not—in contrast to Shakespeare's source—through physical union. In Barnabe Riche's "Apolonius and Silla," Julina becomes pregnant after spending an evening with Silvio. Her pregnancy offers, for Riche, an occasion for warnings and condemnations entirely absent from Shakespeare's adaptation. The contrast is instructive. The human condition, Riche cautions, is one of appetite and humors, longings that are at once damaging and inevitable. Riche writes that on being born we sup "of the Cupp of error, which maketh us, when we come to riper yeres, . . . many tymes to straie from that is right and reason."[56] In actions of love, humans prove besotted and compromised: we "show ourselves to be most drunken with this poisoned cup" when it comes to "our actions of Love . . . for the Lover is so estraunged from that is right, and wandereth so wide from the boundes of reason, that he is not able to deem white from blacke, good from bad, vertue from vice; but onely led by the appetite of his owne affections."[57] Riche's moralizing condemnation of love appears starkly in this language of appetite and drunkenness. Indeed, he parallels lovers and drinkers as those compromised figures lacking reason. Even as he frames his story with a finger-wagging moral, however, the tale nonetheless ends happily for the heroine, suggesting perhaps a less condemning message than Riche overtly advertises.

Riche's view of love as a form of unreasonable appetite, a source of error, transforms in Shakespeare's version into a celebration of love's inspirational force. The reasoned performance of Olivia—as both Petrarchan mocker and dutiful melancholic—yields to a more authentic, if vulnerable and incautious, expression of love. Olivia's willing embrace of dissolution through devotion appears in the race toward, and embrace of, marriage. She views this marriage as a form of service: "What would my lord, but that he may not have, / Wherein Olivia may seem serviceable?" (5.1.97–98). Further, she seeks marriage to help calm her perturbed soul, suggesting her awareness of love as an anxious conversion: as she tells Sebastian, "Plight me the full assurance of your faith, / That my most jealous and too doubtful soul / May live at peace" (4.3.26–28). Of course, on this journey to marriage Olivia exercises a significant degree of agency, asserting her own will in managing the dazzled Sebastian. She insists, for example, that he follow where she leads: "Thou shalt not choose but go. / Do not deny." (4.1.56–57). She further demands, "Be ruled by me" (63), suggesting how her desire for dissolution comes with a keen awareness of her own desires and power in effecting this transformation. Indeed, Sebastian trusts her sanity precisely because she is able to command: considering her potential madness, he reasons, "Yet if 'twere so / She could not sway her house, command her followers" (4.3.16–17). The fall into love thus comes with a continued awareness of position and duty.

Yet Olivia's fall into love, leading to her dissolution within marriage, challenges such an active exercise of her own will and authority by raising fundamental questions about consent. If marital union stands as a contract between two independent, consensual parties, *Twelfth Night* undercuts it. Olivia does not know Sebastian, although she imagines that she does; and Sebastian does not know Olivia, even if he proves willing to join with her and accepts her mistake in calling him "Cesario" (4.1.48).[58] Their marriage thus presses at the boundaries of consent and contract: not only do both parties embrace a kind of unlimited intimacy—the redefining and indeed shattering of oneself through love of a stranger resonant with the theories of Bersani and Dean invoked above—but they also enter into legally dangerous territory by embracing a union marked by deception. The work of the legal scholar Jed Rubenfeld helps illuminate why a marriage such as Olivia's would be so vulnerable on legal, if not spiritual, grounds: if consent is based on misinformation then can it be considered meaningful consent? Such a union is not only untenable, he argues, it is criminal and constitutes what the law problematically deems "rape by deception"—or, as Sebastian describes his encounter with Olivia, "there's something in't / That is deceivable" (4.3.20–21).[59] In consenting to marriage with Cesario but binding herself to Sebastian instead, Olivia finds herself precisely in this position, incapable of offering meaningful consent, even for a union she engineers.

While such a union might be legally problematic, its challenge to consent might be precisely the point. Love and marriage defy conceptions of autonomy that undergird consent, as Rubenfeld goes onto write: "Love wants the other united with the self, and it wants the other to want that same unity. In this way love desires a rupture—indeed it may effect a rupture—in the boundary between self and other. That's why love, for Freud, was so deep a threat to the ego and egotism a threat to love. Bodily integrity, on which individual autonomy depends, is not love's ideal. On the contrary, the *disintegration of individuality* is precisely what love desires."[60] Rubenfeld's analysis speaks to the form of love the early moderns called devoted or addicted love. And such disintegration through devotion, service, transformation, and indeed rupture characterizes Olivia's love in its effect on Illyria itself. Addictive attachment, practiced by Olivia, proves magical: she achieves the seemingly impossible, a marital union between two strangers in love. Earlier in the play Viola dismisses the "vows" of a lover as potentially fickle and untrue. Now the contract between two persons and bodies is similarly interrogated, because this is and is not a valid contract, Olivia is and is not consenting to this match. She marries here a simulacrum of her beloved, a doppelganger who both is and is not Cesario. This union with Sebastian is thus comic in its mingling of characters and lunacies, and yet it offers insight into love by serving logically to extend what loving means into the realm of faith: What does it mean to know the beloved fully? Who would fall in love at first sight? Isn't this the condition of loving generally?

Finding a priest to join her to her imagined beloved, Olivia and Sebastian enter into what the priest describes as

A contract of eternal bond of love,
Confirmed by mutual joinder of your hands,
Attested by the holy close of lips,
Strengthened by interchangement of your rings,
And all the ceremony of this compact
Sealed in my function, by my testimony.
(5.1.152–57)

If the priest emphasizes the contract's physical ties—the joining hands, touching lips, and changing rings—Olivia has insisted on marriage as a salve to her "soul" and a form of "service" to Sebastian; she finds it a source of wonder (5.1.221). What she earlier willed, as a melancholy addict, has magically transformed into the epiphany of love.

Festive Drunkenness

Opening with images of surfeit, drowning, and melancholy to signal forms of love, *Twelfth Night* stages scenes of self-immolation of another kind. Feste wittily posits liquor as a rival sea inviting release into its power, as Orsino claims to do for love. In response to Olivia's question, "What's a drunken man like, fool?" (1.5.126), Feste replies, "Like a drowned man, a fool and a madman: one draught above heat makes him a fool, the second mads him and a third drowns him" (1.5.127–29). The fool's stock accusation insists on drink's transformative powers: the drinker is no longer in his right mind; he is overcome, drowned in liquor until he is unrecognizable in his drunkenness, being foolish and mad. Olivia's reply takes up the image of drowning: "Go thou and seek the crowner, and let him sit o'my coz, for he's in the third degree of drink – he's drowned. Go, look after him" (1.5.130–32). Olivia and Feste lightly satirize these drowned drunkards and in doing so invoke an image resonant throughout the play, from Viola and Sebastian's threatened drowning to Orsino's sea of love.

As the waves of both love and liquor threaten the status quo of Illyria, Malvolio attempts to shore up familiar difference and restore order. Toby and Andrew, he complains, "gabble like tinkers," they make "an alehouse of my lady's house," and their songs are "coziers' catches" (2.3.86–89). In short, as Malvolio asks, "Is there no respect of place, persons nor time in you?" (2.3.89–90). In response to their drinking, Malvolio attacks their behavior as criminal, saying, "If you can separate yourself and your misdemeanors, you are welcome to the house; if not, an it would please you to take leave" (2.3.95–98). According to Malvolio's formula, "misdemeanors" stand apart from oneself. Yet Toby's unbridled festivity is not apart from himself but at the heart of his very character. "Confine?" he responds to Maria. "I'll confine myself no finer than I am" (1.3.9–10). His refusal of limitations, his refusal to restrain himself "within the modest limits of order" (1.3.7–8), invites inversion, festivity, and the carnivalesque. Furthermore, Toby defends his pursuits admirably and in doing so upholds drunkenness as a sign of loyalty and honor: "He's a coward and a coistrel that will not drink to my niece till his brains turn o'th' toe, like a parish top" (1.3.38–40). Drinking, Toby wittily parries, is an expression of devotion.

Indeed, Toby's propensity for excess might seem to resonate with the self-dissolving model of the addict. Shakespeare alternates the scenes of love with scenes of revelry, both sharing a vocabulary of drowning and madness to signal indulgence in the overpowering nature of desire. We watch each of the characters struggle, revel in, or manage the desire they feel—be it for a beloved or for the elusive final glass of canary. Devotion to love, or even drink, requires a willingness to be taken away from reason, rationality, and moderation into a kind of helpless obsession. Yet as the play continues, in the topsy-turvy world of drunken festivity characters retain rather than transform their individual desires. The lovesick drowning that haunts Orsino, the literal drowning that threatens Viola and Sebastian, and the melancholic tide embracing Olivia do not invert but entirely redefine character. By contrast, Sir Toby and Sir Andrew's experiences of drowning are of inversion: toes over brains, madness over reason.

Furthermore, while Toby is besotted, nevertheless he and his fellow drinkers prove resolutely fixed on appetites and quantities. They are materialists, keen to measure their cups and hours even in the name of excess. Orsino invokes appetite figuratively, but Toby's appetites are more realized. "[*Belches*] A plague o'these pickle herring! How now, sot?" (1.5.116–17), he cries, noting that life is made up of "the four elements" (2.3.9), yet "rather consists of eating and drinking" (2.3.10–11). Soon "eat and drink" become "a stoup of wine" (2.3.12–13). Sir Toby's fixations also include, as Toby confesses in his first scene, Sir Andrew's "three thousand ducats a year" (1.3.20). To coax Andrew into remaining, Toby amplifies the wealthy knight's alleged talents. He plays, as Toby informs us, the "viol-de-gamboys," and he speaks "three or four languages word for word without book" (23–25); he's spent time "fencing, dancing, and bear-baiting" (91), he's good at "kickshawses" (111), "the back-trick" (118), and other dances. Such lists of Andrew's talents help flatter the knight into staying in Illyria, where he is encouraged to pursue a most vigorous and physical form of wooing. As Toby counsels Andrew, with Maria, "front her, board her, woo her, assail her" (54–55). Their wooing, too, is based on possession and self-amplification rather than on attachment. Andrew offers little engagement with his alleged beloved, instead wanting to "recover," namely "get" or "gain," a wife. He does not even name Olivia, except in relation to Toby; she is, he claims, "your niece." As Andrew worries, "If I cannot recover your niece, I am a foul way out" (2.3.179–80). In contrast to Olivia, who seeks dissolution of herself in another, these drinkers seek amplification of their tyrannical appetites.

Of course, Toby moves closer to a form of devoted love by the end of the play. Initially he boasts of Maria, "She's a beagle true bred, and one that adores me" (2.3.174–75), which is, perhaps, a testament to her capacity for loyalty, but also an equal statement of Toby's desire to be adored and doted on. Toby does, however, begin to imagine relations differently, upon witnessing Maria's jest: "I could marry this wench for this device . . . and ask no other dowry with her but such another jest" (2.5.176–78). For a knight drawn to others for their cash, Toby's attraction to Maria for her wit over her dowry, even if offered privately and in potential jest, represents a significant shift. As Toby says on her entrance, "Wilt thou set thy foot o'my neck?" (2.5.182), "Shall I play my freedom at tray-trip and become thy bond-slave?" (2.5.184–85). Casting himself as a servant or bond slave to Maria, Toby indicates his growing affection and attachment even as Andrew continues to be reeled into courting Olivia through reminders of "the double gilt of this opportunity" (3.2.23–24) and his need to "redeem" her love, through "some laudable attempt either of valour or policy" (3.2.27–28).

Since the sharp distinction between the two plots of drowning remains, accounting for their relationship requires teasing out their contrasts and distinctions as much as the points of contact. As Lorna Hutson puts it:

> From the analogy developed between drinking and the hazards of navigation (Feste tells Olivia that a drunken man is like a drowned man—1.5.132) there emerges a chiastic narrative of rhetorical *oikonomia,* in which the eloquent and beautiful twins exchange near-drowning for domestic security, while the drunken and inept or

> irresponsible Toby and Aguecheek—initially comfortable with cakes and ale in Olivia's buttery—are finally banished, like the "knaves and fools" they prove to be, to the "wine and the rain" of Feste's song, beyond Olivia's gates. ("On Not Being Deceived," 163)

Two sets of twins thus chiastically pass, as one set moves from comfort to exile, the other from exile to comfort. Or in the terms of this chapter's argument, Shakespeare offers two plots of being overcome, only to expose the elevating potential of addiction to love, and the self-serving nature of drunken good fellowship.

In keeping with the magical economy of Illyria, however, a love triangle turns to two (or even three) happy couples, and Toby's drunkenness leads to neither disease nor exile. In this sense, the chiastic relation between the plots offers a surprisingly optimistic ending. Of course Olivia *threatens* Toby with exile, proclaiming, "Out of my sight!" (4.1.47), when she finds Toby drawing against her beloved Cesario (i.e. Sebastian). And she again drives Toby offstage when he and Andrew return, injured, at the hands of Sebastian. On hearing of Toby's drunken entrance—"Here comes Sir Toby halting" (5.1.187)—Olivia demands, "Away with him! Who hath made this havoc with them?" (5.1.198–99). But even despite these dismissals, and Toby's assaults on her beloved, the play ends not with Toby's banishment but with his further incorporation into the household through marriage to Maria.

Twelfth Night thus offers a gentle critique of drunken fellowship, even as it also ultimately allows such festivity. The play demonstrates the elevation of addicted, devoted love while exposing the limitations of other forms of spirited drowning: Toby and Andrew become more themselves as they ostensibly dissolve their boundaries through liquor. Rather than releasing themselves into their environment, they instead turn to triumphing over their ideological rival. Their revelry-turned-tyranny in their dealings with the play's insecure puritan uncomfortably depends on their dominant social positions and their familial connections; these characters risk little for their desires, or at least nothing of any consequence to them. By contrast, the supposed tyranny of addictive humors such as melancholy—invoked by Montaigne at this chapter's opening—proves an unexpected opportunity for devoted loving promised by early modern lexicographers and embraced by Olivia, Viola, and Orsino at the play's ending.

Endings

Critics might call into question the nature of the final marriages. As Stephen Dickey writes: "Even the central marriages that conclude the play, as they must, seem to represent less a harmony of two different people than a situation in which, for whatever reason, one character agrees to enter wholeheartedly into the private and insular fantasy of another. So Viola accommodates herself utterly to Orsino's 'fancy' . . . whether it be death or marriage, and Sebastian—who merely looks exactly like someone Olivia fell in love with—agrees to wed Olivia with a comic ease so conspicuous as to make us uneasy."[61] Certainly Sebastian enters precisely into Olivia's fantasy. Yet Viola's embrace of Orsino demonstrates—both in itself and transformatively for him—the ways in which loving requires a release of self into devotion of another. This is a radical form of love that alters the affective landscape—for it is the two addicts, Viola and Olivia, who secure the play's final unions. Olivia's devotional love for Cesario, at the expense of herself and her station, transforms Sebastian into a dazzled husband; and Viola's love for Orsino, offered covertly and consistently, finally shifts Orsino's unsuccessful mode of loving into a reciprocal union.

Such a union is surprising since, even to the play's last moments, Orsino remains appetitive and seemingly incapable of merger. On learning that Olivia loves and indeed has, he believes, married Cesario, Orsino responds with his well-known attack:

> Why should I not, had I the heart to do it,
> Like to th'Egyptian thief at point of death,
> Kill what I love—a savage jealousy
> That sometime savours nobly? . . .
> . . . .
> But this your minion, whom I know you love,
> And whom, by heaven I swear, I tender dearly,
> Him will I tear out of that cruel eye
> Where he sits crowned in his master's spite. . . .
> . . . .
> I'll sacrifice the lamb that I do love
> To spite a raven's heart within a dove.
> (5.1.113–27)

These lines at once betray Orsino's love of Cesario (he will kill "what I love," one he "tender[s] dearly," "the lamb that I do love") and his jealous rage at Olivia, his willingness to kill what he "know[s] she love[s]." His response, in other words, indicates affective attachment, but only in the context of cruel and violent action. He still does not give up his own desires and appetites; he instead proves overwhelmed and driven by his own emotions.

It is thus up to Viola, the great transformer, to instruct Orsino toward the more authentic response to and reconciliation of his feelings. In offering to sacrifice herself, Viola models true addicted love, telling Orsino in response to his outburst: "And I most jocund, apt and willingly / To do you rest a thousand deaths would die" (5.1.128–29). When Olivia asks, "Where goes Cesario?" she responds:

> After him I love
> More than I love these eyes, more than my life,
> More by all mores than e'er I shall love wife.
> (5.1.130–32)

This self-sacrifice of the addict, foregoing identity and life itself for devotion and commitment, propels the unions forward, modeling a form of loving so powerful that it transforms the emotional landscape surrounding it. [62]

Of course, in forestalling homoerotic desire, these marriages might seem regressively to reenact the very fixity that opened the play. Thus, as Valerie Traub suggests, the play thoroughly challenges "the ideology of a 'natural' love based on complementary yet oppositional genders," and yet ends, unconvincingly, by upholding heterosexual marriage.[63] As Laurie Shannon writes in response to Traub's reading, however, there is nothing "natural" about heterosexual love in an early modern context in which nature operated "in a homonormative (sometimes, although not consistently, homoerotic) manner. . . . In affective terms, affiliation, affinity, and attraction normally proceed on a basis of likeness, a principle of resemblance strong enough to normalize relations between members of one sex above relations that cross sexual difference."[64] The play indeed depicts relationships built on likeness, affiliation, and affinity. Yet through its staging of love's addiction, the play undermines notions of " 'natural' love" and "normalize[d] relations" in favor of a radical spectacle of loving that erases the boundaries of the self.

Even as my reading of *Twelfth Night* participates in drawing attention to the challenges of Illyrian love, then, I have explored how the play's characters answer this affective challenge by moving beyond individuality, beyond affiliation or affinity, and beyond culturally-scripted unions. Addictive love comes at the expense of other relationships and indeed oneself. Such addictive love is a devotion and gift of oneself, and *Twelfth Night* unexpectedly—given that the play opens with humors and closes with revenge—celebrates such courageous loving. It does so through the generosity of Viola who, like Montaigne's ideal friend, "doth so wholy give himselfe unto his friend, that he hath nothing left to divide else-where."[65] Through the magic of *Twelfth Night*'s twinning, Viola is able to "give himself" entirely, twice: she offers herself to Orsino as Cesario, while a version of her unites with Olivia as Sebastian. But this duplication is no division, for indeed Viola, like Montaigne's friend, "hath nothing left." The play's ending—in its staging of devotional loving—elevates the self by shattering it entirely, into the future of what Orsino provocatively calls, in closing, the "shall be" (5.1.376, 379).

Chapter 3

Addicted Fellowship in *Henry IV*

> All attachment is optimistic, if we describe optimism as the force that moves you out of yourself and into the world.
>
> —Lauren Berlant

Toward the end of *2 Henry IV,* Falstaff famously defends addiction to sack. Rebels surround the king's troops and Falstaff evades the fighting on a nearby battlefield, sucking on his flask instead. The abstemious Prince John chides him for his drinking, and the knight responds with his mock encomium to the dry ("sec") white wine known as sack. Falstaff argues so strongly for the benefits of intoxication that he ends declaring: "If I had a thousand sons, the first human principle I would teach them should be to forswear thin potations and to addict themselves to sack" (*2 Henry IV*, 4.2.120–23).[1] The term "addict" in the final line is crucial to the speech's triumph, crowning the argument's comic hyperbole. To a modern audience this line reads as a cheeky endorsement of drunkenness. Yet, as the previous chapters have illuminated, the term "addict" had a range of resonances beyond what modern audiences hear as the compulsion to drink.

In keeping with the early modern definitions of "addiction" as a "vow" or "dedication," Falstaff's speech is itself a form of sack addiction: a speaking toward, *ad* + *dīcere*, the power of sack. In his extended tribute, the knight expresses his devotion to the drink, offered in the form of a highly structured defense. As with the addictions of Faustus, Olivia, and Viola, the devotion of Falstaff is extraordinary. He gives himself over to a cause and community even at the expense of his own wellbeing. Although clearly fueled by appetite and drunkenness, Falstaff is, like the other addicts studied in this project (and in contrast to Toby and Andrew, invoked in Chapter 2), a transformative figure, demonstrating not a fixity of character but a dizzying dispersal of roles.[2] He offers a theatrical kaleidoscope of devotion, forgoing self-sovereignty in favor of fellowship.

As with *Twelfth Night*'s addicted lovers, Falstaff "gives himself to something," and proves able to "alienate from himself to another," as Thomas Thomas defines "addiction" in the 1587 dictionary analyzed in Chapter 2.[3] Specifically, as this chapter explores, Falstaff gives himself in attachment to Hal, at once choosing the prince and alienating himself within this relation.[4] Devoting himself to their friendship, Falstaff repeatedly demonstrates his unwavering release into intimacy. In the process he defies laws and expectations: sought by the Lord Chief Justice but protected by Hal, he falls asleep; enjoined to fight for the nation but yearning for Hal's companionship, he drinks; rejected by the new king, he nevertheless awaits embrace. In the face of finger-wagging sovereignty, he consistently upholds his extraordinary form of devotion, defying expectations of self-rule in his dogged pursuit of relational release.

Yet, despite Falstaff's devotional persistence, the primary object of his addiction frustrates attachment. Whereas Faustus finds an eager Mephastophilis, and Olivia embraces a baffled but willing Sebastian, Falstaff meets with increasing rejection before his final exile by Hal. In upholding the scene of their attachment in the face of Hal's increasing absence, Falstaff devotes himself to drunken good fellowship, the material condition of their friendship. Yet Falstaff's attachment to sack, even as it helps link him to the prince, comes to control him. As a result, the tavern scenes, so often read by critics as sites of carnivalesque festivity and good fellowship, wither to the singular spectacle of the lone drinker.[5] The knight eats less and less, while continuing to drink. Loyalty to sack overtakes his other commitments.

One might indeed argue, given his trajectory over the course of two plays, that Falstaff's sack addiction comes to appear more compulsive than devotional, a shift in line with addiction's modern definition as a pathology and disease. And this link of drinking and disease would be a potent one for Shakespeare's audience: at the moment of the plays' first performances, puritan tracts began to deem excessive alcohol use an addiction. Falstaff's condition resonates precisely with these writings, from his drinking-related diseases to his poverty, isolation, and social disruption as a result of alcohol abuse. Furthermore, Falstaff's struggle with alcohol reinforces the theological concerns expressed in many such puritan pamphlets: the knight substitutes a devotional aid for the object of devotion itself. The relationship between spirit and matter collapses as the devotee begins to worship the material manifestation of the transcendent spirit, just as Calvin, Foxe, Perkins and other reformist writers warn. For even as these godly writers celebrate spiritual addiction, they warn against the danger of distracting icons (crosses, windows, music and vestments that, in their gorgeousness, might become too attractive, leading the believer not to God but to idolatrous workshop of the object itself). So too with sack, as Falstaff embraces it as a means of securing and then memorializing his friendship with Hal, but he ends by worshipping the drink in itself.

However apt these medical and theological readings might seem, they ultimately pathologize Falstaff's addiction: it is harmful, misguided, mistaken. And attempts to condemn Falstaff have a tendency to fall flat. Even if addiction to alcohol takes its physical toll on the knight and his community, Falstaff's desire for release and connection continues to elicit praise, not censure.[6] Even scholars who argue that the plays condemn Falstaff will themselves recuperate the knight, often turning to a tragic frame: they wish it were otherwise.[7] In seeking an assessment beyond pathology (or mere praise), this chapter turns to Lauren Berlant's concept of cruel optimism as a frame for analyzing Falstaff's devotional addiction. A relation of cruel optimism exists when something a subject desires proves instead an obstacle to his or her flourishing. This concept helps account for Falstaff's dilemma precisely because, unlike the modes of suspended intersubjectivity studied thus far in this book, the knight experiences cruelty in his addiction. If, like Tim Dean and Leo Bersani invoked in Chapter 2, Berlant studies deep attachment in terms of its challenge to self-sovereignty, her approach turns from what she calls Dean and Bersani's "optimism of attachment" to relations she deems "not so buoyant."[8] "What happens," Berlant asks in a question relevant for Falstaff, "when the loss of what's not working is more unbearable than the having of it."[9] The extensions of personhood she examines, the deep relations that she investigates, result in disappointment and exhaustion. And it is this analysis of a relationship at once desirable and taxing that offers a vantage point on the sophisticated density of addictive relations in the *Henriad*. The difficulty for Falstaff is neither his lack of commitment nor his incapacity for release. His challenge comes through the nature of his chosen relation: he attempts to release himself into the figure of sovereignty itself, an impossible gesture in which he persists.

Understanding Falstaff's addiction to Hal and its proxy, sack, as a wearing but enabling relation allows us to appreciate Falstaff's attachment as an expression of devotional intimacy, not as the disease presaged by the early modern medical and religious discourses that are obviously implicated in the plays but are too rigid in their explanatory power. Even as characters condemn Falstaff's addictive propensities, the plays persist in staging the knight's admirable devotion to his relational attachments. Falstaff attempts, in his dedication to Hal, to transform and renegotiate structures of power. If this attachment produces instead a "less buoyant" outcome, the cruel optimism that is an affective yet depleting intimacy, the knight's efforts nonetheless invite admiration. For Falstaff's addiction to sack might be compulsive; it might be mistaken; but it represents the play's most imaginative argument against the isolationist pressures of sovereignty, pressures that pose greater theatrical and affective threats than Falstaff ever does.[10]

A character capable of chameleon-like transformations, Falstaff moves from gigantism to invisibility and back again, transforming himself from sinner to saint, and from fool to Vice to tragic hero. Yet amid such shape-shifting, he retains his singular devotion to Hal and drink, releasing himself repeatedly and faithfully into this relation even as it comes at the expense of living itself. Thus with awareness both of Falstaff's devotion to Hal and, in the face of this friendship's failure, of his resulting attachment to sack, Shakespeare chronicles the drama of addiction's power through his most popular and convivial character.[11]

Addiction as Disease

One does not have to venture far into early modern pamphlet literature to discover the downsides of Falstaff's way of life. Contemporaneous anti-alcohol pamphlets chronicle the diseases attendant on excessive drinking, and Falstaff—perhaps not surprisingly—exhibits all of them. Excessive drinking can provoke, William Fulbecke writes, "a great number of diseases: as Catarres, rewmes, swellinges, goutes, dropsies, doe shake the foundation of our healthe."[12] The author of *The odious, despicable, and dreadfull condition of a drunkard* offer a similar list: "Excessive and intemperate drinking" brings on "a world of diseases and infirmities which shorten [the drinker's] life," including "rhemes, gowtes, dropsies, aches, imposthumes, apoplexies, inflammations, plurisies, consumptions.[13] Thomas Beard, too, links drinking and these precise diseases, including "rheumes, impostumes, gouts, consumptions, apoplexies, and such like." He concludes, "This is one principall cause why men are now so short lived, in respect that they have beene heretofore."[14] Excessive and/or habitual drunkenness is an infirmity accompanied by, and indeed provoking, a whole series of other infirmities.[15]

Nearly all writings on drunkenness, from puritan tracts to medical texts, catalogue precisely this set of diseases linked to alcohol addiction. And corpulence, dropsies, palsies, and consumption characterize Falstaff in just the manner these pamphlets on drunkenness suggest. He is described—in comic yet telling terms—as a "whoreson obscene greasy tallow-keech" (*1 Henry IV*, 2.4.220–21),[16] a "swollen parcel of dropsies" (2.4.438), a "huge bombard of sack" (2.4.439), "fat-kidneyed" (2.2.5), "short-winded" (*2 Henry IV*, 2.2.122), "surfeit-swelled" (5.5.49), "rheumatic" (2.4.56); one who "sweats to death" (*1 Henry IV*, 2.2.105) and who suffers from "diseases" (*2 Henry IV*, 1.2.5, 247), "gout" (1.2.243), "pox" (1.2.243), and "an incurable consumption of the purse" (1.2.235–36). As the doctor claims at the start of *2 Henry IV*, Falstaff's "water itself was a good healthy water, but for the party that owed it, he might have more diseases than he knew for" (1.2.3–5).

Not only does Falstaff exhibit those diseases chronicled by puritan writers, but he also deploys their innovative language of addiction in connection with compulsion to drink. In the proliferating anti-alcohol pamphlets contemporaneous with Shakespeare's plays, writers deploy the term "addiction"—a term whose devotional resonances I have been stressing in this project so far—in a fresh manner: to describe, and deride, the phenomenon of drinking excessively.[17] As seen in the introduction of this project, Downame writes of those who "addict themselves to much drinking, and make it their usual practice to sit at the wine or strong drink."[18] He expands on this point, lamenting how "many of our people of late, are so unmeasurably addicted to this vice . . . as though they did not drink to live, but lived to drink."[19] Downame's formulation is derisory, but it also captures some of the commitment of addiction seen in Falstaff: one lives for one's attachment. Drinking is not merely a necessity of life, to fortify oneself; drinking is a devotion. But even as Downame's formulation leaves a trace of admiration for such attachment, his project lies in attacking drunkenness and drinkers. Addiction to drink, for him, is the wrong kind of addiction: sons should most certainly not, in Downame's world, addict themselves to sack.

Drinking addiction is spiritually as well as physically dangerous, leading, as anti-alcohol pamphlets chronicle, to the enslaved condition of the drinker. The literature on enslaved drinking is repetitious, rehearsing the same medical formula on drinking and habit: "By a long and desperate custome, they [drinkers] have turned delight, and infirmity, into necessity; so that without wine they cannot live," writes the anonymous author of *The odious, despicable, and dreadfull condition of a drunkard.*[20] Samuel Ward's *Woe to Drunkards* draws on this formula of "desperate custom," writing of drunkenness: "if once a Custome, ever necessity."[21] Richard Younge, too, writes in *The Drunkard's Character* that "by a long and desperate custome, they [drunkards] turne delight and infirmity into necessity, and bring upon themselves such an insatiable thirst, that they will as willingly leave to live, as leave their excessive drinking."[22] In *The Trial of Tabacco*, the author and physician Edmund Gardiner invokes this connection of choice turning to necessity as well: "How great the force & power of this cruell tyrant Custome is, that creepeth in by little & little, insinuating and conveighing himself slily into our natures, so that at length he will be so malepart, as to vendicate the whole rule and government of our bodies, prescribing and limiting new lawes, even such as it selfe pleaseth, and abrogating old ancient orders, constitutions, and fashions."[23]

Finally, Downame typifies the conclusions of this sizeable literature, with its emphasis on the cruelty of custom and the tyrannical laws governing the drunkard. He writes,

> The drunkard by his much tipling maketh himself a slave to his vice, and by long custome bringeth superfluity into urgent necessity: for as it is in other sinnes, so in this; before it is admitted, it creepeth and croucheth, flattereth and allureth, like a lowly vassall; but being entertained, it straight sheweth it selfe, not onely a master, but also a Lordly tyrant, which raigneth and ruleth with great insolence. First sinne is committed, then practiced, and often practice bringeth custome, and custome becommeth a second nature, and hath in it the force of a law which must be obeyed, not in courtesie but upon necessity.[24]

The phrase repeated in each of these texts—custom (i.e., habit) turns delight into necessity—comes from Galen: *habitum, alteram naturam*, or "custom alters nature." Writers adopt this Galenic medical formula (one, incidentally, invoked by the medical pioneer Thomas Trotter in the first official study of addiction in 1800) as an explanatory device in accounting for habitual, excessive drunkenness.[25] The phrase helps turn what might otherwise seem to be oppositions into a continuum. One moves from custom, through habit, into necessity, to the point where—as noted above—"without wine [one] cannot live." As a result, drinking takes over the subject, Gardiner and Downame warn, ruling as a "tyrant" and rendering the drinker a "slave."

This language of tyranny and enslavement illuminates the perplexing condition of drunkenness and addiction more broadly: as we have seen in this book so far, the addicted subject at once is and is not himself. The drinker holds some agency—he enslaves himself, just as he addicts himself. The hesitancy on the issue of will and agency is telling. These authors struggle between moralizing and diagnosing, unsure precisely where to credit the will—to the drinker, to the force of custom, or to the power of alcohol itself. Ingesting alcohol begins, much as *I Henry IV* itself, with conviviality: it is, as the pamphlets chronicle above, a "delight," "superfluity," or "entertainment." But the custom of drinking is "sly," "insinuating," and "flattering." Sneaking in as a humble servant, facilitating good times, drink slowly gains power: the formerly "lowly vassall" of drinking becomes the "Lordly tyrant." Then, drinking "raigneth and ruleth with great insolence," establishing its own form of law. "Abrogating" tradition and upsetting the natural order, it has "the force of a law which must be obeyed." Indeed, it proceeds with "prescribing and limiting new lawes." The law of drink is not, of course, the law of nature or of the commonwealth but instead the innovative law of the usurping tyrant. It takes away one's senses and pleases itself instead. Formerly self-sovereign, the habitual drinker becomes overtaken, enslaved, and addicted. "Drunkards" suffer from a "slavish condition," tied to the "tap-house";[26] the drinker, as Downame writes, "maketh himself a slave."[27]

This is the paradoxical position of the addict, willing away his own will. As the writer of *Diet for a Drunkard* poignantly states, "A Drunkard is a man, albeit in his drunkennesse little better than a painted man: as Ambrose said, *What is a Drunkard, but a superfluous creature?*"[28] It is this superfluity of the drinker, the sense of him as emptied out of any purpose beyond drinking, which the play's representation of Falstaff draws on in the final acts of *Henry IV*. Falstaff becomes, Barbara Everitt explains, "an outer man only."[29] He proves both politically and theatrically problematic, as choice has turned, as the Galenic formula cautions, through habit into necessity. Beyond the physical ailments chronicled above, for example, Falstaff becomes preoccupied with cash and his supply of sack, exhausting his friendships with others by preying on their purses to fuel his appetites.[30] Notoriously, Falstaff has made "three hundred and odd pounds" out of those "toasts-and-butter" men who bought their way out of military service. Commanding a regiment of "scarecrows" who are "food for powder," Falstaff increases his purse and supports his drinking (*1 Henry IV*, 4.2.14, 20–1, 37, 65). Uncomfortably turning military service into a fundraising campaign for himself, Falstaff preys on his friends to garner even more funds. He owes Bartolph an angel for his battlefield drinking (4.2.6), Master Shallow a thousand pounds (*2 Henry IV*, 5.5.71), and Mistress Quickly a hundred marks; driven into debt, Quickly is forced to sell her tavern's plate and tapestries to keep from prison. As she claims of Falstaff, he has taken "all I have! He hath eaten me out of house and home. He hath put all my substance into that fat belly of his; but I will have some of it out again" (2.1.71–74). Falstaff convinces Quickly otherwise, and she continues to supply him even to the point of her arrest. Beyond Bartolph, Shallow, and Quickly, Falstaff most obviously draws on Hal, who serves both as the payer of his debts, and as security for the knight's credit.[31]

Falstaff's drinking itself moves from joviality to isolation, from a shared pursuit of fellowship with Hal—the "cups of sack" (*I Henry IV* 1.2.7) for which Hal "hast paid all" (1.2.50)—to singular dedication as the play continues: the devotional aid, sack, becomes the object of devotion itself. By the time *Henry IV* ends, Hal no longer drinks with the knight. Instead, he desires the low-alcohol beverage, small beer, a choice rife with class and national resonances.[32] Falstaff, by contrast, continues imbibing sack, even at the expense of food: "Here, Pistol, I charge [toast] you with a cup of sack" (*2 Henry IV*, 2.4.111–12); "I'll give you a health for that anon" (5.3.24); "Health and long life to you, Master Silence!" (5.3.51–52). Even as he toasts healths (a practice analyzed at length in Chapter 5 of this project), his own deterioration appears in eschewing food—a form of communion—for drink.[33] As Davy reassures him, "What you want in meat we'll have in drink" (5.3.28), and as Falstaff tells Shallow: "Come, I will go drink with you, but I cannot tarry dinner" (3.2.191–92). For a character accused of gluttony throughout, his dogged attachment to drink with a refusal to eat suggests dearth: without Hal he refuses nourishment.[34] Thus by the time Falstaff announces, on learning of Hal's ascension, that "the laws of England are at my commandment" (5.3.136–37), he speaks as if free. But Falstaff is not, at this stage, under his own commandment. The custom of sack drinking, in "prescribing and limiting new lawes" for this character, has arguably overcome him. Falstaff exhibits less, not more, freedom than other characters; he is more constrained, whether financially, physically, or theatrically.

Addiction's Company

Falstaff engages in behaviors deemed—by the pathologizing accounts of addiction, chronicled above—to be shameful: drunkenness, disease, debt, gluttony, lack of control, and cowardice. But Falstaff's relation to his own actions is not one of shame. It is instead one of acceptance, enthusiasm, advocacy, and devotion. He doggedly pursues these behaviors with a sense of how they bring him into community—into a community of knights, drinkers, thieves, and king's men. Indeed, it is notable that, while medical and theological pamphlets condemn the activities linked to Falstaff, the character who most derides Falstaff, and experiences shame by association, is Hal. The prince expresses a sense of embarrassment in his relations, as when he sits with friends but looks down on them, or when he feels disgust at knowing their habits; he expresses frustration—albeit mixed with amusement—at Falstaff's lying about Gadshill; and he's ashamed of his friendships for compromising his ability to mourn his father authentically. Against Hal's sense of shame, and against his politically determined rise, lies Falstaff. For even in the face of Falstaff's mounting challenges—his debt and isolation—he doggedly embraces his relational mode of being. This single-minded devotion helps to determine his popularity and his legacy, even in the face of some of his behaviors that might invite censure.

Further, if anti-alcohol polemic casts addiction as a kind of slavery or tyranny, it is worth acknowledging that such enthralled necessity appears positively in the early modern period as well: not only in accounts of godly devotion, which were surveyed in the introduction, and in definitions of addiction, which were studied in Chapter 1, but also in theories of friendship. Michel de Montaigne, for example, describes friendship in terms resonant precisely with such release, writing of an "amitie" so powerful that friends "entermixe and confound themselves one in the other, with so universall a commixture, that they weare out and can no more finde the seame that hath conjoined them together."[35] Releasing themselves into one another, such friends are boundaryless, unable to find the "seame" between them. Elaborating on his own particular relationship with a friend, Montaigne writes, "I wot not what kinde of quintessence, of all this commixture, which having seized all my will, induced the same to plunge and loose it selfe in his."[36] Montaigne depicts friendship as a version of the addiction studied earlier in this project, in which the will is admirably captured or imprisoned, in which a choice comes to feel like a willed compulsion. His account describes what Falstaff experiences: a seized will, a sense of losing himself in relation. Laurie Shannon deems this Falstaff's "infinite vulnerability," as the knight proves himself an "over-fond partner" who is eventually, and surprisingly, betrayed.[37]

Understanding Falstaff requires recognizing the emotional and what we might call philosophical benefits of his mode of intimate vulnerability. A heroic character who upholds intersubjectivity and relationality, Falstaff persistently releases himself in addicted devotion to Hal, even as he increasingly expresses their intimate relation through drinking sack, the elixir that serves to secure and then memorialize their friendship. From its first moments, the friendship of Hal and Falstaff hinges on the good fellowship of sack drinking. Recognizing this triangulation does not undermine their intimacy but instead admits its foundation in physical and emotional vulnerability. Hal may not embrace such vulnerability ultimately or, even arguably, at all; but Falstaff understands his relation to Hal as a loss of will, a bewitching that takes away even his ability to consent. He has attempted to assert his independence for two decades: "I have forsworn his company hourly any time this two-and-twenty years." Yet his relation to Hal has performed a kind of magic, seizing his will as a consumed potion might: "I am bewitched with the rogue's company. If the rascal have not given me medicines to make me love him, I'll be hanged. It could not be else: I have drunk medicines" (*I Henry IV*, 2.2.15–19).[38] These lines are humorous, of course, given Falstaff's quite active agency in forming friendship with Hal (and his likely exaggeration of the length of their friendship). Yet these lines also speak to his sense of his overcome will, connected both to Hal and liquor. Falstaff understands company and medicines, good fellowship and sack, together: the one—medicine, magic, sack—helps account for the ferocity of his attachment to the other.

The episode at Gadshill in *I Henry IV* stages this triangular relation, namely Falstaff's addicted attachment to Hal and sack. In doing so the scene reveals both the resiliency of Hal and Falstaff's friendship through shared cups of sack, as well as the humiliation Falstaff endures to sustain this intimacy. Hal and Poins wait in the tavern to mock Falstaff, reveling in their own youthful and athletic capacities. On the knight's return, they laugh as he offers exactly the "incomprehensible lies" (*1 Henry IV*, 1.2.176) Poins anticipated. But where such a "jest" might secure distance from Falstaff, as he serves as the butt of their joke, instead Falstaff turns to sack as a unifying tonic. In doing so he asserts his intimacy with Hal even in the face of abuse. With the knightly bombast and heavy-drinking habits typical of soldiers, he enters the tavern as if he were a martial hero: "A plague of all cowards," he cries, calling for sack to restore his body. "Give me a cup of sack, boy. . . . A plague of all cowards.— Give me a cup of sack, rogue" (2.4.110–14). Exposed and mocked, Falstaff transforms the scene away from his humiliation by crying, "Gallants, lads, boys, hearts of gold, all the titles of good fellowship come to you! What, shall we be merry?" (269–71). He speaks as if rallying the troops, not to fight but rather to be "merry," or drunk. Sack offers, for Falstaff, the equalizing liquid that brings together the tavern crew, overcoming potential divisions between himself and his younger, fitter, more aristocratic companions. These men all become, in drinking, "gallants, lads, boys" together.

Falstaff's tenacity in seeking release into friendship with Hal appears on the battlefield at the end of *1 Henry I* as well. Here Falstaff, attempting to assert his intimacy with Hal after a period of separation, is rebuffed and yet persists. Encountering an idle Falstaff, the prince chides him, asks for his sword, and Falstaff responds:

> *Falstaff:* [T]ake my pistol, if thou wilt.
> *Prince:* Give it me. What, is it in the case?
> *Falstaff:* Ay, Hal. 'Tis hot; 'tis hot. There's that will sack a city.
> *The Prince draws it out, and finds it to be a bottle of sack.*
> *Prince:* What, is it is a time to jest and dally now?
> *He throws the bottle at him. Exit.* (*1 Henry IV*, 5.3.53–56)

This exchange betrays Falstaff's desperate desire to uphold the good fellowship he shares with Hal, even beyond the tavern. Attempting to "jest and dally" as they had done at the Boar's Head, Falstaff finds his invitation to humor condemned.[39] For if the knight attempts to recall former alliances, his speech performs a telling substitution: in replacing his gun with a flask, Falstaff announces his greater allegiance to tavern culture than to the nation he has been recruited to defend. His loyalty to Hal persists as friendship, not political allegiance.

Montaigne's model of friendship, which is based in the seamless intimacy of those who willingly release themselves into one another, is helpful for illuminating the strained nature of this scene. Montaigne celebrates the mode of friendship Falstaff seeks with Hal, hoping they might be "rather friends than Citizens, rather friends than enemies of their countrey, or friends of ambition and trouble."[40] Falstaff tries to assert his kinship with Hal, as two drinkers on the battlefield, defying national claims. But of course, they are not equivalent, nor are they like-minded in the way that Montaigne celebrates with friends: "All things being by effect common betweene them; wils, thoughts, judgements, goods, wives, children, honour, and life; and their mutual agreement, being no other than one soule in two bodies, according to the fit definition of Aristotle, they can neither lend or give ought to each other."[41] Reading Montaigne's model of friendship next to the exchanges of Hal and Falstaff exposes how little these men represent "one soule in two bodies," despite the knight's attempts to assert—even to jeers—their companionship.

Even when Hal and Falstaff understand and express their attachment in similar terms, the differences are stark, as when both describe, in separate scenes, their relationship as one of dog and master. When the chief justice accuses Falstaff of having "misled the youthful Prince," Falstaff responds, "The young Prince hath misled me. I am the fellow with the great belly, and he my dog" (*2 Henry IV*, 1.2.145–47). This reply elicits laughs, in part because one can hardly imagine Falstaff physically pulled by anyone. But deeper than this physical joke is Falstaff's inversion of their emotional relationship. It is Falstaff who is led, at least emotionally if not physically; he is entirely vulnerable in relation to Hal, devotedly awaiting his return to Eastcheap. Arguably more apt is Hal's own evocation of their canine-human relationship when he claims in their next scene, "I do allow this wen [Falstaff] to be as familiar with me as my dog" (2.2.104–5). Here Hal confesses both to their intimacy and to the shame he experiences thinking of it, by casting Falstaff as a kind of wart or parasite—"wen"—rather than a sentient being. Since each man references the other as a dog, their twin figurations might speak to their shared understanding of their intimacy. But equally, their opposite evocations of this power relationship attest to their starkly different understandings of, and appreciation for, their friendship: if in both cases the human-dog relation figures attachment, Falstaff's sense of himself as an overcome owner starkly contrasts with Hal's repulsion at an overly familiar beast.

Thus even as Falstaff asserts an intimacy with Hal, as in the scenes at Eastcheap or on the battlefield analyzed above, Hal responds with distance and rejection. This opposition is arguably embedded in the play's convivial lexicon itself. The word "sack," for example, signals both fellowship and ruin in the above exchange with Hal: "There's that will sack a city."[42] Hal sees ruin where Falstaff seeks connection through drink. So, too, does the word "company" reveal both Falstaff's assertion of intimacy and Hal's rejection of it. At the start of *Henry IV*, "company" references Falstaff's addicted relation to Hal: he is "bewitched with the rogue's company" (*1 Henry IV*, 2.2.16–17), and begs Hal to retain him: "Banish not him thy Harry's company, banish not him thy Harry's company" (2.4.465–66). Yet with distance from Hal, company degenerates. Tavern friendship cedes to the impoverished company of wartime: "There's not a shirt and a half in all my company, and the half shirt is two napkins tacked together" (4.2.41–42), Falstaff laments, as the "napkins" for a tavern meal now figure pathetic, inadequate clothing stock. Company is also infectious: "let men take heed of their company," Falstaff warns, for "men take diseases one of another (*2 Henry IV*, 5.1.74–76). Yet even here, voicing his greatest suspicion of company, Falstaff immediately recalls his intimacy with Hal for comfort, saying: "I will devise matter enough out of this Shallow to keep Prince Harry in continual laughter" (*2 Henry IV*, 5.1.76–77). In stark contrast to Falstaff, the prince invokes "company" with a sense of shame. He cannot grieve for his father, he tells Poins, because "keeping such vile company as thou art / hath in reason taken from me all ostentation of sorrow" (*2 Henry IV*, 2.2.47–48). And, of course, his condemnation of Falstaff relies on this loaded term: "I have turned away my former self; / So will I those that kept me company" (5.5.57–58). This last usage of the term drives home its density, with "company"—the very term deployed by Falstaff throughout the plays to mark his attachment to the prince—condemned as a polluting force and a sign of the prince's former corruption.[43]

With both of these words, sack and company, it is the martial requirements of rebellion, and the duties of sovereignty upon Hal, that effect the shifting signification of terms for him: notably, Falstaff persists in his definitions, remaining devoted to the terms' original meanings. Indeed, for a man so playful with words that he turns terms like "honor" and "counterfeit" on their heads, he persistently upholds the relational definition of words, namely the significations that relate to community and, ultimately, to Hal. In his meditation on honor, for example, Falstaff imagines, with a degree of terror, the isolated, senseless world felt by a man who boasts no companionship but honor. He reasons that honor is but "air. A trim reckoning," asking if the recently dead might feel honor: "Do he feel it? No. Doth he hear it? No. 'Tis insensible then? Yea, to the dead. But will it not live with the living? No. Why? Detraction will not suffer it" (*1 Henry IV*, 5.1.135–39). His materialist philosophy, crediting the sensible world against the reputation or honor that cannot be seen, accords with his appetitive life and seems to reinforce the prominence of desire over devotion. But his speech equally invokes life in relation: he wants to follow not honor but a living, breathing being, one who might provide aid in times of trouble ("set to a leg . . . or an arm . . . or take away the grief of a wound," 5.1.131–32). As his first substantial soliloquy in *1 Henry IV*, this meditation on honor confirms what we already know about Falstaff: he exists in relation to others. His scenes consist almost entirely of dialogue to this point; he thinks in relation to others, while Hal meditates alone.

Of course, Falstaff repeatedly claims that he will distance himself from Hal: "I prithee Hal, trouble me no more with vanity" (1.2.78–79), he chides, in bidding the prince not to disrupt him further. He insists of Hal, as noted above, "I have forsworn his company hourly any time this two-and-twenty years" (2.2.15–17). Yet when he confesses, "Thou hast done much harm upon me, Hal; God forgive thee for it. . . . I must give over this life, and I will give it over" (1.2.88–92), no one takes his call to "give it over" seriously; his level of commitment is too strong. In each of these cases, his proclamations evoke laughter precisely because his threat is at once hollow and familiar, the cry of the devoted who attempts to pretend indifference. One might argue that such devotion is fundamentally self-serving: Falstaff fantasizes about his own advancement and detracts from Hal's political obligations. And theirs is, furthermore, a relationship based in libidinal desires: for drink, food, and the money that buys it. But it is worth noting how persistent Falstaff remains in loving Hal, even in the face of the ridicule and rejection charted above. Falstaff tenaciously, and devotedly, clings to their fellowship, upholding their life, awaiting the prince's return, even as fellowship with Hal proves a mere fantasy, a form of playing that the prince offers less and less frequently to his devoted knight.

In this addiction to Hal, Falstaff differs from the figures studied so far in this project in that the play never stages, or even gestures at, Falstaff's independence from his addiction. He might assert that his will has been overtaken by Hal, that he can no longer act independently, but the play does not chronicle any previous independence or clear exercise of the will in this besotted knight. Falstaff's descriptions of his skinny youth, fitting in an "alderman's thumb-ring" (*1 Henry IV,* 2.4.322)—like Shallow's evocations of "the wildness of his youth" (*2 Henry IV,* 3.2.304)—provoke disbelief, a sense of fiction woven through time. It isn't simply disbelief that Falstaff was more virtuous in his youth than now; it's disbelief at imagining Falstaff's youth and self-sovereignty in the first place. It is a testament to this character's theatrical presence that his backstory seems entirely impossible. The man before us—in his size, age, appetites, and affections—represents the knight as he exists through time. And when his connection to Hal dissipates, he dies.

Defending Sack

In his last, great soliloquy of the *Henriad*, when Falstaff offers his addict's pledge to sack, it is worth noting how much of Falstaff's argument expresses a yearning for the embrace of friendship that he alone upholds, even in the face of both condemnation by the Lord Chief Justice, and absence from Hal. For if the speech overtly counsels addiction to sack, its argument hinges not on drink itself but on the communal relations, the friendships, that drink fosters. Falstaff might open the speech with a medical defense of sack, arguing in Galenic terms that alcohol helps in "warming the blood, which before, cold and settled, left the liver white and pale" (*2 Henry IV*, 4.3.101–3), but this defense is clearly tongue and cheek, not least because Falstaff defies the moderation prescribed in the medical literature he references. His more vigorous assertion trumpets sack as the birth of wit and sociability: "A good sherris sack" ascends to the brain and "makes it apprehensive, quick, forgetive, full of nimble, fiery, and delectable shapes, which, delivered o'er to the voice, the tongue, which is the birth, becomes excellent wit" (*2 Henry IV*, 4.2.94–100). Here Falstaff frames sack addiction not as a form of individual transformation (offering only health) but as a sociable enterprise, bolstering communities both of witty drinkers and, as he goes on to claim, fighting soldiers: sack fuels, he argues, the birth of the fighting spirit, for it "illumineth the face, which as a beacon gives warning to all the rest of this little kingdom, man, to arm" (4.2.106–7).[44]

The speech's argumentative crescendo—and its commitment to community—sounds in its conclusion: Falstaff ends in praising not sack but Prince Hal. The prince is Falstaff's primary evidence for his link of drinking, wit, and valor. For Hal has turned his dour family stock into witty, warm sociability through the drinking of sack: "Hereof comes it that Prince Harry is valiant; for the cold blood he did naturally inherit of his father he hath like lean, sterile, and bare land manured, husbanded, and tilled with excellent endeavor of drinking good and good store of fertile sherris, that he is become very hot and valiant" (*2 Henry IV*, 4.2.115–20). For Falstaff, Hal demonstrates courage both on the battlefield and in the tavern, where he returns witty barbs with vigor. Furthermore, like sack itself, Hal appears as the great "captain, the heart" who inspires "all the rest of this little kingdom" (4.2.107–9) and commands the "vital commoners and inland petty spirits" (4.2.108) in battle. In a speech dedicated to sack, then, the devotional object securing Falstaff's argument—the object of his addiction and the end of his reasoning—is not drink but Hal. The knight uses his encomium to announce his own attachment to and admiration of the prince. The speech tellingly ends with Falstaff's image of himself as father to a thousand Hals, the "thousand sons" who "addict themselves to sack" (4.2.120–23). Falstaff posits a family army of his own, a grouping of men like—or as—Hal.[45]

This argument about the sack speech as a sign of Falstaff's addiction to Hal might seem perverse, given the depiction of Falstaff's gluttony and drunkenness in the play. Yet when viewed through the triangulated relation of Falstaff, Hal and sack, this final soliloquy can be seen as one more assertion of the knight's devotion to friendship. This issue for Falstaff, at this late, impoverished, and isolated moment on the battlefield, is not the error of his speech's logic. It is that the ground has moved beneath him: he struggles to maintain a mode of relational being in the face of the isolationist sovereignty that governs Henry IV's, and then Henry V's, England. And this tug of war between intersubjectivity and sovereignty becomes even more acute at the play's end. Here, Falstaff reasons that his dirty clothing at the coronation march will signal his devotion even more powerfully to the prince than a clean shirt: "This poor show doth better; this doth infer the zeal I had to see him" (*2 Henry IV*, 5.5.13–14); "it shows my earnestness of affection" (5.5.16), he assures himself. It displays, he claims, "my devotion" (5.5.18). He elaborates that he will "ride day and night" and will "stand stained with travel and sweating with desire to see him [Hal], thinking of nothing else, putting all affairs else in oblivion, as if there were nothing else to be done but to see him" (5.5.20–27). His description insists upon his "devotion" to Hal and his "zeal." He has lost himself in "earnestness," "affection," and "desire" for the prince. His terms are telling, as he chronicles his single-minded dedication to Hal, "thinking of nothing else" and casting "all affairs else into oblivion." In Falstaff's struggle, and in his excitement at this final ride to London, the play stages how relational modes of being, even at their least productive and even in the face of their failure, invite admiration for their dogged pursuit of a connected life.[46]

Half-Faced Fellowship

Falstaff offers the play's philosophical and political challenge to sovereignty through a practice of addiction. And such a challenge may be precisely why Hal must reject Falstaff in order to establish his polity, as critics from C. L. Barber through Laurie Shannon have suggested: Hal must banish the "Vice" figure, Falstaff, or reframe the terms of *mignonnerie* to ascend as sovereign. But this political logic sits uneasily with the plays' affective, ethical landscape, as generations of viewers have noted: the rational, ambitious, and even Machiavellian prince betrays the life-affirming Falstaff. Hal and Falstaff thus form a chiastic partnership when viewed from the vantage point of attachment: Hal's anti-addiction, his incapacity or unwillingness to develop strong bonds, allows him to rise as an effective ruler, whereas Falstaff's deep devotion to forms of fellowship, which initially buoy him, sink him into solitary drinking even as he maintains his connection to audiences within and outside of the play.

Over the course of the *Henriad*, Hal's love (or performance of love) wanes when attachment no longer serves, a form of fickle intimacy first practiced by his father, Hal's relational model.[47] Examination of King Henry IV helps reinforce the link of sovereignty to isolation in these plays, exposing why Hal, too, must reject—or arguably revels in rejecting—addiction. The king palpably frustrates loyalty and devotion, not only in committing the treason that gained him the crown but also through a series of disloyalties that mark the *Henriad*. *1 Henry IV* begins, notably, with broken pledges. The first concerns the voyage to Jerusalem, invoked from the end of *Richard II* and now rehearsed again. If addiction is a form of speaking toward a desire, goal, or object—pledging oneself, *ad* + *dīcere*—King Henry perpetually gestures toward, only to fall away from, his commitment: "this our purpose now is twelve month old, / And bootless 'tis to tell you we will go" (*I Henry IV*, 1.1.28–29). The voyage, never realized, is mentioned yet again at the end of *2 Henry IV*, as a reminder of the king's failure to reach Jerusalem beyond the named room in his own castle. Opening on a failed mission, the play's first scene further rehearses broken commitments, from the northern and western countrymen who revolt against the king to his own truant son. Henry's lament for his son's character introduces one of the play's central conflicts in the opposition of Hal and Hotspur. More tellingly in terms of addiction, the king's speech alerts the audience to his own fickle character, his willingness—albeit imaginary—to part with his child: "O, that it could be proved," he speculates, "That some night-tripping fairy had exchanged / In cradle clothes our children where they lay, / And called mine 'Percy,' his 'Plantagenet'; / Then would I have his Harry, and he mine" (1.1.85–89). Sadly, and ironically obscured from his knowledge, Henry has taught Hal well, for his son practices the same mode of abandonment, rejecting both biological and adoptive fathers as the plays proceed.

Both king and prince actively refuse attachment in the name of isolationist sovereignty. The king, distanced and condemning in his dealings with his former allies (as Worcester's description of Henry suggests: "see already how he doth begin / To make us strangers to his looks of love" (*1 Henry IV*, 1.3.284–85)), finds a mirror image in a son who plots the estrangement of his current allies: "I know you all, and will awhile uphold / The unyoked humour of your idleness" (1.2.185–86). In distancing themselves from friends, King Henry and the prince also prove forgetful. When Hotspur calls Henry IV a "forgetful man" (1.3.160), he draws attention to the self-serving mode of rule practiced by Henry and anticipated by Hal: forgetfulness masks the calculation of a king who remembers precisely how he gained the throne and now suspects those who helped him to it, just as it frames Hal's equally purposeful dismissal of Falstaff at the end of *2 Henry IV*: "I know thee not, old man" (5.5.46). In the end, the rebels and Falstaff are condemned, precisely as Hotspur predicts at the start of the *Henry IV* plays: allies are "fooled, discarded, and shook off" (*1 Henry IV*, 1.3.177). Hotspur bluntly calls this mode "half-faced fellowship" (1.3.207), a description that might stand as much for Hal as for King Henry.

By *Henry V* the new king's isolation from his former friends is complete, with Henry having disavowed his "addiction to courses vain" (*Henry V*, 1.1.54), as the archbishop of Canterbury and the bishop of Ely put it. But the play's audience knows what Ely and Canterbury do not and what Falstaff learns only belatedly: Hal never suffered from such "addiction" in the first place. The attachment Canterbury and Ely posit was only an act. Hal proves, even from the start of *1 Henry IV*, to be every bit as calculating as his father, free from the devotions that his friends and allies imagine for him. This is part of the tension throughout *Henry IV*. The audience remains aware of the gap between Falstaff's deep attachment to Hal and Hal's distance from the knight. Falstaff's attachment is so strong that he remains rooted in tavern culture regardless of his locale, just as Hotspur makes battlefields of great halls and bedrooms. Hal, by contrast, isolates himself—or skillfully molds himself to place and circumstance—to serve his own self-interest. He might take on the languages surrounding him, speaking of familial loyalty with his father and of drunken thieving with Falstaff, but he does so for pragmatic rather than transformative or expansive purposes, proving himself self-possessed rather than addicted.[48] He can speak with a tinker in his own language, but this multilingual facility is arguably to command his subjects more effectively: "The Prince but studies his companions / Like a strange tongue, wherein, to gain the language, / 'Tis needful that the most immodest word / But looked upon and learnt" (*2 Henry IV*, 4.3.68–71), Warwick reassures the King.

Nowhere is Hal's contempt for devotion more obvious than in his vow to triumph over Hotspur. The prince hopes to fell his opponent in one economical gesture, and this hope betrays his transactional, rather than devotional, understanding of honor. But the sustained dedication of Hotspur, developed over years of battle feats, cannot be transferred in an instant, despite Hal's fantasy of exchange. Hotspur leads "ancient lords and reverend bishops on / To bloody battles and to bruising arms" (*1 Henry IV*, 3.2.104–5), admired, even in his treasons, by church and court alike. Hal might boast that he, too, will wear a "garment all of blood," and a "bloody mask" on a "glorious day," filled with "glorious deeds," and further "glorious deeds" (3.2.135, 136, 133, 146, 148), taking on Hotspur's language of honor and bloodshed as a rhetorical plea to his father: I can, he seems to reassure the king, at least imaginatively inhabit Hotspur's ethos of war. But more than bloody images, transactional language dominates his speech, and it is this language that most proves Hal to be his father's son.[49] Drawing on images of purchasing and pawning, Hal boasts, "I will redeem all this on Percy's head" (3.2.132); "I shall make this northern youth exchange / His glorious deeds for my indignities" (3.2.145–46); "Percy is but my factor, good my lord, / To engross up glorious deeds on my behalf" (3.2.147–48). Hal invokes commercial images of pawning one's reputation, exchanging it, or purchasing it through an agent. He moves away from the language of chivalry or glory—language that evokes an entire ethos of commitment and community—to the economic language of a singular event, one "glorious day," "the day," "the time" (3.2.133, 138, 144) when one transaction undoes history: Hotspur will instantly "exchange" or "render" his former honors to Hal, giving him his reputation and indeed identity in the process. In his triumph against Hotspur, Hal imagines he can instantly accrue all the honors his opponent had gleaned in his years of fighting: "I will call him [Hotspur] to so strict account / That he shall render every glory up, / Yea, even the slightest worship of his time, / Or I will tear the reckoning from his heart" (3.2.149–52). Arguing that reputation and honor are transferable, Hal proves every bit his father's son: he, too, hopes that one son might substitute for another.

The structure of the *Henriad* might rely on Hal's triumph over rebels and his dismissal of his tavern allies for the battlefield, but these plays also shape a counter trajectory, an alternate and affective arc that upholds addiction over calculation. The play's most devoted characters, Falstaff and Hotspur, serve as the energetic foils for royal ascension while Hal, like his father before him, forms prudent but disposable alliances. Indeed, in contrast to Hal, Falstaff and Hotspur remain loyal to allies even to the point of self-destruction. They demonstrate a resolve at odds with their own good health: these men respectively drink and fight themselves into the grave, challenging the political environment of self-serving courtiers in the process. Inspired and spirited—indeed, Falstaff's boasts about the power of sack as "nimble" and "fiery" could describe Hotspur himself—both men view their rebellious loyalty through language of friendship, a language increasingly outmoded in a mercantilist and pragmatic England: "Our plot," Hotspur cries, "is a good plot as ever was laid, our friends true and constant; a good plot, good friends, and full of expectation; an excellent plot, very good friends" (*1 Henry IV*, 2.3.15–18).

It seems Henry IV's court is one that draws on the addictive capabilities of firmly set believers only to reject such fortitude and commitment as the price of rulership. By contrast, those dedicated addicts, Hotspur and Falstaff, appear as imaginative dreamers, unmoored from the real politick that surrounds them: "Imagination of some great exploit / Drives him beyond the bounds of patience" (*1 Henry IV*, 1.3.198–99), Northumberland claims of his son; "he apprehends a world of figures here / But not the form of what he should attend" (1.3.208–9), his uncle Worcester replies. Imagination and figuration are Falstaff's bread and butter: Hal accuses Falstaff of an imagination gone wild—"These lies are like their father that begets them, gross as a mountain, open, palpable" (2.4.218–19)—and Falstaff ultimately responds with a further gesture toward fiction: "Shall we have a play extempore?" (2.4.271). If Hotspur at times appears distracted and enraged—"What, drunk with choler?" (1.3.128)—he is also single-minded: "All studies here I solemnly defy, / Save how to gall and pinch this Bolingbroke" (1.3.226–27). He pronounces his commitment to opposition and remains true to his pledge, even in the face of desertions and isolation, as Northumberland and Glendower leave him to fight an impossible battle unaided. So, too, with Falstaff, who insists upon his intimacy with Hal even beyond his own banishment, a single-minded devotion painful in its staging: "Do not you grieve at this: I shall be sent for in private to him" (*2 Henry IV*, 5.5.75–76).

In their fancy, Hotspur and Falstaff make the mistake of imagining themselves equal to royalty. If Hotspur's dreaming drives him to commit actual treason, Falstaff's imagination deceives him into thinking that he is Hal's partner in greatness and in the process to challenge Hal's sovereignty. But Hal reasons instead that Falstaff is "the tutor and the feeder of my riots" (*2 Henry IV*, 5.5.61), banished for his "surfeit" and "gormandizing," as the prince puts it in his final speech to Falstaff (5.5.49, 52). If the political outcome of such addictive capacity, as Hal argues, is oppositional and deeply problematic for the crown, such capacities also speak to the ability to feel and move others.[50] Hal's moderate position appears, from this angle of attachment, to be a weak and fickle one. Standing between two devoted addicts who uphold fellowship on the one hand and honor on the other, Hal places himself awkwardly and insecurely between them.

Theatrical Transformations

If the play's political plot requires isolationist sovereignty, the play's affective plot does not. Looking at the play from a relational vantage point highlights the ingenuity of Falstaff's attempts to reframe his environment in addictive terms: How to remain devoted to a rejecting idol/God? How to maintain connection in the face of abuse? Falstaff provides a study of the strains and costs as well as the beauty of such an attempt. He reveals the danger of pathologizing addictive relations: even in their diseased form, such devotions offer a way out of the isolationist impasse of sovereignty, which requires distance between father and son, son and friend, court and county, peer and peer. Such sovereign isolation destroys every relation in its wake and locates shame at the heart of connection. The plays may well chronicle, as Laurie Shannon carefully argues, "polity's founding moment," but the resulting "relocation" of Falstaff is more heartbreaking than triumphant.[51] For even in his most overcome state Falstaff continues to shine, in contrast to the characters who are incapable of intimate relation.

As optimistic, indeed as buoyant, as this chapter's reading of Falstaff's addiction might be, it is hard not to notice the degenerating quality of Falstaff's transformations over the course of the plays. Falstaff might promise to swear off sack at the end of *1 Henry IV*—as he claims in his last lines of that play: "If I do grow great, I'll grow less, for I'll purge and leave sack and live cleanly, as a nobleman should do" (*1 Henry IV*, 5.4.163–65)—but his claim is a humorous one. And the reports of his demise acknowledge, even as they attempt to suppress, the knight's constraining habits. He died, Nym notes, from "bad humours" (*Henry V*, 2.1.121), a reference to the prince's ill humor but one that also evokes the humorological excesses of Falstaff leading to death.[52] Furthermore, if Falstaff dies of a broken heart as Mistress Quickly claims, he nevertheless spends his final moments thinking not of Hal but of sack: "They say he cried out of sack," Nym reports—and the hostess confirms, "Ay, that a' did" (2.3.26–27). While these lines provoke laughter, nevertheless Falstaff dies acknowledging what others might not: drinking—connected to Hal, to rejection, and to his addicted heart—overcame him.

Falstaff's end might have the unfortunate effect of fulfilling godly polemic, a surprising outcome given Shakespeare's presumed hostility toward his puritan detractors. Why might Shakespeare offer such an unexpectedly moralizing portrait of a drinker condemned by his friend and felled by sack? Richard Strier provides one persuasive and indeed moving answer: Shakespeare stages, in the rejection of Falstaff, a moral position he does not necessarily embrace. Even if the plays demand the rejection of Falstaff to facilitate the rise of Hal, Shakespeare does not take such a moral position comfortably. In the *Henriad*, "prudence, order, and morality had to prevail, and Shakespeare never forgave himself for that. He never again put himself in a position of seeming to favor (as Falstaff puts it) 'Pharaoh's lean kine.' "[53] As a result, Strier argues, in a subsequent play like *King Lear*, the foolish old man becomes the hero, and those daughters who chide him with foolishness are the villains.

Closer, however, to *2 Henry IV* than *King Lear* is *Twelfth Night*, a play embracing transformation in all its theatrical possibilities. If *Twelfth Night* stages, as seen in Chapter 2, the transformative potential of love addiction and the role of theater in facilitating this, *2 Henry IV* offers a stark contrast. For it is the theatrical cost of Falstaff's addiction, finally, that might account for why Shakespeare brings his popular knight to such an apparently moralizing end. The performance of Falstaff and Hal's good fellowship is initially full of possibility. The capers, gags, and witty exchanges of the tavern foreground the playful imagination of the playwright, who is freed from the more constraining historical narrative of the plays' other plot lines. Falstaff triumphs in his rhetorical and theatrical range, adopting Cambysean sovereignty and Euphuestic exuberance to the delight of Quickly and tavern patrons: "O Jesu, he doth it as like one of these harlotry players as ever I see!" (*1 Henry IV*, 2.4.385–86). But the compulsion to present Hal's rise and the knight's drunkenness perhaps wears thin as it extends over the course of two plays. What began as flexibility turns to fixity, as Falstaff's devotional attachment to Hal and to sack makes him unable to shift, change, or adapt. Such devotional capacity might stand as a welcome form of resistance to the changeable, mercurial politics of the royal court, but it ultimately proves a theatrical liability.

This theatrical liability seems especially acute when noting how, at the start of *1 Henry IV*, Falstaff serves as one of Shakespeare's most metatheatrical characters. Coming after Shakespeare's masterful and widely popular shaping of Richard III as the comic Vice character, Falstaff draws on the same theatrical tradition—he is the comic tempter of the medieval morality plays, in league with the devil but also intimate with the audience. He is, as Hal rehearses in *1 Henry IV*, "that reverend Vice" (2.4.441) and "That villainous, abominable misleader of youth, / Falstaff, that old white-bearded Satan" (2.4.450–51). Shakespeare modifies, expands, and exponentially magnifies the possibilities of this role, inflecting the stock character of medieval drama with other theatrical traditions embedded in the sub-roles that Falstaff plays. For, as Marjorie Garber reminds us, while Falstaff functions as a Vice figure, he is also "a Lord of Misrule . . . an ordinary man temporarily raised to high estate."[54] Mikhail Bakhtin analyzes the political and social power of this rebellious figure, an embodiment of Carnival, triumphing temporarily against the official culture's forces of Lent; and Shakespeare returns to this character type in *Twelfth Night* with Toby's hijinks against Malvolio. Shakespeare's capacious knight is also a *miles gloriosus*, the braggart soldier figure familiar from Plautine comedy; finally, Falstaff embodies the theatrical clown, that fool with license to mock authority figures, poking fun at their hypocrisy and lampooning their elevated speeches. Probably played by Will Kemp, Shakespeare's company's most famous clown, the character of Falstaff would be a physical one, complete with Kemp's famous dance moves. Ultimately, Falstaff is a transformative shape shifter, taking on a dazzling number of roles: he's an honorable thief, a good fellow, a cowardly lion, a euphuistic rhetorician, and an old melancholy man; he famously adopts the roles of prince and king.[55]

Shakespeare combines a dizzying range of theatrical traditions in creating Falstaff. It is a testament to the character's larger-than-life status that he has so many character types informing him. He isn't just one stock character: he's too big for that. As a consummate actor, indeed, an actor's actor, Falstaff begins the *Henriad* able to transform himself through imaginative feats. This transformative ability proves crucial not only for Falstaff as a character but also for the man playing him onstage: an actor was deemed capable of changing not merely his speech and costume but also, as Joseph Roach establishes, his humorological body, "precisely controlling the instantaneous transitions between passions."[56] Figures like Richard Burbage and Edward Alleyn were famous for their ability to shape-shift on stage. Exhibiting "Ovidian alterations of bodily state," these players demonstrated a range of passions through manipulation of their humours, a practice that allowed transformations from role to role and at different moments within a single role. As Thomas

Heywood writes of Alleyn, he is a "Proteus for shapes, and Roscius for a tonge / So could he speak, so vary."[57]

In his attachment to sack, Falstaff proves not only resistant as a political subject—a welcome position in Henry's corrupt England—but also, increasingly over the course of two plays, debilitated as a transformative actor. Even the most convivial activities become dangerous when compulsory, a point Hal makes at the start of the plays: "If all the year were playing holidays, / To sport would be as tedious as to work" (*1 Henry IV*, 1.2.194–95). So Shakespeare faced a theatrical choice: he could remain trapped in a form of compulsory conviviality, representing his devoted knight in his form of good fellowship, night after night, through multiple plays. Or Shakespeare could—and did—find a way of deepening Falstaff's character: through his fall into a contemporary, topical concern surrounding habitual drinking and disease, presenting such addiction through an awareness of its dangers and an appreciation of its devotional commitments. The banishment of Falstaff is potentially moralizing, but it also emerges out of the play's dogged portrait of the knight's fidelity. The plays illuminate the costs of addictive fellowship within rejection, the tenacious attachment to Hal at the expense of self. Falstaff thus develops from the comic Vice, the braggart soldier, the carnivalesque drinker, and the fool into a tragic figure: he is the isolated drinker, a type increasingly familiar from medical pamphlets, puritan tracts, and state laws; and he is the abandoned addict, longing for his devotional object. In facing the choices between conviviality and tragedy, the playwright ultimately chose his own craft, upholding theatrical range even at the expense of his most popular character.

As a result of such dramatic transformation Shakespeare banishes, even as he generated, one of the theater's best roles and one of its most novelistic characters, both in terms of tragic trajectory and comic range. Falstaff becomes Shakespeare's legacy, a characterological testament to his dramatic talents. The play's end thus stages a heartbreaking spectacle: Falstaff—the man who "foreswear[s] thin potations and addict[s himself] to sack"—is condemned. The audience, of course, desperately wishes it were otherwise. But in depicting Hal's rise as sovereign, Shakespeare represents the point at which deep attachment—living in unerring, persistent devotion to another—proves too threatening to endure, and that point is Falstaff's addiction.

Chapter 4

Addiction and Possession in *Othello*

> The locus of addictiveness cannot be the substance itself, and can scarcely even be the body *it*self, but must be some overarching abstraction that governs the narrative relations between them.
>
> —Eve Kosofsky Sedgwick

With the Venetian troops in Cyprus and the Turkish fleet destroyed at sea, the Herald reads Othello's general announcement, inviting "each man to what sport and revels his addiction leads him" (2.2.5–6).[1] This addiction, as analyzed in this project thus far, might be to study or to God; it might be to love or to friendship. Or it might also be, as the scene in *Othello* proceeds to reveal and this chapter explores, addiction to drinking. Regardless of what "revel" the addict choses, the Herald's phrasing insists that he or she will be "lead," drawn out by the agentive force of addiction itself. Even if the choice of one's addiction signals liberty, as the Herald's invitation seems to suggest, it also betrays subservience before one's own desires or leanings. Compressed into this short invitation lies one of the play's more pressing questions: if a character is led to drunkenness, violence, or murder, where does responsibility for such action lie? Does it lie with the individual who carries out the action? Or does it come from the force that "leads him" and overtakes him—whether it be in the form of alcohol, love, or some other passion?

Such questions about addiction and responsibility have remained unaddressed in this book so far. The first chapters of this project considered largely reflexive addictions: the devotions in *Doctor Faustus*, *Twelfth Night*, and the *Henriad* were actively chosen and embraced. But as noted in Chapter 1 of this project, addiction—and the other key terms associated with addiction, like conviction, commitment, and compulsion—might represent choice, or they might signal an external dictate. One might be committed to a spouse, or to an asylum; one might have a conviction as a viewpoint, or as a criminal record; and one might be compelled by an idea, or by the law. In turning to imperative addiction, this chapter explores individuals controlled, or at least guided by, an external force.[2] Such imperative addictions, in their transformative power, result in criminal outcomes, at once determined by the play's tragic genre and the compulsive nature of the addictions themselves. The Herald might invite soldiers to "sport and revel," but addiction within tragedy concentrates on overpowered choices leading to criminal acts. The tragic will proves, as Mariana Valverde puts it in her study of addiction, "diseased" or misled, compromising the choice to act in the first place.[3] The result is that an addict is at once agentive and subjected, thus frustrating the essential task of assessing responsibility.

Othello stages criminal action mitigated by the incapacity of addiction. Othello is prompted to murder in the name of love and loyalty, compelled yet also choosing to act. Possessed by another, Othello commissions or attempts to execute the murder of Cassio, Desdemona, Iago, and himself. He understands his actions as his own, and yet as doubly foreign: he names the external prompter as Iago, but he also externalizes his own criminality, as the turbaned Turk or a possessive madness. In its broad consideration of agentive selfhood, *Othello* engages the early modern conception of addiction as a form of possessive devotion that overcomes the consenting subject. Othello's love for Desdemona demonstrates the admirable, extraordinary capacity of the addict to dedicate himself to another. Yet other forms of addiction come to challenge Othello's primary devotion to his new wife. One might call his addiction jealousy, and indeed, such an addiction appears prominently in one of Shakespeare's potential sources for the play, John Pory's translation of *A geographical historie of Africa*, by Hasan ibn Muhammad al-Wazzan, christened Johannes Leo Africanus.[4] Africanus describes people exceedingly "addicted to jelousie," as well as those "addicted to witchcraft."[5] For some critics, a predisposition to jealousy—or what Africanus calls an addiction to jealousy—is exactly what Othello displays: he becomes most himself when jealous. Othello could also be called, in keeping with Africanus, addicted to magic or witchcraft in his association with the charmed handkerchief, a family heirloom.[6] Yet Shakespeare shapes Othello as startlingly independent from social and characterological compulsion: a newlywed allegedly free from sexual desire, a Venetian Moor initially self-assured and calm in the face of his city's hysterical racism, and a former bondslave who secured his own freedom, Othello excites precisely the expectation of escape from tyranny and compulsion, whether physical or environmental.

In contrast to the *Historie*, then, Othello's addiction is not to jealousy or magic; his devotion is distinct from humoral predisposition. As elaborated in Chapter 2, a humoral predisposition amplifies a preexisting state. Addiction, by contrast, challenges agentive selfhood. Othello, in loving Desdemona, addicts and shatters himself. He begins the play in the state of devoted love, already redefined and transformed even if he only reluctantly admits this. Attacked for his marriage, conscripted through love into what Ian Smith has described as "an internal 'race war' initiated by the play's resident racist, Iago," Othello is eventually convinced that his devotion is misplaced and counterfeit.[7] As a result he shifts his pledge of love from Desdemona to the ensign he mistakenly imagines to be honest and true. Ultimately, in shifting his addiction from his wife to Iago, Othello destroys the lives of those he loves, at once compelled and yet seemingly free to act otherwise.

The extraordinary devotional capacity of Othello—his predisposition as an addict—proves, in a corrupt environment, his undoing. The play heightens Othello's drama by staging a form of addictive undoing not once but twice, anticipating Othello's tragedy by staging Cassio's earlier on. The play's early scene of drinking, in which Cassio's intoxication leads to brawling, explores how excess compromises the will and draws authority into question. Why offer such a meditation on drunkenness, stretching from the scene of Cassio's health drinking through the remainder of the play? Why feature drunken excess as a source of criminal wrongdoing in a play about much deeper and more troubling forms of villainy? Through unexpected discussions on the origins of drunken habit, the play investigates the tangled relation of addiction to responsibility and capacity posed in the play's more dominant plotline. The surprising outcome of the drunken brawl—even in the absence of clear resolution—lies in the sympathy shown toward Cassio as an excessive drinker, one who experiences the double bind of the addict, who is at once taken to be strictly responsible for his guilt and also figured as *non compos mentis,* or incapacitated. Cassio may well be, as James Siemon argues, derelict in his duty and supported by "intrinsically unqualified female agents" who trespass codes of military masculinity.[8] Yet nonetheless the play invites us to consider his innocence, due at least in part to his incapacity from both drink and the plotting of Iago.

This sympathy toward—and interpretive sophistication in presenting—

Cassio's plight is in sharp contrast to the contemporaneous conversations on drunken responsibility within early modern criminal law. Early modern legal theorists, as this chapter's first section will explore, insist upon strict responsibility when it comes to addictive action. Specifically, criminal acts undertaken by drunken defendants do not excuse but exacerbate guilt. Shakespeare's more complex understanding of the addict's conundrum—at once actor and acted upon—pushes against these contemporary legal views, asserting the transformative power of addictions against the law's insistence on strict responsibility: how can an addict be deemed guilty when he is no longer himself, when his actions have been determined by another? Staging drunken criminality allows exploration of these questions by putting pressure on conceptions of individual will and *mens rea* at stake in early modern (and modern) understandings of addiction. And questions of *mens rea* are, of course, exactly what haunt the tragedy, from Cassio's drinking to Othello's murderous act.

Cassio's condition, being both exonerated from guilt because of his compromised ability to choose and condemned due to his culpable and predictable actions, anticipates in small measure the complex position of Othello. Othello, too, acts with compromised freedom. But his ability to consent—in defiance of his own training and profession—to being overtaken by another, and to experience the deep passion that results, remains admirable. Addictive propensities allow both Cassio and Othello to open themselves to others, to relate to the world around them. Being possessed is to be a participant in community. From possession of the handkerchief to possession of a spouse, the play highlights the vulnerability of such attachments to loss, manipulation, and deceit.[9] Yet even as the play stages a tragic result emerging from such possession, it equally upholds addiction over an excessive exercise of the will. The oppositional relation of Iago and Othello—one driven by the will and the other releasing himself into addiction—reveals how the exercise of the will, so lauded in legal rulings, comes at the expense of connection, commitment, and devotion. Even, then, as *Othello* draws on contemporary concerns about drinking addiction in its portrait of Cassio, the play uses the drinking scene to think through, with sympathy, the play's broader meditation on the ability to be possessed, to open oneself up to love and community even at the risk of oneself.

Drunken Incapacity in English Law

If a drunken man staggers blindly across a tavern, gets in a fight, and kills someone, what role does his drunkenness play in his guilt? He was, by virtue of alcohol, incapacitated. But the legal definition of incapacity, by which a defendant is deemed incapable of *mens rea*, or willful action, most often references mental or physical illness, often due to aging or disability. While it can also stem, as this example suggests, from drunkenness, such drunken incapacity poses difficulties for legislators. Unlike the incapacity of illness and aging, which are brought on involuntarily, defendants might actively choose to get drunk. And because the capacity for willful choice is fundamental to the legal understanding of a natural person (namely a human being who exercises rights and obligations), the implications of pardoning criminal defendants on the grounds of drunkenness, a chosen incapacity, are troubling. As the legal historian Jeremy Horder writes, "The possibility that defendants may plead that, through no fault of their own, they lacked the capacity for free agency at the time of their alleged offence, raises the most fundamental issues of principle in the criminal law."[10] Horder underscores how legislation on drunkenness reveals the dilemma of legal personhood more generally: incapacity challenges legal understandings of the individual as a "natural person" who acts under the law. The law's assumed link of personhood with self-possession, autonomy, liberty, and rights runs up against modes of human life that deviate, whether by choice or compulsion, from such sovereignty.[11]

Today the legal solution to this conundrum of intoxication and *mens rea* lies in the manslaughter charge, which admits the compromised will of the drunken defendant while still prosecuting him as a natural person who has killed someone.[12] The early modern solution is quite different, primarily because early modern legislators did not consider drunkenness as a form of incapacity. On the contrary, it was deemed in the landmark case of *Reniger v. Feogossa* (1551)—the first ruling on intoxication in English law—to be a voluntary condition, a sign of an active and errant will: "If a person that is drunk kills another, this shall be felony, and he shall be hanged for it, and yet he did it through ignorance, for when he was drunk he had no understanding nor memory; but inasmuch as that ignorance was occasioned by his own act and folly, and he might have avoided it, he shall not be privileged thereby."[13] Since an individual elects to drink he is granted no special consideration.

Indeed, as a subsequent case goes on to establish, the defendant might even be further penalized for drunkenness. According to *Beverley's Case* (1603), "Although he who is drunk is for the time *non compos mentis*, yet his drunkenness does not extenuate his act or offence . . . but it is a great offence in itself, and therefore aggravates his offence, and doth not derogate from the act which he did during that time."[14] Prominent jurists and legal theorists of the early modern period recapitulate this point. "A drunkard," Sir Edward Coke states, "is *voluntarius daemon* . . . ; what hurt or ill soever he doth, his drunkenness doth aggravate it."[15] William Blackstone writes, in a section on deficiencies of the will caused by madness, chance, and drunkenness, that "as to artificial, voluntarily contracted madness, by *drunkenness* or intoxication, which, depriving men of their reason, puts them in a temporary phrenzy: our law looks upon this as an aggravation of the offence, rather than an excuse for any criminal misbehaviour."[16] Punishment for drunkenness has classical precedent: Blackstone cites the Greek enactment "that he who committed a crime when drunk should receive double punishment," and Aristotle writes in *Nichomachean Ethics* that "we punish a man for his very ignorance, if he is thought responsible for the ignorance, as when penalties are doubled in the case of drunkenness; for the moving principle is in the man himself, since he had the power of not getting drunk and his getting drunk was the cause of his ignorance."[17]

Not all early modern lawyers pressed for higher sentencing, but the understanding of drunkenness as the defendant's choice and therefore personal responsibility was nearly universal. A person who committed a criminal offence should not be excused when, as Aristotle says, "the moving principle is in the man himself." Thus Francis Bacon and Richard Hooker both agree that even if the law should not punish involuntary action, it should nevertheless punish actions committed while drunk. As Bacon puts it, "If a mad man commit a felonie, he shall not lose his life for it, because his infirmity came by the act of God; but if a drunken man commit a felonie, he shall not be excused because his imperfection came by his owne default."[18] Similarly, Hooker—in his discussion of how humans might transgress their own natures—asserts how drunkenness might cause a man to commit a crime against nature. Yet such a crime is a matter of choice, not compulsion, and thus he is responsible for it: "It is no excuse . . . unto him, who being drunk committeth incest, and allegeth that his wits were not his own, inasmuch as himself might have chosen whether his wits should by that mean have been taken from him."[19] Of course, this is not to say that the courts always meted out higher sentences to drunken offenders.[20] Drunkenness does appear, at times, as a mitigating circumstance in criminal cases. For example, the case of Thomas Baynard vs. Georg Haythorne and his wife Anne featured the "purchase of unspecified property at unfair price after making plaintiff drunk."[21] Nevertheless, the key legal theorists of the period, and the landmark cases, uphold a strict view of drunkenness as a choice and as a sign of voluntary madness.

In tandem with these rulings on incapacity, legislators pressed to criminalize drunkenness itself. One of the period's most significant and understudied legal struggles concerns this long and ultimately successful parliamentary battle to deem drunkenness itself a crime.[22] Prior to the sixteenth century, legislation targeted not the drinker but the ale sellers, as Peter Clark, Judith Hunter, and others demonstrate in their surveys of the history of tavern legislation.[23] Yet this control of alehouses was as much an economic as a social or criminal issue: "The magistrates were far more concerned with the economic aspects of running alehouses than with the number of those who got drunk," F. G. Emmison writes.[24] Over the course of the sixteenth century, however, this economic interest was met by a social and moral one: controlling the heavy drinker. No longer content to regulate alehouses, legislators now sought to create yet another, more stringent measure to control drinking.[25] Thus a series of bills on drunkenness, presented in Parliament between 1566 and 1606, began to posit excess drinking itself as a crime. These bills do not concern—as modern law does—drunken *action*. Rather than targeting those who are drunk and disorderly, these bills instead criminalize drunkenness itself, deeming it an overthrow of the subject, a form of tyranny, sin, and waste that endangers both the drinker and the community.

The 1606 "Act for the Repressing of the odious and loathsom Sin of Drunkenness" (4. Jacobus 5) begins the official history of criminalizing drunkenness in England. The culmination of forty years of legislative efforts, the act targeted drunkenness alone, claiming that drink overthrows, disables, impoverishes, and abuses otherwise loyal subjects: "The loathsome and odious Sin of Drunkennesse," the act argues, "is of late grown into common use within this Realm" and leads to "the overthrow of many good Arts and Manuall Trades," as well as "the disabling of divers Work-men; and the generall Impoverishing of many good Subjects."[26] The act had significant staying power, with all subsequent attempts at legislation, including in 1614, 1620–21, and 1623, repeating this act's sense of drunkenness as an acute problem.

Since the history of this legislation on drunkenness is largely unfamiliar—in contrast to the extensive studies on the history of tavern regulation—it is worth piecing together the forty-year battle to pass this 1606 act. Doing so reveals precisely the concerns, and at times the sympathy, shown by parliamentarians and monarchs toward the power of drunkenness. This history also reveals a set of shifting conceptions of and accommodations toward drinking. Notably, the 1606 legislative effort shifts from earlier attempts to regulate the "drunkard" to instead attacking "drunkenness." This may account for the success of the act's passage after four decades of effort: condemning a so-called drunkard proved a losing battle. By locating the agency in drinking rather than in the individual who drinks, legislators shifted blame away from a person and onto an activity—one that, as the legislation indicates, should be controlled and prevented. If the earlier drafts insisted on the damage a drinker causes the nation, the final act blames drinking itself for abusing individuals.

The legislative attempts to pass a bill on drinking began in 1566, with the bill "on swearing and drunkenness" to appear in Parliament.[27] Linking drunken excess to swearing, gaming, and other allegedly sinful pursuits, the bill failed. Two decades later, the more substantial 1584 bill, labeled "An Act against excessive and common drunkenness," more aggressively justifies the need for such legislation. Even as it echoes the 1566 version in condemning drinking a spiritual problem, a "sin" that brings on the "displeasure of almyghtie God," the 1584 bill's language more immediately insists on the antisocial effects of drunkenness, its "poornes of life." Habitual ("common") or "excessive" drunkenness represents a misuse of resources: "A few in excesse" consume that which "moderately used would nourish and satisfie many." Further, drinkers commit crimes, including "swearing and blasphemynge . . . quarrelless fyghtinge bloodshede, manslaughter," as drunkards are driven to "unlawfull shifte, and become more like brute beastes than resonable creatures." As a result, the drinker should be punished "as a comon Barretor," namely a quarrelsome person given to brawling or riot, a "rowdy" or hired bully.[28]

The 1584 bill exposes the conceptual tension at stake in regulating drunkenness. The heavy drinker actively abuses the nation and wastes its resources. Indeed, he is a bully and a "barretor." Yet drinkers, in becoming "brute beastes," by "falling into the saide vice," appear enslaved and transformed by drinking, suggesting ways in which they lack legal agency. In this regard the term "barretor" is especially telling, because it describes a particular kind of bully, one hired out as a "rowdy." Rather than acting on his own behalf, the barretor quarrels for others. This conundrum concerns a criminal who is at once free and bound: the drinker both exercises free will in his drunkenness and seems bound in it, behaving as alcohol dictates. If this condition is a familiar one, it is still worth noting how even the era's strictest legislation hesitates about the drinker's agency. Indeed, subsequent drafts of legislation move between views of the drinker as an agent and as a slave.

When the 1584 bill failed, Parliament raised the issue again in 1601, this time targeting the drinker even more directly. Heavily amending the 1584 draft into a "bill against drunkards and common haunters of alehouses or taverns," the 1601 bill targets neither singular drunkenness nor alehouse activities, but instead excessive, habitual drinking: "drunkards" and "common haunters."[29] Indeed, the revised title of the bill reinforces the shift: the 1584 bill regulated "excessive and common drunkenness," namely an activity; now the bill regulates "drunkards" and "haunters," that is to say, individuals defined by their relationship to alcohol. 1601 was a year that also saw the introduction of "a bill for reformation of abuses in alehouses," and "a bill against victualling houses and taverns." This was revised to "a bill for supressing alehouses and tippling houses" read on December 2 in the House of Lords, and on December 4 in the House of Commons.[30] The "drunkards" bill thus appears in tandem with regulation of alehouses, rather than in place of such regulation.

The 1601 bill, like the two before it, did not pass. It took another five years before legislation on drunkenness received approval. The resulting 1606 act condenses much of the language of the prior bills while also shifting the focus—as noted above—to the drinker victimized by alcohol (Figure 1). It reads:

> A Statute against Drunkennesse [An Act for the Repressing of the odious and loathsom Sin of Drunkenness] (4. Jac 1, c.5).
>
> WHereas the loathsome and odious Sin of Drunkennesse is of late grown into common use within this Realm, being the Root and Foundation of many other enormous Sins, as Bloodshed, Stabbing, Murder, Swearing, Fornication, Adultery, and such like; to the great dishonour of God, and of our Nation; the overthrow of many good Arts and Manuall Trades; the disabling of divers Work-men; and the generall Impoverishing of many good Subjects, abusively wasting the good Creatures of God:
>
> Be it therefore enacted by the Kings most Excellent Majestie, the Lords and Commons in this present Parliament Assembled, and by the Authority of the same, That all and every person or persons, which shall be Drunk, and of the same offence of Drunkennesse shall be lawfully convicted, shall for every such offence, forfeit, and

> loose five shillings of lawfully Money of *England,* to be paid within one week next after his, her, or their conviction thereof, to the hands of the Church-Wardens of that Parish, where the offence shall be committed, who shall be accomptable therefore to the use of the poor of the same Parish: And if the said person or persons so convicted, shall refuse, or neglect to pay the said forfeiture, as aforesaid, then the same shall be from time to time, levyed of the Goods of every such person or persons so refusing or neglecting to pay the same, by Warrant or Precept from the same Court, Judge or Justices, before whom the same conviction shall be: And if the offender or offenders be not able to pay the said sum of five shillings, then the offender or offenders shall be committed to the Stocks for every offence, there to remain by the space of six houres.
>
> And it is further enacted by the Authority aforesaid, That if any person or persons, being once lawfully convicted of the said offence of Drunkennesse, shall after that be again lawfully convicted of the like offence of Drunkennesse; That then every person and persons so secondly convicted of the said offence of Drunkennesse, shall be bounden with two sureties to our Soveraign Lord the Kings Majestie, His Heirs, and Successors, in one Recognisance or Obligation of ten pounds, with condition to be from thence forth of good behaviour.[31]

Whereas earlier versions had a much more elaborate definition of drunkenness based in an identity (namely the 1601 "drunkard or comon haunter of alehouses for needless excessive drinking"), here the legislation condemns the act of "drunkenness." Whereas earlier versions specified two justices of the peace as the arresting officers or witnesses, here the legislation omits reference to the precise mechanisms of arrest. Perhaps this act resolved some of the (now lost) debate that occurred around the earlier drafts of the legislation: a parliamentary exchange on swearing in 1601 reveals that certain members of Parliament were concerned about the opportunities for corruption offered to justices of the peace who might make arrests as a means of lining their own pockets.[32] A later parliamentary debate in 1620 saw the committee revising the 1606 legislation to "define a drunkard."[33] This discussion suggests how members might not share consensus on what precisely constituted a "common drunkard" or "common haunter," despite the appearance of these phrases in local records and in the titles of legislation. Finally, the 1606 bill underwent heavy revision. Its introduction on December 5 provoked dissent: "Sir Rob. Johnson offereth a reformed Bill, for repressing the odious and loathsome Sin of Drunkenness—the first Reading, and hissed."[34] Then the bill, engrossed in committee, re-appeared on March 3: "Upon a third Reading, much disputed; Two Provisoes read thrice, and, upon several Questions, added; and the Bill, with the Provisoes, upon another Question, passed."[35]

This law received lavish praise, at least among those puritan ministers who vigorously supported criminalizing drunkenness and who, in increasing numbers, held parliamentary seats. Richard Rawlidge, for example, praises the legislation in a pamphlet advertising all parliamentary acts against drunkenness (Figure 2). John Downame, the puritan minister and brother to King James I's chaplain George Downame—and familiar from this project's earlier chapters—also praises Parliament for the 1606 act. He does so in terms that highlight the stakes of the debate, calling habitual drunkenness an addiction

A Statute againſt Drunkenneſſe, Viz.

4. Jacobus 5. *The penaltie of a Drunkard, &c.*

WHereas the loathſome and odious Sin of Drunkenneſſe is of late grown into common uſe within this Realm, being the Root and Foundation of many other enormous Sins, as Bloodſhed, Stabbing, Murder, Swearing, Fornication, Adultery, and ſuch like; to the great diſhonour of God, and of our Nation; the overthrow of many good Arts and Manuall Trades; the diſabling of divers Work-men; and the generall Impoveriſhing of many good Subjects, abuſively waſting the good Creatures of God:

Be it therefore enacted by the Kings moſt Excellent Majeſtie, the Lords and Commons in this preſent Parliament Aſſembled, and by the Authority of the ſame, That all and every perſon or perſons, which ſhall be Drunk, and of the ſame offence of Drunkenneſſe ſhall be lawfully convicted, ſhall for every ſuch offence, forfeit, and looſe five ſhillings of lawfully Money of *England*, to be paid within one week next after his, her, or their conviction thereof, to the hands of the Church-Wardens of that Pariſh, where the offence ſhall be committed, who ſhall be accomptable therefore to the uſe of the poor of the ſame Pariſh: And if the ſaid perſon or perſons ſo convicted, ſhall refuſe, or neglect to pay the ſaid forfeiture, as aforeſaid, then the ſame ſhall be from time to time, levyed of the Goods of every ſuch perſon or perſons ſo refuſing or neglecting to pay the ſame, by Warrant or Precept from the ſame Court, Judge or Juſtices, before whom the ſame conviction ſhall be: And if the offender or offenders be not able to pay the ſaid ſum of five ſhillings, then the offender or offenders ſhall be committed to the Stocks for every offence, there to remain by the ſpace of ſix houres.

And it is further enacted by the Authority aforeſaid, That if any perſon or perſons, being once lawfully convicted of the ſaid offence of Drunkenneſſe, ſhall after that be again lawfully convicted of the like offence of Drunkenneſſe; That then every perſon and perſons ſo ſecondly convicted of the ſaid offence of Drunkenneſſe, ſhall be bounden with two ſureties to our Soveraign Lord the Kings Majeſtie, His Heirs, and Succeſſors, in one Recogniſance or Obligation of ten pounds, with condition to be from thence forth of good behaviour.

Figure 1. The 1606 Statute against Drunkennesse (4. Jac 1, c.5).
© The British Library Board. (669.f.7.(70)).

undermining the nation: "Our wise Statesmen thought it necessary in Parliament, to inact a law for the suppressing of this sin; for *Ex malis moribus bonae nascuntur leges*; evill maners occasion good lawes. And indeed not without good cause is the sword of the Magistrate joyned with the sword of the spirit . . . For who seeth not that many of our people of late, are so unmeasurablie addicted to this vice."[36] Downame's treatise echoes the arguments of the legislation. First, like the legislation, which distinguishes between the single and repeat offender, Downame targets the "addicted" drinker over those who "drink to live." Second, if drafts of the legislation complain of the few who "waste in excess" what might "nourish and satisfie many," Downame also criticizes the drinker for consuming precious resources: he is a "devourer of the fat of the land, in which respect he is more pernicious to a State, & more fit to bring a comon dearth, than either canker-worme or caterpillar."[37] Like the "caterpillars of the commonwealth" (Shakespeare's term for bad counselors), these habitual drinkers corrupt the state from within.[38]

Finally, just as the legislation notes the "overthrow" of trades and the "disabling" of workmen, so too does Downame show how drunkenness robs the commonwealth: the drinker "disableth himselfe that he cannot performe any good service to his countrie."[39] Although Downame invokes the language of festivity—the drinker "carouses"—he goes on to define the problem in more political terms. Unruly subjects are tyrannized by the power of drink: "A drunkard can neither bee good Magistrate, nor good subject, seeing hee cannot rule others that cannot rule himself."[40] This legislative attack on drunkenness, and its attendant condemnation in godly polemic, reinforces the legal theory cited above: drinkers are doubly responsible, both for their drinking and for the criminal actions committed while drunk. This 1606 act thus works in tandem with theories of drunken responsibility to condemn the drinker for his errant but active will. At the same time, this act embeds the fissures evident in broader theories of addiction: the drinker is at once entirely responsible, and abused and overthrown.

Given the difficulty in establishing responsibility—even the strictest legislation hesitates about the drinker versus the drink as culprit—perhaps unsurprisingly this legal insistence that drunkenness is a sign of the defendant's active will in choosing errancy over health is short-lived. With the growing view of excessive, habitual drunkenness as a kind of permanent madness, indeed as an addiction in modern terms, jurists swing the pendulum of legal reform toward the opposite extreme, no longer blaming but entirely exonerating drunken defendants. What happens, the law begins to ask, when the

A MONSTER LATE FOVND OVT AND DISCOVERED.

Or

The ſcourging of Tiplers, the ruine of *Bacchus*, and the bane of Tapſters.

wherein is plainly ſet forth all the lawes of the Kingdome, that be now in force againſt Ale-houſe keepers, Drunkards, and haunters of Ale-houſes, with all the paines and penalties in the ſame lawes.

With ſundry of their cunning inventions, hatched out of the Divells ſtore-houſe, and daily practiſed by Ale-houſe-keepers, Tapſters, &c. With an eaſie way to reforme all ſuch diſorders.

Compiled by *R.R.*

Iſa: 5.11. *Wo be to them that riſe vp early to follow drunkenneſſe, &c.*

Imprinted at *Amſterdam* Anno 1628.

Figure 2. Title page of Richard Rawlidge, *A Monster Late Found Out and Discovered* (1628). Rare Book 51778. The Huntington Library, San Marino, California.

defendant is an addict? Sir Matthew Hale considered this question as early as the 1670s, but it was in the 1820s that jurists came to rule on what Hale calls the "habitual or fixed phrenzy" of drunkenness.[41] After the "discovery" of the more familiar notion of addiction as compulsion and disease in 1800 by the physicians Benjamin Rush and Thomas Trotter, legislators began to consider habitual drunkenness as a mitigating circumstance. In cases including *Rex v. Grindley* (1819) and *Regina v. Cruse* (1822), the court ruled that drunkenness diminishes criminal intent. *Burrow's Case* and *Rennie's Case* further established that "fixed, habitual, and permanent" madness as a result of intoxication reduced a man to a state of "being without reason or mind," and therefore "not accountable or responsible for his actions."[42] As Horder writes of incapacity in general, "Our system of law is not based . . . on principles of brute deterrence, and presupposes that to become criminals people must be responsible for harm, and not just cause it."[43] This Enlightenment view of incapacity yielded, by the twentieth century, to a more flexible understanding of how a drunken defendant might be held criminally liable for his or her actions, while also retaining possible protection from the charge of specific intent to harm, a charge required for the prosecution of crimes such as first-degree murder. Hence, after the early modern views of responsibility and the Enlightenment notion of guiltlessness, modern jurists introduce a *via media* by using the manslaughter charge for drunken defendants.

This survey of views on drunkenness exposes the contrast between early modern and modern conceptions of the drunken will. To look at rulings such as *Feogossa* and *Beverley*, one might imagine that sixteenth-century legislators and citizens deemed drunkenness as a matter resting entirely in an individual's control. Furthermore, one might imagine that it is only with the "invention" or "discovery" of addiction that the concept of the incapacitated will of the habitual drinker enters legal and lay understandings. The legislative record even supports this assumption. As suggested above, Rush and Trotter, the two physicians credited with the discovery of alcoholism, published their findings in 1800; twenty years later the first rulings on the incapacitated will of the habitual drinker would appear.

Even as early modern legislators engage with will so strictly in cases of drunkenness, however, contemporary nonlegal writings acknowledge the tangle of incapacity and responsibility at stake in drunkenness and addiction more broadly. Precisely at the moment of *Beverley* in 1603, Shakespeare explores the ways in which the addict is often the victim of either his own uncontrolled impulses or the criminality of others. In *Othello*'s broad investigation of incapacity, both Cassio and Othello commit crimes engendered through their "own act and folly," to use *Feogossa*'s language. They are both "voluntarily contracted" and compelled in their incapacity, being, in legal terms, "*voluntarius daemon*" and displaying "imperfection" of their "owne default."[44] The play invokes the legal link of human personhood with agentive choice, showing Cassio punished for criminal action and Othello arrested. But the play more dramatically stages the incapacity of these victims and links agentive choice to Iago's villainous hyperexercise of the will.

The incapacity of Cassio and Othello might invoke a counterfactual response: What if each had been strong enough to stay fixed in their duties, against Iago's goading? What if they had maintained strong boundaries against attack? But such a thought experiment denies the ways in which incapacity does not, in itself, make these characters culpable. Instead their addicted incapacity emerges from the desire to join fellow soldiers, in the case of Cassio, or one's beloved, with Othello. Such desire for union might, in itself, be called admirable. These characters represent a potentially troubling but also expansive notion of self, a notion that extends beyond the legal insistence on self-possession. The legal preoccupation with such self-possession or sovereignty ironically secures only isolation as the fundamental human right. The legal theorist Jennifer Nedelsky critiques the law on precisely these grounds, as a "(misguided) attempt to protect individual autonomy." She explains: "The perverse quality of this conception [of the law] is clearest when taken to its extreme: the most perfectly autonomous man is the most perfectly isolated."[45] Othello's heroics reveal in contrast, even at tragic cost, a mode of living based in addiction's transformative and connecting power.

Cassio's Infirmity

Shortly after the Herald invites "each man to what sport and revels his addiction leads him," Iago encourages Cassio to join the soldiers and drink: "Come, lieutenant, I have a stoup of wine, and here without are a brace of Cyprus gallants that would fain have a measure to the health of black Othello" (2.3.26–29). But Cassio avers, prompting the following exchange:

> *Cassio:* Not tonight, good Iago, I have very poor and unhappy
> brains for drinking. I could well wish courtesy would invent
> some other custom of entertainment.

> *Iago:* O, they are our friends. But one cup, I'll drink for you.
> *Cassio:* I have drunk but one cup tonight, and that was craftily qualified too, and behold what innovation it makes here! I am unfortunate in the infirmity, and dare not task my weakness with any more.
> *Iago:* What, man, 'tis a night of revels, the gallants desire it.
> *Cassio:* Where are they?
> *Iago:* Here, at the door, I pray you, call them in.
> *Cassio:* I'll do't, but it dislikes me. (2.3.26–44)

This exchange hinges on the encounter of opposites: the frank Cassio versus the crafty Iago, the honorable lieutenant versus the villainous ensign. The audience watches Cassio snared by his enemy, since Iago has announced in the scene before that he is plotting the lieutenant's ruin, and he reiterates this intention as Cassio goes in (2.1.270–75).

But the scene is about more than the opposition of these two characters. It is also about the opposition of two incompatible views of drinking. Iago represents merriment and sociability: he tells Cassio that "gallants" are drinking to "the health" of Othello. It is a "night of revels," and these drinkers are "our friends." This custom of health drinking—the focus of Chapter 5—was a prevalent one, associated with male communities bound in political-military unity. Drinkers would pledge the health of their superior or absent friends, either passing the glass around the table or draining it entirely. Thus Iago's invitation might be in Cassio's best interest, since drinking rituals were ways of cementing alliances and demonstrating fortitude. For Cassio to refuse to drink with the other soldiers might betray his unfitness for his job (which Iago alleges from the start). The lieutenant could appear, in abstaining, as unsociable, haughty, or puritanical, as well as disloyal and unmanly, a particularly potent charge for a lieutenant accused of inexperience in the field.[46]

Against Iago's view of festive drinking lies Cassio's own. He knows that health drinking is a "courtesy" and "custom"; he knows that he should participate, and he fully accepts Iago's depiction of drinking. Indeed, he capitulates in the end, attesting to the strength of the social pressure to drink and to his own desire to join friends. But at the same time, Cassio offers an entirely different picture of alcohol's effects. For him, it is not about good fellowship but about disease. He is open about his own troubles with drinking: he has "poor and unhappy brains for drinking"; it is "an infirmity," "a weakness." He speaks of the physical effects that drinking alcohol has on him. Looking at his drunken body, he claims, "What innovation it makes here!" While on the one hand, Cassio's language might merely suggest he has no tolerance—he gets quickly drunk because he is a lightweight—on the other hand, his description invokes the notion of infirmity familiar from religious and legal discourses. He draws attention to the addicted brain, or the "unhappy brain," which is not unaffected by alcohol but indeed is too much affected by it. In this sense, Cassio stands in contrast to other characters affected by toxins in the play, most notably Othello, who little knows how the poison of Iago's jealousy will work on his system and who proves unable to protect himself. Cassio knows about the dangers of alcohol and tries to prevent its effects. So he has been pretending to drink more than he actually has to keep up appearances, knowing that he should drink no more.

The scene thus stages a confrontation between two alternative views of drinking. The men speak in opposite languages (social vs. physical). These languages are considered equally legitimate, and both press upon the loaded term "addiction" deployed by the Herald. Iago's call on behalf of the merry gallants is a familiar one; Cassio knows it well. Part of Iago's mastery is that he takes stock arguments and twists them to his own ends, as in his use of the timeworn cuckoldry story to bait Othello. But even as the play rehearses this customary drinking practice, it also stages the opposite, but equally true, notion of drinking: it is a "custom" that overcomes participants, as the previous chapter revealed. Custom, as the Galenic formula puts it, alters nature, a maxim that addiction theorists deploy in later centuries as an explanatory device in accounting for habitual, excessive drunkenness: *habitum, alteram naturam.*[47] Cassio performs this maxim as he moves from externalized custom—the call to drink—to an alteration of his nature, the innovation his body experiences.[48] Cassio's "unhappy brains" recall these contemporary accounts of daily bouts affecting the drinker's brains: wine gives "the braine a blow, that like a subtil wrastler, it may supplant the feet afterwards," or "wine takes away the heart, and spoyles the braine."[49] Cassio's defense of abstinence, then, has to be taken seriously. Indeed, the audience recognizes that Cassio should take his own argument more seriously than he does, for in capitulating, he facilitates Iago's vengeful rise.

The rest of the scene continues to juxtapose Cassio's disease theory of drinking with Iago's conviviality thesis. But in doing so it tips the scales toward disease, not merriment, as the truer position, for it is precisely the soldiers' weakness for drink that allows sober Iago to become stage manager of events. Plying Roderigo and three Cyprian soldiers "with flowing cups"

(2.3.55) to ensure their drunkenness, Iago repeatedly calls for more drink throughout the scene: "Some wine ho!" (2.3.64), "Some wine, boys!" (2.3.70) "Some wine ho!" (2.3.93). Iago punctuates his calls for wine with drinking songs. Of his first song he claims, "I learned it in England, where indeed they are most potent in potting. Your Dane, your German, and your swag-bellied Hollander—drink, ho!—are nothing to your English" (2.3.72–75). The drunken Cassio asks, "Is your Englishman so exquisite in his drinking?" (2.3.76), to which Iago responds, "Why, he drinks you with facility your Dane dead drunk; he sweats not to overthrow your Almain; he gives your Hollander a vomit ere the next pottle can be filled . . . O sweet England!" (2.3.77–84). This laudatory inset on the strength of the English drinker resonates with any number of ballads circulated in the decades surrounding the play's performance, ballads portraying national pride in both English ale and English drinkers—particularly against the Dutch and Danes, their rivals in beer production and consumption.[50]

To have such a ballad sung by Iago, in Cyprus, is peculiar. How might the audience receive this bit of pro-England rhetoric? Does this scene of comic relief help promote nationalist drinking in the face of the famed abstinence of Muslim opponents? To enjoy this drinking scene *as* a drinking scene—indeed, to enjoy the scene as comic relief at all—requires viewers to suspend awareness of the unfolding plot and the role of Iago in generating it. And, were the drinking episode to achieve this affect, it would highlight even more powerfully the dangers of drink. If all it takes is a drinking song and the call of "sweet England" to get the audience to forget what Iago announces before the scene—he is not drunk, but merely acting so to entangle others—then they are just as likely as Cassio, Roderigo, Montano, and others to be the dupes of a villain. So even as an audience of early modern Londoners might take a bit of national pride in Iago's claims—the English can outdrink even the most notoriously drunken foreigners—the play also reminds viewers to suspect such rousing words: they feed into precisely the type of national competition that Iago hopes to take advantage of on this "warlike isle."

What is the origin of Cassio's infirmity? Why is he so "rash and very sudden in choler" (2.1.270) when drunk? To Cassio himself, such an infirmity comes from his "unhappy brains," his baffling weakness as a drinker. But this infirmity might equally come, as Cassio's arresting, drunken non sequitur might indicate, from reprobation, from failed election: in the midst of his drunken ranting he claims, "There be souls must be saved, and there be souls must not be saved" (2.3.99–100). The scene, if not the play itself, bears out this stark judgment. While all the soldiers drink, only Cassio is dismissed as a drunkard, ending the scene bloody, incapacitated, and mute. He imbibes like the other characters, but he alone receives severe punishment, a sanctioning that seems both extreme and inevitable. In this sense, as Cassio himself claims, he is among those unsaved. He becomes, in terms of his office under Othello, the unelected soul: he is excised from employment, marriage, and friendship.

Yet this fall from grace is complicated, of course, by Iago's role in engineering it, and thus Cassio's infirmity is and is not his own. Iago convinces Montano that Cassio is habitually drunk: " 'Tis evermore the prologue to his sleep: / He'll watch the horologe a double set / If drink rock not his cradle" (2.3.125–27). While drink, like tobacco and opium, was often prescribed to help cure insomnia, Iago's lines are more damning.[51] Cassio, in his view, is compelled by alcohol, habitually imbibing it to the point of necessity. And this possession by alcohol poses political risks, as Iago elaborates: 'Tis pity of him: / I fear the trust Othello puts him in, / On some odd time of his infirmity / Will shake this island" (2.3.121–24). As Cassio himself had claimed, drinking is an infirmity. And Montano quickly adopts this language in response: "And 'tis a great pity that the noble Moor / Should hazard such a place as his own second / With one of an ingraft infirmity" (2.3.134–36). The term "infirmity" has migrated in resonance: when deployed initially by Cassio, it meant weakness or inability for drinking, but now it means the weakness of *habitual* drinking. Furthermore, Montano's description indicates that what might have been an external, or foreign, practice has become incorporated into Cassio's character: the infirmity is now "ingraft[ed]" and permanent.

The scene ends with Cassio dismissed from his post as a drunkard. What begins as a single instance of drinking now stands for Cassio's identity writ large. Cassio is convinced he cannot sue for his place because, as he puts it, "I will ask him for my place again, he shall tell me I am a drunkard: had I as many mouths as Hydra, such an answer would stop them all. To be now a sensible man, by and by a fool, and presently a beast! O strange! Every inordinate cup is unblest, and the ingredience is a devil" (2.3.298–304). There is a strong dramatic irony accompanying Cassio's lament—the audience knows, as he does not, that the "devil" that torments him is not wine but Iago.[52] Yet there is also logic to Cassio's speech—it was truly wine that made him easy prey for Iago and Roderigo. Wine has indeed "stolen" from him; it has "transformed" him.[53] Where does responsibility for Cassio's fall lie? He is slandered, with his drinking bout exaggerated as habitual and excessive. He is also vulnerable to the social custom of health drinking, a common and nearly compulsory practice among groups of soldiers. Cassio nevertheless deems himself a drunkard and understands himself to be responsible for his choice to drink. As with the 1606 act analyzed above, Cassio is at once entirely responsible—and abused and overthrown.

Othello's Addiction

Iago preys upon Cassio's vulnerability, deploying customary arguments and social pressure against him while meting out precisely enough poison to ensure the lieutenant's incapacity. Cassio consents to drinking and thus subsequently holds himself responsible for the events, not realizing he has been manipulated by another. This dynamic repeats itself in Iago's dealings with Othello. The move from Cassio to Othello is a move from the custom of health drinking to that of cuckoldry, from the passion of wrath to jealousy, from the poison of liquor to fantastical speech. But the practice is the same, and the resulting questions—about capacity and responsibility—remain. In both cases Iago preys upon the predispositions and vulnerabilities of his victims. Cassio begins the play predisposed to drunken brawling. Othello begins the play possessed by intimate relation to another, and it is precisely this predisposition—this addiction to love—that Iago exploits.

Recalling the addiction to love as a devotional, not humoral, state helps illuminate the condition of Othello in the play's opening. Love addiction, as explored in Chapter 2's analysis of *Twelfth Night*, is a release of control that requires commitment beyond a singular act; it is a sustained process of giving up oneself again and again. To addict, to recall Thomas Thomas's formulation, is to "bequeath or give himselfe to something: to saie: to avow: to alienate from himselfe to another, and permit, graunt, & apponit the same to another person: to condemne: to approoue or alow a thing to be done, to deliver, depute or destinate to; to judge, to constraine, to pronounce and declare."[54] Thomas's definition resonates with that of other lexicographers in its gloss of addiction as bequeathing, giving, or devoting. But he puts this process in starker terms, calling it a form of alienation from oneself in favor of another; indeed, to addict is to "deliver, depute or destinate" oneself to another. And it is this process of delivering oneself to another that is at stake in *Othello*.

Othello's addicted state—his willingness to be overcome by devotion to love at the risk of his own identity—unfolds over the course of the play's first act, in spite of his vigorous attempts to reassure the Venetian state otherwise.

Indeed, what is most obvious about Othello's love in the first act is his attempt to downplay it. He has, he admits, put himself into "circumscription" for love: "But that I love the gentle Desdemona / I would not my unhoused free condition / Put into circumscription and confine / For the sea's worth" (1.2.25–28). His formula implies that deep relation to another requires loss of freedom and indeed a kind of entrapping possession, even as such love also represents a form of awakening. It is a familiar and problematically misogynist formulation to see marriage simultaneously as entrapment and the possession of another. But Othello's statement tellingly exposes how he feels not ownership of Desdemona but confinement of himself: he is the one possessed—by Desdemona, by the institution of marriage, by the domestication it implies, or perhaps most obviously, by love itself. And he chooses this relation. *He* loves, *he* has "put" himself into this circumscribed condition, consenting to what might otherwise seem uncomfortable or undesirable because his heart has transformed.

A military hero who marries privately, Othello—in his union—challenges the singular nature of his commitment to the state. The play's first act brings Othello before the senate on a military matter, where he endures instead a trial of sorts, asked to defend his course of private wooing. Condemned by the very man, Brabantio, who previously sought his companionship, Othello finds that Venice, his home, has become the site of interrogation and persecution as a result of his marriage. He maintains composure in the face of Venetian slander and refuses to be reactive even when attacked: "Let him do his spite," "my perfect soul / Shall manifest me rightly" (1.2.17, 31–32). Through the interrogation Othello asserts himself more vigorously as a trained soldier, comfortable in the theater of war, than as a newlywed. Indeed he seems unclear on his relation to what he calls Cupid's realm, hesitating on the agency behind his union: he has "ta'en away" Desdemona, he has "married her" (1.3.79–80), but he has also "won" her (1.3.94), as if granted her by merit. He did "draw from her a prayer" (1.3.153) but he also "did consent" (1.3.156) to her. This dispersed agency, moving between Othello and Desdemona as the agentive force behind the marriage, speaks affirmatively of their equal participation in wooing. But it also illuminates Othello's hesitancy to claim his will as his own, a hesitancy perhaps born of the "trial" scene that confronts him: in fielding racialized accusations of witchcraft and sorcery, he might well highlight Desdemona's agency as a means of placating his interrogators and protecting himself and their marriage.

In contrast to Othello's muted account of his own love, Desdemona's account of marriage in the opening scenes insists on its addictive, transformative power. "Desdemona," Ayanna Thompson writes, "makes it clear that she has been made wholly new through her marriage to Othello."[55] Roderigo claims she has made "a gross revolt" (1.1.132), and even in challenging such a view, Desdemona concurs that her marriage occasions "downright violence and scorn of fortunes" (1.3.250). Indeed, she must admit, as Othello does not, the shattering and revolutionary potential of their union from the start: she leaves her home, she redefines her heart as "subdued / Even to the very quality of my lord" (1.3.251–52), she takes to the transforming seas. Desdemona loves so deeply that she challenges Othello's attempts to maintain his separateness, a point Paul Cefalu makes in his analysis of the play: "Othello becomes discontented when he is compelled to leave the comforts of his relative mind-blindness. This is a process of losing the self in the other."[56] If deep connection in *Twelfth Night,* the *Henriad* and even in *Doctor Faustus* provides a model of expansion beyond egoistic boundaries, for Othello such expansive bonds are challenging, drawing him into conflict and potentially confining him.[57] Addiction in love is achieved, but the opening scenes stage its cost to Othello's sense of independence and his self-possession as a military commander.

As a sign of his chaffing against the addiction he experiences in love, Othello claims that he can separate his professional and emotional duties, and he furthermore reassures the senate of the primacy of his military commitments over his new marriage. When granted the commission to Cyprus, he says, "I do agnize / A natural and prompt alacrity / I find in hardness" (1.3.232–34). He is a warrior first and foremost, he tells them; he submits to their request, he is still "bending to [their] state" (1.3.235). He is, he claims, their servant, despite the fact he is also now a spouse. Even in asking for Desdemona's presence on Cyprus, he insists on the primacy of his military commitment over his interest in her, for the Cyprus mission is "serious and great business" (1.3.268), against the "light-winged toys / Of feathered Cupid" (1.3.269–70). He goes to Cyprus, he tells the duke, "with all my heart" (1.3.279), as if none of it belongs to Desdemona.

Yet Othello, too, experiences the "divided duty" (1.3.181) that Desdemona admits. At the end of the council table scene the strain between Othello's devoted love and his louder claims of committed service to Venice appears clearly. This is where, in planning the journey to Cyprus, Othello attempts to make, publically, private accommodations for his wife. Having assured the state it has his full heart, he nevertheless exhibits concern and preoccupation with Desdemona's condition, arguably to the frustration of the duke. Witnessing the intramarital planning of Othello and Desdemona over her journey to Cyprus, the duke responds with a reply at once supportive and dismissive: "Be it as you shall privately determine, / Either for her stay or going; th'affair cries haste / And speed must answer it" (1.3.276–78). The duke tries to reestablish the private and secondary nature of the marriage in the face of the geopolitical threat of the Turks; the couple's time-consuming desire to determine Desdemona's new home is, his comments indicate, beside the point.

Othello nonetheless lobbies for his love, acknowledging his strong affection even as he downplays it. He wishes Desdemona's company, he admits, not for physical reasons but "to be free and bounteous to her mind' (1.3.266). This sentence pops out from its speech precisely because it is framed through negatives: he does not wish to comply with his appetite, because he claims he does not have one; he does not value the toys of Cupid, because they are anathema to him. He just wants, he claims, to commune with her mind. If some critics find his protestations naïve for dismissing bodily desires, others point to the prurient, prejudicial environment surrounding him. Yet in approaching his love as transformative and devotional, his statement about feeling "free" with her "mind" offers a powerful testament to their connection: he begs for her presence on the island and offers a reluctant admission of his own deep attachment in the process. Freedom now comes in relation to Desdemona, he reasons. At once confined and circumscribed, he nonetheless experiences freedom with, or for, her.

As the play proceeds, Othello's addictive love for Desdemona comes into greater focus. Reunited after the tempest, Othello tells her, "it gives me wonder as great as my content / To see you here before me!" (2.1.181–82). Othello's repetition of the word "content" three times in this one scene demands critical attention, with the valences of the word shifting, in E. A. J. Honigmann's analysis, from pleasure and contentment to self-indulgent satisfaction. But the appearance of "content" in Othello's formulations is also about its destabilization through "wonder" and the "absolute." The homely feeling of contentment—what might be familiar and moderate—is here yoked to uncommon states, of wonderousness, surprise, and newness: "I fear / My soul hath her content so absolute / That not another comfort like to this / Succeeds in unknown fate" (2.1.188–91).[58] The feeling of contentment becomes extraordinary, exceeding anything in the known world. Othello's equations suggest precisely how transformative his love for Desdemona proves: it renders him mute, as when he tells her, "I cannot speak enough of this content, / It stops me here, it is too much of joy" (2.1.194–95). The excess of love, its unexpected power, overcomes him. He dwells in exclusive relation to Desdemona, despite his attempts to uphold his professional role above or against his marriage.

Othello's love provokes a transformation parallel to Cassio's "innovation" from drinking. He watches himself reunite with Desdemona after the tempest with some disbelief: "I prattle out of fashion, and I dote / In mine own comforts" (2.1.205–6). Mocking himself in love, his statement betrays—in a man who claims more comfort with weapons than words—the changes he experiences, even a kind of emasculation, with the words "prattle" and "dote" being overdetermined as female. But his lines betray curiosity about this state, not concern. For even as loving Desdemona might destabilize him, it also transforms him through attachment. He experiences, like Desdemona, the simultaneity of rulership and service in love; he may command, but so does she. Indeed, she is powerful enough to command an emperor, Othello claims, in an image notable for its tangle of equity and hierarchy: "O, the world hath not a sweeter creature," he claims of Desdemona; "she might lie by an emperor's side and command him tasks" (4.1.180–82). Lying side by side in bed, lover and beloved both experience sovereignty and vulnerability at once.

In loving, Othello feels the interconnection at the base of ethical, if not legal, definitions of personhood. Indeed, his state challenges legal notions of the autonomous self. He is, in legal terms, "voluntarily contracted" in his love, a phrase invoked above in the rulings on drunkenness, but one equally appropriate for the condition of married love in which Othello now engages. He is also, as Coke claims of the drinker, "*voluntarius daemon*," intoxicated in his condition of loving, surprised at himself. To early modern jurists, such compromised will is criminal; it is a condition to be policed and condemned, as suggested above. In sharp contrast, the play does not uphold strict models of individual action and isolated will. Othello's willingness to embrace incapacity distinguishes him as a sympathetic hero transformed by love.[59] While Iago's claim to uphold his interests at the expense of others makes him culpable, and ultimately criminal—his attack on "obsequious bondage" (1.1.45), and his desire "to serve [his] turn upon" (1.1.41) Othello, is sinister—in opening himself to love of another, Othello experiences an acute form of such connectedness. In possessing each other, Desdemona and Othello love with the shattering power of true union, the kind of addictive relation analyzed in earlier chapters of this project. They challenge and overturn the relations that formerly held them—to family, to nation, to identity. No longer self-possessed, they are connected to, devoted to, each other. Iago, by contrast, celebrates those isolated actors who "keep yet their hearts attending on themselves" (1.1.50), a formula that ultimately demonstrates the dangers of strict responsibility, a legal defense of the hyper will.

Other characters comment on Othello's devotion and his diminished self-possession. Cassio calls Desdemona "our great captain's captain" (2.1.74), and cynical Iago proclaims a version of this as well: "Our general's wife is now the general . . . for he hath devoted and given up himself to the contemplation, mark and denotement of her parts and graces" (2.3.309–13). Iago's description of a devotion so overpowering as to undo someone, to unseat them from government of others and themselves, offers a precise account of the transformative, ravishing addiction this project has traced so far. He expands, in soliloquy, saying, "His soul is so enfettered to her love / That she may make, unmake, do what she list" (2.3.340–41). Iago's characterizations are not laudatory; he does not admire Othello's ability to transform himself. Yet part of the reason for Iago's sinister success, as Paul Cefalu's analysis illuminates, lies in his uncanny ability to read other people's minds, in this case Othello's experience of a transporting addiction that he will soon transform to a liability.

In loss Othello most fully articulates the addictive nature of his love. Lamenting Desdemona's alleged betrayal, Othello cries that he could have endured "captivity," living in "poverty," and even shame and humiliation, standing as the "fixed figure for a time of scorn" (4.2.51, 52, 55), but it is torture and death to be tormented "there where I have garnered up my heart, / Where either I must live or bear no life" (4.2.58–59). Connection to his wife becomes life and death: he no longer exists as an independent being, he cannot serve the state with "all [his] heart" as he claims in the opening. It is being wrenched away from her, and experiencing the pain of being "discarded" (4.2.61), that undoes him. His devotion to her and commitment to their marriage has transformed him to a degree that he cannot exist without it. He defines himself in relation to her; his love for her is "the fountain from the which my current runs / Or else dries up" (4.2.60–61). Indeed, as he claims at the play's ending, Desdemona has become the world. He defines value entirely in relation to her, a point he expresses through a troubling image of trafficking: "Had she been true, / If heaven would make me another world / Of one entire and perfect chrysolite, / I'd not have sold her for it" (5.2.139–42). This image on her impassable value is also an image of possession—he imagines owning his wife and therefore being free to sell her. As striking as this image might be—not least because Othello has experienced, as other Venetians have not, the human trafficking of being "sold to slavery" and of "redemption hence" (1.3.139)—it deeply contrasts with his view of their relationship elsewhere in the play.

Othello might attempt, as he does here, to assert possessive control over Desdemona ("O curse of marriage / That we can call these delicate creatures ours"; 3.3.272–73). But he more frequently and convincingly speaks of their mutual dependency or his own possession by her: "perdition catch my soul / But I do love thee!" (3.3.90–91). He, like Olivia, is caught in love, giving voice to the strong feelings hinted at but not fully realized from the play's opening scenes, when Othello was more reluctant to admit their presence. This vulnerability in love, this staging of the addict's deep attachment, shapes Othello as a hero, but it also, as the next section explores, proves his undoing.

Othello's Possession

Possessing another, and being possessed in turn, can prove a violent process. *Othello* stages this violence by demonstrating possession to be both ownership and entrapment.[60] The very openness to attachment that fuels addictive love also leaves Othello vulnerable—in the horrifying course of act 3, scene 3—to control by Iago. As the play reveals through the circulation of the handkerchief, possessions—and possessive relations—are vulnerable to manipulation and loss. If this handkerchief had been woven by a "charmer" as a gift to "subdue" another (4.3.59, 61), it becomes, as Paul Yachnin argues, "a possession that possesses the possessor."[61] This formulation helps illuminate the peculiar condition of Othello: far from offering the hero magical power, possessions—a handkerchief, a love, a military unit—instead serve as a means for others to control him.

If the previous section tracked Othello's transformation through love—the release of agentive selfhood in marriage—this section turns to his much more obvious transformation in the play, as he changes in relation to Iago. As with Falstaff, when faced with the alleged failure of his primary addicted relation, Othello develops another addiction in its place. He moves from his devoted attachment to Desdemona toward an irrational, shattering, destructive attachment to Iago, pledging himself and in so doing, opening himself to Iago's emotional world. They become partners, exchanging oaths: "I am bound to thee forever" (3.3.217) he tells Iago, prompted by Iago's pronouncement of love: "I humbly do beseech you of your pardon / For too much loving you" (3.3.215–16).[62]

How is it Othello is so vulnerable to Iago? How does he become "perplexed in the extreme" (5.2.344), as he puts it? This question haunts us, as critics, viewers, and readers. One answer lies in the poisonous theater that Othello imbibes over the course of act 3, scene 3. As with Cassio's "one cup," so too with Othello: what had been drink for Cassio is strong emotion prompted by insinuations and allegations with Othello.[63] After one short exchange, in which Iago insinuates Desdemona's infidelity, Iago boasts, "The Moor already changes with my poison" (3.3.328); his "poisons" will "burn like the mines of sulphur" (3.3.329–32) and engender sleeplessness: "Nor poppy nor mandragora / Nor all the drowsy syrups of the world / Shall ever medicine thee to that sweet sleep / Which thou owedst yesterday" (3.3.334–37). "Work on, / My medicine, work!" (4.1.44–45), Iago urges as Othello falls into a trance. This poison is, certainly, jealousy, and here Leo Africanus's *A geographical historie of Africa* offers one context for understanding Othello's potential responsiveness to such a toxin. In now-familiar passages cited for their relevance for Shakespeare's play, Africanus chronicles the addiction of jealousy prominent in regions of the continent. Residents of Delgumuha are, for example, exceedingly "addicted to jelousie."[64] "No nation," he writes in his general introduction, "in the world is so subject unto jealousie; for they will rather leese their lives, then put up any disgrace in the behalfe of their women."[65]

Yet if this poison is jealousy, it is not Othello but Iago who is dominated by the emotion, as Mary Floyd-Wilson has argued. Jealousy is, initially, anathema to Othello, as Desdemona herself says: "my noble Moor / Is true of mind, and made of no such baseness / As jealous creatures are" (3.4.26–28); "the sun where he was born," she claims, "drew all such humours from him" (3.4.30–31). It is only with extraordinary baiting, with outright invention and lies, that Othello comes to feel jealousy at all. His entry into jealousy thus challenges Emilia's theory, offered to the dumbfounded Desdemona as she reels from Othello's insistent call for the handkerchief: "jealous souls," she says, are "jealous for they're jealous. It is a monster / Begot upon itself, born on itself" (3.4.159–62). The circuit of Othello's jealousy is not as precisely hermetic as Emilia suggests, since her own husband helped with the begetting of it. But the possession of jealousy, like Yachnin's formulation of the handkerchief, is nonetheless circular: it possesses the possessor. This is, of course, not the case with Othello, as the audience knows. Jealousy was not born of itself but born of Iago, becoming Othello's dominant, defining emotion only after he transfers his loyalty to this newfound partner.

If jealousy is the motivating emotion and outcome of Othello's union with Iago, jealousy does not explain why Othello transfers his loyalty to Iago in the first place. Instead, I would argue, it is Iago's attack on the nature of

Othello's addiction that unravels him. Iago exposes Othello's marriage to be a form of cruel optimism, whereby "the object that draws your attachment actively impedes the aim that brought you initially," in Berlant's formulation rehearsed earlier in this project.[66] For it is precisely when Iago insinuates Desdemona's deceitfulness and insincerity that Othello's addiction transforms. Asserting one last strong defense against jealousy ("I'll see before I doubt"; 3.3.193), Othello tells Iago twenty-five lines later, "I am bound to thee for ever" (3.3.217). In the intervening exchange, Iago contrasts Othello's "free and noble nature" filled with "self-bounty" (3.3.202–3) to Desdemona's "pranks," which she and other Venetian women "dare not show their husbands" (3.3.205–6). Such women keep, he claims, their actions "unknown" (3.3.207). Desdemona, he argues, has already proved insincere: "she did deceive her father, marrying you" (3.3.209).

Iago's nationalistic argument, carefully analyzed by Emily Bartels and others, preys upon Othello as an outsider, unfamiliar with the "country disposition" (3.3.204) of Venetian women. Iago insinuates, and goes on to elaborate, an incendiary critique of their marriage, describing as "unnatural" (3.3.237) Desdemona's failure to marry men of "her own clime, complexion and degree" (3.3.234). If Othello successfully defended himself against such arguments in Venice, the difference now is that Iago attacks not Othello but Desdemona: it is her fidelity to Othello, not just physically but emotionally and spiritually, that Iago calls into question. "Iago compels Othello," Andrew Sisson writes, "to become aware that his marriage depends upon his partnership with a virtue that cannot be known, displayed, judged, or valued in a way that would satisfy him of its reality."[67] In reframing her as deceitful, claiming that she hides her actions and even her feelings, Iago troubles and undoes the basis of Othello's attachment, a devotional relationship that the hero embraced.

In deeming Desdemona a false idol, Iago shatters Othello's faith. He insinuates that the match that Othello willingly entered, at the expense of his liberty, never existed. It is a sign of Othello's addictive capacity that he transforms himself so violently and utterly in relation to Iago—it is a sign, precisely, of his relational, devotional abilities, in sharp contrast to Iago himself, the unbending force of will. Iago distains devotion and service; he never will, he claims, "wear my heart upon my sleeve / For daws to peck at" (1.1.63–64). He shuns intimate relations, mocking those figures who experience love and savoring their vulnerability to his machinations. He is, as many critics have noted, an actor—he is a theatrical Vice figure, a Machiavel combining staging and imaginative fancy as a theater director might. If this project began by teasing out the intimate relation of the actor and the addict, as two figures who pledge and transform themselves in relation to another, *Othello* pushes on this connection to the point of undoing it, staging instead their opposition. For even as Iago pledges himself to Othello, taking an oath to serve him faithfully, we know his pledge is an act. A manipulation, it stands as the worrying form of theater as deception and counterfeiting asserted by writers like Prynne. Othello, by contrast, is the addict: he takes a sincere vow, pledging himself and speaking to his commitments: "Now by yond marble heaven / In the due reverence of a sacred vow / I here engage my words" (3.3.463–65). Othello promises Iago to maintain his "compulsive course" of "bloody thoughts" (3.3.457, 460), securing their shared pledge through an inverted marriage ceremony.[68] This newly established marital relation compels Othello: even as it appears consensual, the audience recognizes it as deception. This second devotional relation—to Iago, to the dominant emotion of jealousy within their relationship—overcomes Othello to the point of undoing all previous attachments. He claims he turns to stone, admitting and yet also seemingly unaware of the stakes of his own transformation: "My heart is turned to stone: I strike it, and it hurts my hand" (4.1.179–80), a phrase that, like Cassio's deterministic "souls must not be saved," evokes Calvin and Luther's writings on the hardened heart untouched by grace.

The very capacities that elevated Othello at the start of the play—the capacity to love, to temper his military life with marital attachment—are now what open him to Iago's machinations. As a result of this possession—by love, wrath, jealousy, Iago—Othello transforms utterly. He changes to the point that Desdemona claims, "My lord is not my lord, nor should I know him / Were he in favour as in humour altered" (3.4.125–26), while Lodovico exclaims of Othello's behavior in hitting his wife, "My lord, this would not be believed in Venice / Though I should swear I saw't" (4.1.241–42). He asks, "Is this the noble Moor whom our full senate / Call all in all sufficient? This the nature / Whom passion could not shake?" (4.1.264–66). This project's earlier chapters traced the admirable process of addictive change, as a hero emerges out of him- or herself. Theorizing the laudable ability to will away one's own will, this project highlights devotional aspects of service to another. But this chapter instead exposes the environmental strains and threats of addiction: its release leaves one vulnerable to deception and villainy. With Othello, the "speaking to" of addiction, the pledging, occurs within an environment of white Christians who may or may not pledge back. As Ian Smith asks: "Among his white, Christian auditors, whom can he trust to tell his story or speak of him in a balanced way?"[69] Pulling away from the addictive relation that had offered him life outside of public definitions and expectations, Othello commits himself, pledges himself, to Iago—but his supposed exercise of will is compromised. Poisoned and possessed, Othello lacks the ability to offer meaningful consent. Indeed, he no longer demonstrates *mens rea* required to be fully responsible. In the end, even Othello does not recognize himself, ending the play in the past tense, speaking as "he that was Othello? Here I am" (5.2.281).

Diminished Responsibility

In *Othello*, who is responsible? Desdemona says "nobody. I myself" (5.2.122). But her answer claims excessive, inappropriate responsibility, even as she also protests "a guiltless death I die" (5.2.121). In her refusal to mete out blame, Desdemona brings the question of responsibility to the fore. For the most obvious answer to Emilia's question—"Who hath done / This deed?" (5.2.121–22)—is also an inadequate one: if it is obviously not Desdemona, neither is it clearly Othello. Through diminished capacity, Othello's act of murder, like Cassio's drunken brawling, simultaneously is and is not his own. In the play's early scene of drunkenness, Othello attempts to assert a model of strict responsibility, condemning Cassio entirely. But it is precisely this strict resolution—in upholding radical responsibility for Cassio, who is depicted, in part, as vulnerable to forces greater than himself—that helps produce the play's tragedy. Dismissing his lieutenant, failing to recognize the mitigating circumstances and compromised will at issue in the episode, Othello leaves himself open to the same process of manipulation that felled Cassio. The play thus doubles the audience's awareness of the law's strict responsibility as an inadequate response: the audience, throughout, sees that the charge of drunken brawling unjustly falls on an incapacitated man; so too with Othello, who is possessed at the prompting and design of Iago.

Othello wrestles with this insight on mitigated responsibility. He understands himself as both externally and internally compelled, but he initially claims his action as purely his own: "'Twas I that killed her" (5.2.128), "I did proceed upon just grounds" (5.2.136). He takes full responsibility and even deems himself a fool—"O fool, fool, fool!" (5.2.321)—thus echoing Cassio, who claims of himself, "to be now a sensible man, by and by a fool, and presently a beast! O strange!" (2.3.300–302). In their admitted guilt, both men reinforce the link of folly with criminality in the legal ruling of *Reniger v. Feogossa* (1551): because the drinker's "ignorance was occasioned by his own act and folly, and he might have avoided it, he shall not be privileged thereby."[70] In such folly, Cassio and Othello were for "a time *non compos mentis*," yet as with the drunken man in *Beverley's Case* (1603), such a condition "does not extenuate his act or offence."[71] Othello, like Cassio, should have been able to choose differently, for as Aristotle reasons, "the moving principle is in the man himself, since he had the power of not getting drunk," or impassioned, or jealous.[72]

Yet the play's final recognition scene insists not on guilt but on diminished responsibility for both Cassio and Othello. Insight into their mitigating circumstances comes in tandem. First Othello learns of his manipulation at the hands of Iago, turning to the heavens for retribution against the perpetrator of events: "are there no stones in heaven / But what serves for thunder?" (5.2.232–33). He finds himself immediately undone, unrecognizable, losing all of the qualities formerly precious to his sense of self: "Let it go all," he proclaims, in lines that signal the belatedness of his own death. "Who can control his fate?" (5.2.244, 263), he asks, turning to the heavens again, now not for intervention but explanation. He understands himself as ultimately possessed, owned by another—be it Iago, fate, Desdemona, or the passionate part of himself: "O cursed, cursed slave!"(5.2.274). Then Cassio learns of his own manipulation: he was led into drinking, and goaded into fighting, by Iago and Roderigo. Cassio's drunken criminality now emerges as Iago's. Far from enduring the law's strict sentence, Cassio ends ruling Cyprus, absolved of any charges and promoted in the process.

Othello and Cassio thus stand in externalized relation to their own deeds, the agency behind their criminal action dispersed. When Cassio recounts Roderigo's letter, stating how Iago "made him [Roderigo] / Brave me upon the watch, whereon it came / That I was cast" (5.2.323–25), he draws attention to precisely such dispersed agency, as the action moves from Iago to Roderigo to Othello as the party responsible for Cassio's fall. So, too, with Othello, who has "fallen into the practice of a cursed slave" (5.2.289), as Ludovico puts it; Othello can only ask the "demi-devil" why he "hath thus ensnared [his] soul and body" (5.2.298–99). Finally, meditating on the events, Othello deems them "unlucky deeds" (5.2.339); he imagines he has been "wrought" upon by another, and thereby "perplexed in the extreme" (5.2.343–44). Cast out, wrought upon, both Cassio and Othello find their consent compromised, even as Othello seeks punishment and retribution: "Whip me," "blow me," "wash me" he cries to the "devils" he attempts but fails to conjure (5.2.275–78).

How, then, to understand addicted action? If the Herald's invitation initially frames addiction as a choice, taken up as a temporary inclination, ultimately the play reveals addiction as a form of possession. One is overcome and ravished. Othello's solution to the question of responsibility, a powerful one, lies in imagining himself as a partial agent, a hand independent of a body or soul: he is "one whose hand, / Like the base Indian, threw a pearl away" (5.2.344–45). "These hands have newly stopped" the life of Desdemona, he claims (5.2.199). He ultimately understands his own actions as self-divided. A portion of his body—his hands—are responsible, and yet even they were led by another. In famous lines, he deems himself both Turk and Venetian, both responsible and vulnerable: "in Aleppo once, / Where a malignant and a turbanned Turk / Beat a Venetian and traduced the state, / I took by th' throat the circumcised dog / And smote him—thus!" (5.2.350–54). Here Othello condenses the play's dilemmas around criminal responsibility by externalizing a part of himself in calling it foreign—a Turk—but also malignant, resonant with Iago.[73] Othello deems himself culpable for a crime he committed while incapacitated, but he insists that addictive possessions—to the predisposition of jealousy, and to poisonous magic of Iago—overcame him. Thus even in assuming responsibility he also insists, tragically, on his vulnerability, if not innocence.

It is worth recognizing, in ending, how Desdemona displays the same transformative capability as Othello, redefining herself in love, shattering prior attachments, and remaining devoted even to the point of audience frustration at her supposedly meek response to Othello's violence. She displays what Lauren Berlant calls "nonsovereign relationality," putting connection above individuality.[74] Part of the play's tragedy lies in the demise of two characters with such equal capacity as addicts. For considered from the vantage point of addictive attachment, Desdemona's response is anything but passive. Othello's addiction to Desdemona collapses upon attack; hers never does. If Othello transfers his addiction to another in the face of assaults, she does not. Of course, her devotion might seem pathological, in that it supports the abuse that might accompany such a deep structural commitment. But the lens of early modern addiction invites us to parse, cautiously, the distinction between devotion and pathology. For it is not, I would argue, Desdemona's loyalty or devotion that's diseased. The disease comes from Iago and from Othello's violent, inverse addiction to him. Both heroes, Othello and Desdemona, are compelled as addicts, both are possessed in devotional relation, and this possessive devotional relation might have been to each other. But Othello, as insecure in his attachment as Faustus had been in his, chooses the wrong relationship, and he does so at the expense of the beloved who perfectly mirrors—and in fact sustains, in the face of his failure—his own heroic abilities as an addict.

This argument on Desdemona's propensity as an addict, in the face of Othello's failure, might serve as the limit case of addiction. When devotion and attachment lead to violent death, can addiction continue to be called—as I have argued through this project—laudable, heroic, and extraordinary? Applauding the strength of Desdemona's attachment could resonate with dangerous and familiar arguments counseling loyalty in the face of abuse, self-sacrifice in the name of marital ideologies. Desdemona anticipates, one might argue, the more familiar modern view of addiction as diseased and destructive. But before denouncing such attachment, it is worth pausing to distinguish admirable devotion from its gruesome outcome. In the case of Desdemona, does her death diminish her right to love? In the case of Othello, does his ultimate fall to Iago cheapen his earlier devotion to his new wife? Their devotional attachments offer the play's strongest challenge to Iago's hyper-exercise of the will. Demonstrating "nonsovereign relationality as the foundational quality of being in common," they expose a mode of loving worthy of admiration, even in its demise.[75] Yoking incapacity and sympathy, devotion and humanity, the play challenges early modern legal emphasis on strict responsibility. More than this, the play upholds—even in its deadly outcome—a model of addiction. Othello and Desdemona meet heartbreaking, tragic ends, but their capacities to release themselves into love, and their willingness to relinquish the will, hold the potential to end otherwise. It is Iago who—in his counterfeit attachments, and his failure to connect—speaks to the inhumanity of those who strive never to addict themselves at all. Iago and Othello, in their oppositional stances to attachment, expose how the exercise of the will—the autonomy and self-possession so lauded in legal rulings—can come at the expense of another, more related model of living, one based in devotion and addiction.

Chapter 5

Addictive Pledging from Shakespeare and Jonson to Cavalier Verse

> Mould all our Healths in your immortal *Rythme*,
> Who cannot sing, shall drink in time.
> We'll be one Harmony, one Mirth, one Voice,
> One Love, one Loyalty, one Noise,
> Of Wit, and Joy, one Mind, and that as free
> As if we all one Man could be.
>
> —Charles Cotton

> Is not a Health now become a signal of a battle, in which many lose their precious lives basely and inhumanely; others their senses, their clothes, their modesty; and they who escape with their lives, lose themselves for the time?
>
> —*The Great Evil of Health-Drinking*

Both Charles Cotton and the author of *The Great Evil of Health-Drinking* invoke the ravishing effects of health drinking, namely the practice of toasting to the health of another. Cotton celebrates the loss of self in drink and community. Unified as "one Harmony" and "one Mind," health drinkers experience, he claims, a form of liberation unavailable to the isolated individual. Being overcome, Cotton's speaker enjoys freedom, "as free / As if we all one Man could be." Finding liberty in liquid erasure, the speaker embraces self-shattering through drinking, love, and loyalty combined. In *Health-Drinking* (alternately attributed to Charles Morton or William Prynne), health drinking is indeed a ravishment of self and senses, but to base ends: health drinking fells participants in a compulsive ritual, losing mind and body in the process. Both writers invoke this drinking, that is to say, as the addictive, transformative phenomenon seen in earlier chapters of this project. Their opposite interpretations of the same ritual in fact mirror precisely the tension within discourses of addiction, as the overcome addict invites either admiration for his or her extraordinary devotion, or concern and diagnosis for a compromised will and compelled body.

This chapter examines these precisely oppositional views of health drinking to expose the interpretive challenge of addiction. Certain addictive practices bolster individuals and communities in times of hardship, transforming loss into freedom and providing an occasion for expressions of loyalty and devotion to faith. These same practices can also invite condemnation. Compulsive healthing puts such tensions into high relief. What might it have meant to raise a glass to another's health in early modern England? To drink to a friend's health was the most popular and chronicled drinking ritual of the period. It was, as Joshua Scodel writes, "the central communal ritual of early modern drinking culture."[1] Drinking a health represented, as Angela McShane notes, "a mark of loyalty and fealty in both courtly and civic circles."[2] A group of companions would pledge to the health of one another, as well as to absent friends, loved ones, or superiors, often while kneeling or doffing a cap. Each drinker would pledge in turn. Obliged to respond, the pledger's companions would raise their drinks and either drain their glasses in unison or pass around a healthing bowl, from which everyone would take a gulp. Such health drinking signified "a public display" of "close social bonds."[3] To drink to another's health restored "harmony and concord." Healthing "served to oil . . . networks of credit and community."[4]

The above description of health drinking resonates with forms of addiction already addressed in this project. First, health drinking reinforces our definition of addiction as a verbal pledge: health drinkers willingly pledge themselves to a community, proving themselves addicts who speak toward (*ad* + *dīcere*) their commitments. Second, health drinking attests to addiction's devotional applications. Indeed, the devotional stakes of health drinking are particularly high: the practice of healthing arguably resonates with—and at times stands in for—the Christian Eucharist, a point that Angela McShane reveals in her comparison of these practices over the course of the seventeenth century.[5] The communion cup might serve, or have once served, as a healthing vessel, and in both cases the ingestion of alcohol with its attendant pledge signals devotional, indeed spiritual, connection. Finally, health drinking speaks to the paradox of willing away one's will: health drinkers are actors on their own behalf, choosing to drink as a sign of their loyalty, and yet they also enter into a ritual known to be compulsive, one that overcomes them and redefines them in relation to the drinking community. The practice is at once reflexive and imperative, addicting oneself and being overcome in the process.

It is this last aspect of healthing that invites censure as well as celebration: drinkers are compelled to remain in the drinking community until all toasts have been discharged. Failure to participate leads to brawling and sometimes death. If a range of scholars highlight health drinking as a communal ritual, the practice invited virulent attacks as well: drinkers are no longer free to exercise discretion and moderation, but are participants in a practice that exceeds their control. The second epigraph highlights this bullying: "Many lose," the author writes, "their precious lives . . . their senses, their clothes, their modesty." Indeed, they "lose themselves for the time." This process appears tyrannous and violent—it is a "battle." Despite sustained celebration of health drinking as loyalist pledging, then, it was a controversial, heavily debated, and at times illegal practice in early modern England. The Protectorate banned it, and Charles II issued a royal proclamation against it.[6] Furthermore, a host of writers depict health drinking as divisive and downright sinister. What is the custom more honored in the breach than the observance? Health drinking. Trumpets "bray out / The triumph of [Claudius's] pledge" even as Hamlet derides it.[7] What fells Cassio in *Othello*? Iago's call to health drink: "Here without are a brace of Cyprus gallants that would fain have a measure to the health of black Othello," Iago claims. "O, they are our friends; but one cup."[8]

Through such cross currents in its representation, health drinking offers a case study in addiction as simultaneously devotional and dangerous, reinforcing arguments made in earlier chapters. Yet in its study of a specific addictive practice this concluding chapter's methodology is distinct. Earlier chapters concentrated on addiction through the study of one exemplary play, while this chapter opens with a range of drama before moving to lyric poetry. If earlier chapters remained closely trained on historical moments in the sixteenth and early seventeenth centuries, this final chapter moves from the 1580s through the 1660s. Finally, earlier chapters illuminated shared attitudes to certain forms of addiction, tracking the celebration of addiction to God, study and love, and the suspicion of addiction to drink and jealousy. This conclusion instead concentrates on disparate attitudes to one practice. I justify such a methodological shift in this final chapter because analyzing one addictive practice over this broad historical and generic range allows me to demonstrate both the long reach of addiction as devotion and the variability of attitudes toward addictive practices themselves. If some forms of addiction invite celebration and others condemnation, as previous chapters have argued, this chapter reveals how attitudes toward a single addictive practice can shift radically, even in a short space of time. Health drinking appears, in the 1580s and 1590s, as a sinister, foreign, deplorable activity. But by the 1630s and 1640s, health drinking is invoked as a devotional practice uniting the careworn faithful. This addiction becomes, after decades of suspicion, suddenly laudable, desirable, and compelling: it serves, in turbulent times, as a way to forge community, both through a verbal pledge of loyalty and through the shared action of communal drinking.[9]

This book thus ends with a chapter that frustrates what might appear a historical trajectory on addiction embedded in this book's argument. In uncovering a link between Reformation theology and perceptions of addiction, this project's first chapters demonstrated how Calvin and his English followers embraced addiction as a desirable achievement, even as subsequent chapters have illuminated increasing concerns about idolatrous, material addiction to the wrong spirit, whether in the form of false idols, false friends, or cups of liquor. But to read a historical narrative onto these shifting, alternate views on addiction would be reductive, for it would ignore the persistent gestures toward addiction as devotion through the seventeenth century. Despite the tragic turn charted in this project's last two chapters, a buoyant embrace of addiction remains, as this project's final chapter reveals.

I end this project, then, with a methodologically distinct chapter, but one that encapsulates many of addiction's countercurrents illuminated throughout. Furthermore, this final chapter, on health drinking, allows me to emphasize the longevity of addiction's mobility and density, across the voluminous early modern literature on health drinking as it extends over an eighty-year period, from pamphleteers and playwrights of the 1580s–1610s through Cavalier poets in the 1640s and 1650s up to the period of Charles II's restoration.[10] Beginning with an analysis of devotional drinking in the poetry of Robert Herrick, this chapter then turns to earlier evocations of healthing, tracking the practice as it appears in chronicles, state papers, satires, and the sectarian polemic. Here healthing appears as a form of compulsory sociality and drunken excess, an addiction that resonates with the habitual, excessive ingestion studied in Chapters 3 and 4. Depictions of health drinking in plays from the period are surprisingly consonant with puritan tracts: Jonson, Shakespeare, and others illuminate pledging as compulsory and violent. In sharp contrast to forty years of attacks on healthing, the drinking verses by Caroline poets, studied in this chapter's last section, establish the dominant critical view of healthing as a pledge of loyalty toward the community, countryside, and king. Within the ethos of wine, women, and song characteristic of such verse, this poetry ingeniously rewrites the cultural suspicion of health drinking so evident in the literatures of the prior decades, reframing pledging as a country custom and a bonding ritual.[11] These poets claimed the practice as a laudable addiction and transformative devotion at a moment when their opponents condemned and even outlawed healthing as a problematic pastime. Speaking and acting, commanding others and compelled themselves, these poets—in their rich, playful, and dark verse—attest to the complexity of addiction's legacy.[12]

Devotional Drinking

In the "The Hock-cart, or Harvest home," Robert Herrick invokes healthing as a signal of devotion to his country alliances. The speaker encourages country workers to enjoy their customary holiday and to express their attachments—to the country estate and the commonwealth—through the healthing ritual:

> freely drink to your Lords health,
> Then to the Plough, (the Common-wealth)
> Next to your Flailes, your Fanes, your Fatts;
> Then to the Maids with Wheaten Hats:
> To the rough Sickle, and crookt Sythe,
> Drink frollick boyes, till all be blythe.[13]

Of this poem Cedric C. Brown writes, "the greatest symbol of unity of minds is the health, here offered to lord, plough, and commonwealth."[14] As with the Cotton verse cited at the opening of this chapter, a group of drinkers, unified in their loyalty pledges, constructs their own celebratory "commonwealth," losing themselves in the process. The "rough" and "crookt" features of life—a nod to the hard work of country labor as well as to the corruption of government—dissipate in drink, so that those exhausted workers become "blythe," a term signifying their jocund, merry bodies and spirits. "Frollick" here serves both as an adjective modifying "boyes" (these men have become youthful, joyous boys) and as a verb linked to drinking (these men drink and make merry), suggesting the way in which the action of drinking reshapes these men.

Healthing connects poets to loyalist audiences while also providing poetic inspiration: breathed into by the spirit of the health, the poet reaches new heights. In "To live merrily, and to trust to Good Verses," Herrick's speaker deploys healthing as such a devotional aid, pledging loyalty to his favorite authors, beginning with a toast to Homer:

> *Homer*, this Health to thee,
> In Sack of such a kind,
> That it wo'd make thee see,
> Though thou wert ne'r so blind.
> (*CP*, 76.13–16)

Herrick's speaker proceeds through Virgil, Ovid, Catullus, and Propertius, draining a full glass with each pledge. By stanza 9 the speaker's head is spinning: "Round, round, the roof do's run" (77.33). Yet he offers a few more health pledges, consuming larger and larger portions as the poem continues: as he says to Tibullus, "This flood I drink to thee" (77.38). Through health drinking the speaker signals his devotion to the classical forbearers who shape his verse. More specifically, the speaker becomes a devotee of Bacchus, offering an appropriately erotic image of himself heated with desire and ingesting the overdetermined symbols of the god as a result: "O *Bacchus*! coole thy Raies! / Or frantick I shall eate / Thy *Thyrse*, and bite the *Bayes*" (77.30–32).[15] Continuing on and on in his drinking, despite his weak plea for "cool[ing]," the speaker does indeed become frantic, spinning around. He is, he tells us, "ravisht thus" (77.34). Like Cotton or indeed like Faustus, Herrick's speaker finds self-erasure and embrace in a greater community, one most obviously composed of drinkers and poets, but gesturing outward to all men, as the final image of the poem suggests: "all Bodies meet / In *Lethe* to be drown'd" (77.49–50). Bodies, minds, spirits, reputations, verses: all unite in the liquid spirit of alcohol.

Health drinking as a devotional pledge brings Cavalier poets and audiences together with their poetic forbearers, including not only Anacreon and Horace but also Jonson. In Herrick's "A Bacchanalian Verse," the speaker drinks a health to Jonson, echoing the poet's *Leges Convivales*:

Fill me a mighty Bowle
  Up to the brink:
  That I may drink
Unto my *Johnsons* soule.

Crowne it agen agen;
  And thrice repeat
  That happy heat;
To drink to Thee my *Ben*.

Well I can quaffe, I see,
  To th'number five,
  Or nine; but thrive
In frenzie ne'er like thee.
(*CP*, 215)[16]

The tongue-in-cheek excess of the poem suggests both poetic debt and forgetfulness. Herrick at once celebrates (and gently mocks) the "frenzie" of Jonson, chronicling the earlier poet's role in inspiring his verse. Indeed healthing brings the speaker into union with Jonson: through drunkenness they both reach inspired heights, even if the speaker fails to reach the level of Jonson's poetic (and drunken) achievement ("thrive . . . ne'er like thee"). Here the spirit of alcohol and inspiration unite in a kind of metempsychosis as Jonson's soul migrates into Herrick's. Of course Herrick, in drawing on the biographical lore surrounding Jonson's excessive drinking, overlooks the moderation upheld in the verse of his poetic forbearer. In "Inviting a Friend to Supper," Jonson might praise "rich Canary-wine" (*WBJ*, 5.29) for its inspiration, but he also reminds his guests to "sup free, but moderately" (l.35). Similarly in *Underwood* he contrasts "the Tribe of Ben" with those "that live in the wild Anarchie of Drinke."[17] Most notably in "To Celia," perhaps the best known poem on health drinking (despite its disavowal of healthing), Jonson's speaker counsels, "Drink to me only with thine eyes, / And I will pledge with mine" (*WBJ* 5.227). As Joshua Scodel's scholarship on moderation has established, Jonson "reveals his concern for self-restraint in his ambivalent treatment of . . . the health."[18]

While Jonson's speaker replaces wine in the cup with glances and pledges, Herrick offers no such substitution. His speaker celebrates "the wild Anarchie of Drinke," which is dismissed by Jonson, as a sign of his community's bountiful fortune. Yet if Herrick's excess appears to contrast with Jonson's, his hard drinking may well be a poetic construction, as much as Jonson's moderation was. As a county curate who complained of isolation, Herrick may not have enjoyed the "wild Anarchie" celebrated in his verse—or he may indeed have enjoyed drinking alone, imaginatively surrounded by the classical figures he invokes. Poetic fancy allows Herrick to express a form of heavy, communal drinking in which he likely did not indulge, at least in a conventional sense. As Katharine Eisaman Maus puts it, "What replaces the striving for conventional forms of success, in Herrick's representation, are the pleasures of the imagination, often an imagination divorced from action."[19] Health drinking thus offers an opportunity for addictive transformation through pledging beyond the physical practice itself, into its imaginative possibilities. If, as Scodel argues, "royalist civil war and Interregnum drinking poems defiantly respond to Puritan-Interregnum values with calculated excess," for Herrick health drinking fosters this excess through a devotional mode uniting him, despite physical isolation, with the greater spirit of the nation.[20]

Compulsive Healths

Herrick celebrates healthing as a devotional ritual, pledging himself to literary precursors and country communities, finding inspiration through drink. But such an embrace of healthing is entirely innovative. Not only was healthing derided for decades before a poet like Herrick came to uphold it, it was also condemned as a foreign import. Tracing the history of health drinking in England reveals how this addictive ritual, embraced and lauded by Herrick and other Cavalier poets as a devotional pledge, attracted scorn and satire as a form of compulsive binge drinking, leading to disease and death. Healthing thus condenses the oppositional modes of addiction as deep transformative attachment and compulsive excess, and in doing so surprisingly inverts the broad historical shift from early modern devotional addiction to the more familiar modern notion, by celebrating healthing against earlier attacks.

According to the early modern chronicle of William Camden, soldiers returning from the Dutch wars introduced the practice of compulsive health drinking into England. In his *Annales of the reign of Elizabeth,* Camden writes: "Yet here wee must not omit to observe, that our Englishmen who of all the Northerne Nations have beene most commended for sobrietie, have learned since these Low-Country warres so well to fill their cups, and to wash themselves with Wine, that whilest they at this day drinke others healths, they little regard their owne."[21] Camden's view of the English as "commended for sobrietie" before the 1580s is of course fantasy, and health drinking has medieval precedent in England, appearing in both *Beowulf* and Geoffrey of Monmouth's *Chronicle*.[22] But in surveying the social practice of healthing, the ritual (or perhaps more accurately, concerns about the ritual) spiked dramatically in the 1580s and 1590s. What emerges during this period is the repeated attempt to locate health drinking as a foreign practice. Playwrights, satirists, and sectarians all understood, or pretended to understand, its rise as a foreign phenomenon coming out of the Low Countries, which had both a customary practice of health drinking, as well as a reputation for excessive drinking. Simon Schama notes that "the Dutch reputation for hard drinking went back at least to the early sixteenth century, when Lodovico Giuccardini noted it as 'abnormal.'"[23]

English writers, drawing on the stereotype of the drinking Dutch, blamed the health drinking of the 1580s and 1590s on the Low Countries. Like Camden, Thomas Nashe writes of returning soldiers in *Pierce Penniless* (1592): "From gluttony of meats, let me descend to superfluity in drink: a sin that, ever since we have mixed ourselves with the Low Countries is counted honorable, but, before we knew their lingering wars, was held in the highest degree of hatred that might be."[24] The English association of healthing with the Dutch appears in writings such as *Hans Beer-Pot: acted in the Low Countries by an Honest Company of Health-Drinkers* (1618), featuring three men who enjoy multiple healths together.[25] More generally Thomas Heywood condemns health drinking as an unwelcome foreign practice in *Philocothonista*, his drinking pamphlet. The frontispiece to the pamphlet reads:

> Calves, Goates, Swine, Asses, at a Banquet set,
> To graspe Health's in their Hooff's, thou seest here met;
> . . . like Cyrcean Cups, Wine doth surprise
> Thy senses, and thy reason stupifies,
>   Which Foe, would Warre-like Brittaine quite expell,
>   No nation like it, could bee said to excell.[26]

Heywood opposes "Warre-like Brittaine" to this stupefying "foe," which the nation should "expel," terms of national overthrow that echo the legislation against drunkenness studied in Chapter 4. Like Odysseus's men, English fighters have turned to drunken swine entrapped by Cyrcean sack, as figured in the woodcut on the pamphlet's title page (Figure 3).

Philocothoniſta,

OR, THE

DRVNKARD,

Opened, Diſſected, and Anatomized.

LONDON,

Printed by *Robert Raworth*; and are to be ſold at his houſe neere the *White-Hart* Taverne in *Smithfield*. 1635.

Figure 3. Title page of Thomas Heywood, *Philocothonista or the Drunkard, Opened, Dissected, and Anatomized* (1635). Rare Book 61481. The Huntington Library, San Marino, California.

Camden and Nashe are writing of the period when England officially entered the Dutch wars. The first English campaign, a notorious disaster led by the Earl of Leicester in 1585, featured Robert Devereux, the second Earl of Essex, making his military debut. Military efforts ended in the battle of Zutphen, which resulted in the death of Sir Philip Sidney, and the elevation of Essex to knighthood. Reports of health drinking, preserved in state papers, appear in correspondence from Holland during this first campaign. A 1586 report, for example, of a quarrel between Captain Edward Norris and Count Hollock, the commander of the Dutch troops, features health drinking before the battle of Zutphen:

> Supper time being come, everybody set, and drinking beginning, Capt. Edw. Norris perceived himself more than ordinarily pressed; and after many carouses, the Lord Marshal [Count Hollock] took a great glass and drank to the health of the Lord Norris and my lady. Captain Norreys desired his lordship to take a lesser [glass] . . . But being urged, he pledged him, and drank it. . . . Immediately my Lord Marshal took the like glass again, and drank to Capt. Norris, who musing much at it . . . took the glass and set it by him as the manner is. But the Lord Marshal, reaching over the board . . . said "Capt. Norris, take your glass, and if you have any mind to play, seek other companions, for I will not be played withal; therefore pledge me." Capt. Norris said I trust your lordship will not force me to drink more than I list. . . . [but he nevertheless consented to drink]. He was drinking to Mr. Sydney when the Count Hollock . . . took the cover of a great bowl and threw it violently at Capt. Norris's head, . . . and cut him a great gash to the bone, the blood running down his face and eyes. The Count presently rose to have stabbed him, but was stayed by Sir Philip Sydney and others and so carried away.[27]

Hollock represents healthing gone awry, turning to force and violence when the Englishman Norris attempts to withdraw from the ritual. The scene stages the compromised consent explored in the discussion of Cassio: Norris at once wants to withdraw and yet does not, finding himself overcome by drink and compelled by violent healthers. As Camden puts it in the citation above, "whilest [the soldiers] at this day drinke others healthes, they little regard their owne."

As figures like Norris and Essex returned home, such wary references to health drinking proliferated. Queen Elizabeth expressed her suspicion of health drinking, at least if this anecdote from Sir Nicholas L'Estrange is to be believed: "In comes one of her peeres into the Presence: she observing more jollity than usuall in his fashion and discourse, askt him where he had been? 'Y'faith, madame,' sayes hee, 'drinking your health.' 'So I thought,' sayes she, 'and I am sorry for't; for I have observed I never fare worse than when my health is drunk.' "[28] The queen plays on the increasingly familiar association of healthing with sickness or poor fortune. The sorry spectacle of healthing, whether at Zutphen or at Elizabeth's court, continues at Essex House where, after the second earl's fall, his followers and allies ran into trouble because of such drinking.[29]

Even witty drinking pamphlets warn against the compulsory nature of healthing. Health drinking had "*Customes to be observ'd*," since rules insist "to keep the first man, and to know to whom you drink. To have a care to see your selfe pledg'd"; most importantly, one must "see the health go round."[30] Gina Bloom explores such rules as a form of what she calls "disciplined play."[31] To put a bit of pressure on her link of drinking and discipline, it is worth noting that these pamphlets openly acknowledge the danger of tyrannical drinking laws, suggesting an awareness that "play" might turn deadly, and that laws might serve to damage rather than discipline eager drinkers. Such customs could prove dangerous, for example, in forcing drinkers to imbibe well beyond their own limits, as Richard Brathwaite explains in *A Solemne Joviall Disputation . . . shadowing the law of drinking* (1617): "*But what if wee be injoyned* [to health drink] *upon tearmes of honour?* Neither then, do I say, are we bound unto them. For such things as prejudice piety, our healths safety, and are directly against the rules of civility, we are not to imagine them to be done by us."[32] Brathwaite might suggest that the laws of honor or the tavern are suspended for the "rules of civility," but contemporaries repeatedly draw attention to healths as compulsory, precisely the kind of practice that enacts the troubling Galenic formula of custom altering nature: compelled by companions to drink, the healther might soon be bound by drink itself.

George Gascoigne equally reminds drinkers not to force others. One of the earliest discussions of health drinking from this period, his *Delicate Diet for daintie mouthde Droonkardes* (1576) ponders what happens when one "compels" or "bindes" a man to drink: "[If] by this curtesy, and friendly entertainement of yours, a friend which is constrayned thus to pledge you, doo chance to surfeyte, and to fal thereby into such distemper, that he dye thereof:

what kind of curtesie shall we then accoumpt it?"[33] Gascoigne depicts the dominant practice within health culture: the ritual depends precisely on binding a group of militant drinkers, thereby provoking a dilemma for drinkers forced into and felled by compulsory sociality. In the language of these texts, the drinker feels bound and constrained. William Hornby writes of the reception awaiting an abstaining man in his otherwise humorous poem *The Scourge of Drunkenness* (1618):

> But if to pledge a flash hee doth refuse;
> They'l take the pot, and throw drinke in's face,
> And with broad scoffs, most grosely him abuse,
> Thus will they urge him to his great disgrace.[34]

At risk of "abuse," "disgrace," and indeed violence, the health drinker might stay in the ritual even against his own will and to the point of endangering his health.

As writers from Gascoigne to Hornby suggest, health drinking is a compulsive and compulsory practice. It endangers health through both sickness and violence. Participants hurt their bodies through a ritual that is a "surfeit." The English who were formerly "commended for sobrietie" and who held healthing in the "highest degree of hatred" now give "little concern" to their own health and "dye" under constraint. Overcome by drink, participants are equally overwhelmed by fellow drinkers, threatened with violence if they attempt to withdraw from the ritual. If addicts are bound to service for someone or something, then critics of healthing expose such bondage as tyrannous. The pamphlets, through a language of being "injoyned," and "constrained," "compel[led]," and "bound" "by oaths," depict healthing as a compelled commitment and express a degree of concern taken up even more vehemently by puritan critics.

For if Gascoigne, Brathwaite, and others satirize healthing as a violent ritual, attacks on the practice become more pointed in the work of sectarian preachers during the next decades. Over twenty separate publications on health drinking appeared throughout the course of the seventeenth century, beginning with John Downame's 1609 *Foure Treatises* up to the compilation on health drinking by Samuel Clark, *A Warning Piece to all Drunkards and Health-Drinkers* (1680). Often taking the form of a polemical sermon, the publications on health drinking come largely from puritan ministers who offer arguments resonant with earlier critiques of the practice cited above.

Such sectarian literature on health drinking went through multiple editions, as was the case with the puritan pamphleteer William Prynne's one-hundred-page polemic against healthing, *Healthes Sickness* (1628). In it he amplifies the attack on health drinking as compulsory: "Our ordinary drinking of Healthes, doeth take away Christian libertie and freedome, and puts a *kinde of Law and necessitie upon men*, in the use of Gods good creatures: For it confines and restraines . . . mens drinking, to the will and pleasure of such as begin the Health: and puts a kinde of Law and necessitie upon all the company that are present, both in the matter, manner, measure, time, and end of drinking."[35] Using the same language of confinement that appears in Prynne—as well as in Gascoigne, Brathwaite, and Hornby—the author of *The Great Evil of Health-Drinking* warns against the compulsion of healthing: "our young masters, who swear you shall drink, or swear they'll run you through, they'll see through you, they'll pin you to the wall, or fasten you to the ground. . . . It is dangerous to drinke, and it is deadly to refuse."[36] For all these writers, the drinker has no "Christian libertie" or "freedome" at all, compelled by a new "Law and necessitie." Furthermore, the drinker finds himself under the "will and pleasure" of another, since he must continue to imbibe as long as pledges are raised and spoken.

This compulsory view of healthing—"it is deadly to refuse"—resonates with other Stuart pamphlets about pushy drinkers bullying friends into excess. The author of *A looking glasse for drunkards* (1627) notes how men "take it in hye disgrace, and take occasions to quarrell if the healthes bee not observed."[37] Richard Younge notes in *The Drunkard's Character* (1638) that "it is an unexcusable fault, or, as may say, an *unpardonable crime* to *refuse* an health, or not to *drink equall* with the rest."[38] To these religious writers health drinking represents an alternate (and illegitimate) form of government, whereby abstainers are called criminals and punished for transgressing a set of tavern laws. The drinker cannot extract himself from this drinking commonwealth without fear of physical retribution. As Downame puts it in *Foure Treatises* (1609), healthing is "a Lordly tyrant, which raigneth and ruleth with great insolence."[39]

Attacking health drinking as compulsory, these writers also view it as physically dangerous. *A looking glasse for drunkards* (1627) notes the physical dangers of healthing, since a drinker is forced to consume even when his body flags: "When a man pleads . . . that he can drinke no more, without great hurt to himselfe, then the drunken rout will wind him in, by drinking healths to one great personage or another."[40] Downame concurs, writing that health drinkers "draw one another to excessive quaffing, by making challenges, who can expresse most love to their absent friends by largest drinking, not caring to bring themselves, through their intemperancy into grievous diseases, by drinking healthes to other men."[41] "Grievous diseases" attend health drinking. These writers repeatedly highlight the ironic link of "health" to disease, claiming that drinkers ruin rather than preserve their well-being by such drinking.

To reinforce the dangers of health drinking, many authors offer examples of unfortunate drinkers, killed or maimed by their excessive quaffing. The puritan preacher Samuel Ward illustrates such a danger (as seen briefly in Chapter 4) in *Woe to Drunkards* (1622): "At a Taverne in Breadstreet in London certaine Gentlemen drinking healthes to their Lords, on whom they had dependence, one desperate wretch steps to the Tables end, layes hold one a pottle-pot full of Canarie Sack, sweares a deepe oath; What will none heere drink an health to my noble Lord and Master? And so setting the Pottle-pot to his mouth, drinkes it off to the bottom, was not able to rise up, or to speak when hee had done, but fell into a deepe snoring sleepe, and . . . was within the space of two hours irrecoverably dead."[42] Ward offers several other stories, complete with pictures of fatal health drinking. The "desperate wretch" noted above lies on the ground, dying, with a text issuing from his mouth: "Oh cursed health drinking" (Figure 4).

Of course, these sermonizing writers warn against the practice of health drinking for their own highly invested purposes. The godly focused particular energy on the damage that drunkenness posed for church services. Sabbatarian debates about church attendance and pious living persistently raised the specter of the drunkenness unfolding in taverns and alehouses, not only after the services but also during them. Healthing, as one of the most popular and notorious drinking rituals, posed a special risk. It threatened to offer an alternative government, complete with tavern laws and a unified commonwealth. This is not to say that these writers uphold monarchical loyalty against drinking. Instead they trumpet the Christian commonwealth, answering the lures of alcoholic spirits with a rival community, one based in the spirit of God. The Christian liberty of such godly community stands in stark opposition, these writers argue, to the compulsive and tyrannical rituals of health drinking: binding potential churchgoers to the tavern rather than the sacristy, the healthing ritual compelled men into participation (and thereby into excessive drunkenness). Godly writers thus emphasize the horrifying specter of compulsory conviviality, and their tales of dying health drinkers might be read as part of their attempt to scare drinkers straight. While a health drinker might

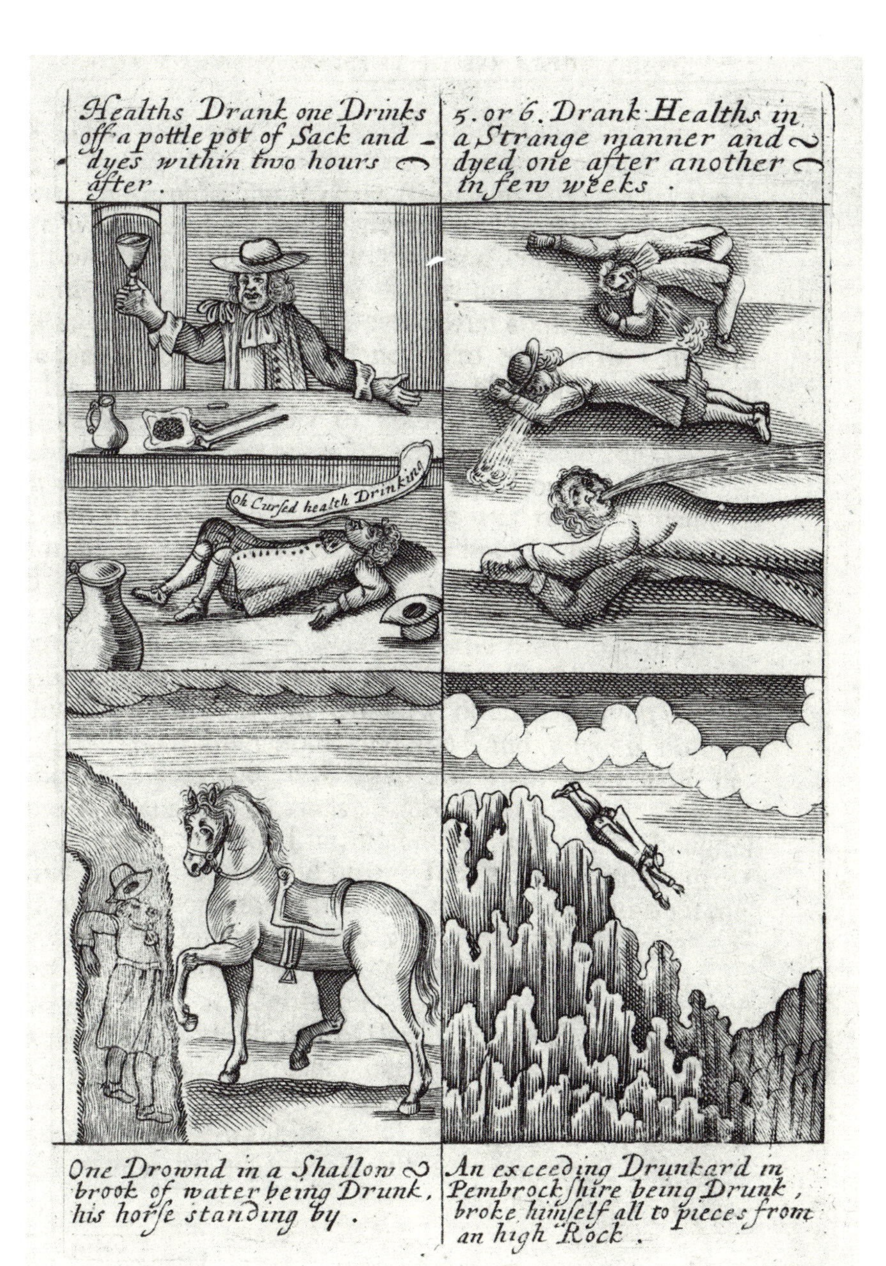

Figure 4. Effects of health drinking as shown in Samuel Ward, *Woe to Drunkards* (1622). Rare Book 42944. The Huntington Library, San Marino, California.

not attend to the state of his soul, he might respond to stories of rotting bodies instead.

Even as sectarian writers have a particular agenda in their attack on health drinking, nevertheless their arguments against it—as compulsory and dangerous—resonate with the earlier, secular depictions of the practice, as suggested in the examples from Camden, Norris, Gascoigne, Brathwaite, and Hornby above. The conversation about healthing as a social practice is indeed rather one-sided. Health drinking is rarely if ever praised in pamphlet literature, despite its potential to construct community around shared oaths and transformative spirits. Pamphlets trumpet instead "safety," "civility," and "piety," while expressing concern over "surfeyte" and "distemper." Such hesitation may well be prudent—but it also stands in stark contrast to Faustus's heroic embrace of dangerous impiety, or Olivia's acceptance of the "plague" of love. Even when imaginative writers offer dramatic health-drinking scenes on stage, as they do for comic effect or to propel their plots forward, the tangle of compulsion and violence quite often remains, as the next section reveals.

Drama of Health Drinking, 1590–1620

If the sectarian ministers underline the dangers of healthing, for Ben Jonson the ritual serves as a joke. One of the most humorous and extended portraits of healthing in early modern drama appears in his *Every Man out of his Humor*.[43] The character of Carlo Buffone begins the play by serving as prologue, drinking healths to all the audience members. Continuing to imbibe as the play proceeds, his drinking climaxes with a healthing scene between himself and two inanimate cups. As the stage directions tell us, "*He sets the two cups asunder, and first drinks with the one, and pledges with the other.*" Watched by amazed onstage characters standing in the wings, Buffone moves between the cups in the following fashion:

> *1 Cup*: Now sir, here's to you, and I present you with so much of my love.
> *2 Cup*: I take it kindly from you, sir, (*Drinks*) and will return you the like proportion.[44]

Buffone continues the routine, with each cup offering the other a pledge. But eventually the cups fall to blows when 2 Cup is convinced that 1 Cup did not drink his full pledge, recalling Count Hollock and Edward Norris. Indeed, the Zutphen episode resonates strongly with the scene's satire, its violence now overlaid with humor: overturning tables, cups, and chairs, Buffone is restrained by the alarmed drawer just as Hollock had to be restrained by Sidney after throwing a drinking bowl at Norris.

The health drinker is a swaggerer who senselessly attacks himself, overtaken by manic lunacy. Furthermore, this ritual is so evacuated of meaning that the cups fight over nothing; and its narrative arc—from drunkenness, to quarrelling, to destruction—is so formulaic that mere cups, rather than men, can enact the drama. In its joke, the scene reinforces precisely the critique against healthing familiar from pamphlet literature. Indeed its compulsory nature is part of the humor, as inanimate objects are conjoined to drink. Carlo Buffone perfectly encapsulates the razor's edge of healthing. He is the play's wittiest character, and he initially embodies the theater's conviviality as he welcomes the audience in the play's opening scenes. But in the drinking scene, he becomes the buffoon his name promises, leaving the audience in stitches as he relinquishes his authority to cups, compelled (by his own imagination) to continue drinking or risk violence if he abstains. Health drinking here stands in simultaneously for compulsion and inspiration. It is a mode of drinking that propels Buffone forward into an imaginary narrative, a comic scene that illuminates how health drinking at once exposes him and creatively prompts his comic artistry and invention.[45] Health drinking is, at least here, a clownish pursuit serving as an "ironic burlesque of dramatic 'realism.'"[46]

Jonson chronicles buffoonish, swaggering health drinkers in several other comedies. In *Epicene*, Captain Tom Otter, a dedicated health drinker with his own carousing cups named the "bull," the "bear," and the "horse" according to size, is overcome with drink and condemns his social-climbing wife to his audience (4.2.3). "Wives are nasty, sluttish animals," he begins—and ends being beaten by his wife who had been positioned to overhear his diatribe (4.2.45).[47] In *Bartholomew Fair*, too, healthing proves, like the game of vapours itself, a peculiar social ritual that overtakes its participants, freeing them from their own social habits to revel in the unexpected antics of the fair space. Jonson's Quarlous, disguised as Trouble-All, initially claims, "I may not drink without a warrant" (4.6.4) a mock-fastidiousness easily overcome when Knockum supplies a cooked-up warrant. Trouble-all begins to pledge vigorously, trumpeting, "In Justice Overdo's name, I drink to you, and here's my warrant" (4.6.117), eliciting laughs for his eager imbibing.[48]

While comedic in tone, Jonson's plays stage the cultural link of health drinking and bullying: practiced drinkers entice new members, who find themselves humiliated by having let their guards down long enough for the plot to proceed without them.[49] Yet his comedy also exposes the freedom granted characters through drunkenness. Health drinking transforms previously antisocial figures—Quarlous, Morose, and in *Eastward Ho*, Security (indeed, the very names of these abstaining characters insist upon the barriers between themselves and others)—into members of a comic community. Thus even as Jonson invokes health drinking as a laughable practice, he depicts its transformative potential in shifting otherwise oppositional, isolated characters into more open, indeed inspired, figures. In each of their cases the compulsion to drink brings addiction, voluminous speech, and expansion.[50]

While Jonson both satirizes and lightly celebrates health drinking, Shakespeare repeatedly invokes the practice to sinister ends, moving beyond satire to outright villainy. If *Twelfth Night*'s Sir Toby Belch provides one of the era's most humorous defenses of health drinking—he insists on "drinking healths to my niece . . . as long as there is a passage in my throat and drink in Illyria" (1.3.36–38)—the practice more frequently appears as a substitute or smokescreen for individual ambition: Claudius, Macbeth, Iago, Lucio, Simonides, Westmoreland, Prince John of Lancaster, Cardinal Wolsey, and Pompey all invoke health drinking to entrap their enemies or opponents, or to create false community in the midst of their criminal plotting.

In *Hamlet*, Claudius's health drinking bookends the play, providing one example of how a devotional drinking ritual turns compulsive and violent. In his first scene, the king counsels Hamlet to remain in Denmark and celebrates the prince's acquiescence with health drinking: "This gentle and unforced accord of Hamlet / Sits smiling to my heart, in grace whereof / No jocund health that Denmark drinks today / But the great cannon to the clouds shall tell / And the King's rouse the heaven shall bruit again" (*Hamlet*, 1.2.123–27). Such allegedly celebratory health drinking is precisely what Hamlet and Horatio hear from the battlements of the castle two scenes later; as noted above, the prince deems the custom more honored in the breach than in the observance. The king "takes his rouse, / Keeps wassail" (1.4.8–9) and, as Hamlet puts it in his description to Horatio, "as he drains his draughts of Rhenish down / The kettledrum and trumpet thus bray out / The triumph of his pledge" (1.4.10–12). Hamlet mocks this alleged "triumph" of drinking; instead of bringing honor to Denmark, Claudius brings infamy. Now other nations, Hamlet claims, "clepe us drunkards" (1.4.19). Packed into his short reference to health drinking lies a salient critique of Claudius's character and his practice of kingship. Health drinking exposes him as a King of Misrule rather than a legitimate sovereign. Honor in battle turns to honor in pledges, while stamina and courage support drunken excess, not wartime valor. If the narratives on health drinking and Dutch wars from this chapter's first section establish a link between drinking and fighting, Hamlet wryly undercuts it. Claudius is no soldier, and his alleged triumph in drinking only "takes / From our achievements" (1.4.20–21) and undermines the nation from within.

Opening with a reference to health drinking, the play ends with a poisoned cup in a scene reinforcing contemporary concerns about healthing and death. Claudius offers a series of compulsive toasts to the prince, attempting to goad him into drinking poison: "The King shall drink to Hamlet's better breath" (5.2.248), "Now the King drinks to Hamlet" (5.2.255), and "Stay, give me the drink. Hamlet, this pearl is thine: / Here's to thy health" (5.2.264–65). In his various pledges, Claudius orders a servant to "give him the cup" (5.2.265), but Hamlet refuses it. These pledges are highly dramatic: they might serve as a sign of Claudius's anxiety, his appetite for drinking, or his overeagerness to snare Hamlet. But the repetition of the gesture draws attention to it, not least because Claudius's pledging mania prompts Gertrude to drink: "The queen carouses to thy fortune, Hamlet" (5.2.271). On hearing Gertrude's response to this pledge—"No, no, the drink, the drink, O my dear Hamlet, / The drink, the drink—I am poisoned" (5.2.294–5)—Hamlet finally moves against the king. He forces Claudius to drink the health: "Drink of this potion" (5.2.310). By contrast, the prince begs Horatio to refrain: "As thou'rt a man / Give me the cup. Let go! By heaven, I'll ha't!" (5.2.326–27). Here Hamlet interrupts the round of healths initiated by Claudius and continued through Gertrude, and he does so because the survival of Horatio depends entirely on his abstinence. The audience palpably feels the two critiques of healthing rehearsed above: it is unhealthy, indeed poisonous, and it leads to violence. In a line of reasoning surprisingly evocative of godly complaints against health drinking, the play's final scene teaches us to refuse a pledge in the name of salvation, or at the very least survival.

What difference does it make to note that health drinking begins and ends *Hamlet*? While Claudius pledges loyalty to his nation through health drinking, the play invites its audience to admire Hamlet's singularity: he refrains from and indeed derides the practice. Hamlet is no godly moralist; he does not offer an extended, Prynne-style, sermon on the topic. Rather, as the above survey of contemporary views on health drinking in both pamphlets and drama helps illuminate, Hamlet need only invoke the familiar and largely ridiculed social practice to establish a flaw in Claudius's character: the audience would know precisely what it meant for a king to be drunk on healths. In contrast to this illegitimate king, Hamlet upholds Danish moderation against innovative, unwelcome drinking practices, much as Camden and Nashe attempted to remind the English of their former health and sobriety before the (alleged) introduction of health drinking from foreign shores.[51]

Having drawn attention to the range of healthing scenes in early modern drama, it is worth noting, in this section's final point, that healthing often appears as merely a background distraction in these plays or as dramatic shorthand. In comedies, characters like Buffone, Otter, Quarlous, and Sir Toby imbibe for our amusement. Health drinking appears laughable, a stock excuse for drunkenness and distraction. In *Hamlet*, health drinking appears as white noise, a diversion that allows characters to plot without arousing suspicion. But the ubiquity of the white noise of health drinking speaks to an extended, if at times critically invisible, conversation on a simultaneously devotional and compulsive practice. Indeed, the paradox of addiction itself is compressed in this practice: the addict is lauded and condemned, agentive and overcome. Primarily, though, if earlier chapters featured figures striving to uphold laudable attachments in addicted relation, here dramatic characters attempt to secure community through healthing, but succeed only in compelling others with hollow pledges. This process appears especially starkly in tragedies, which take the threat of violence attendant on healthing in comedies a step further, by actively linking health drinking and attempted murder. And both genres insist on healthing as a ritual allegedly constructing, but in practice eroding, community. Finally it is this continuity in representations of healthing that warrants attention. Whether in comedy or tragedy, and whether played for laughs or suspense, health drinking appears onstage as a gesture of false loyalty upheld by self-serving, swaggering roisterers. Some of these roisterers divert and amuse us, of course. But the sustained satire of health drinking suggests that early modern dramatists largely mocked the ritual: healthing, in their invocations, appears an empty gesture, failing to bind questionable communities of desperate drinkers. Despite its claim to addict—through pledging and fellowship—healthing is only mocked and reviled.

Health Drinking as an English Devotional Custom, 1630–1660

While Shakespeare and Jonson mock communities of health drinkers, poets such as Cotton and Herrick instead construct "one mind" through inspired drinking, ravishing themselves in addicted union. Cotton's speaker imagines himself merging with Anacreon and Horace, "as if we all one Man could be" (l.53).[52] Herrick also trumpets healthing as an ancient, historical, and literary rite, a form of "crowning" as the bowl goes around. As this final section reveals, from the 1630s through the 1650s, poets such as Herrick, Waller, Cotton, and Lovelace uphold health drinking as a devotional practice, a form of pledging that transforms the individual into an anonymous member of a loyalist group. In doing so they posit what we can now see as an innovative classical and pastoral origin for health drinking. Further, they represent healthing as a country ritual, akin to other traditional festivities including Whitsun ales, Morris dances, and maypoles. In their ingenuity, these poets not only reframe the representations of the ritual as villainous and compulsive by their playwriting predecessors, then, but they also actively challenge the condemnation of healthing in the contemporaneous pamphlets of their puritan opponents. By the 1630s, the discourse of health drinking thus fractures. The godly have launched their sustained attack on health drinking while Cavalier poets begin to celebrate it as a loyalty oath, a means of addicting oneself to one's political allies.

Health drinking becomes an expression of freedom and defiance expressed through loyalist pledges to community and country. Lovelace's "To Althea, *From Prison.* Song" exemplifies these celebratory views of healthing. As "flowing Cups run swiftly round," the poem's prisoners unite their "hearts with Loyall Flames." The men celebrate their drinking:

> When thirsty griefe in Wine we steepe,
> When Healths and draughts go free,
> Fishes that tipple in the Deepe,
> Know no such Libertie.[53]

Lovelace's poem rehearses a view of health drinking conventional in Cavalier verse. Such drinking is about loyalty ("our hearts with loyall flames"), "libertie," and release from care. What has seemed, in earlier invocations of healthing, to be the mutually exclusive freedom of abstaining and the compulsion of loyal participation here appears resolved: loyalty and liberty go hand in hand.

In "The Royalist," Alexander Brome amplifies and extends Cavalier engagement with health drinking as an expression simultaneously of celebration and compulsion. He writes,

> Come, pass about the *bowl* to me,
> A health to our distressed *King*;
> Though we're in hold, let cups go free,
> Birds in a *cage* may freely sing.[54]

Brome offers a familiar notion of the royalist supporters as caged birds who nonetheless sing for joy: as Lovelace writes in "To Althea, *From Prison.* Song," "stone walls do not a prison make / nor iron bars a cage" (79, lines 9–10). Both poems link healthing and liberty, with the latter signifying both liberality—of the generous host or abundant cups—and political freedom.

If the pamphleteers and dramatists above satirize or condemn healthing for *denying* the drinker liberty, Brome, like Lovelace and others, celebrates health drinking for *providing* such liberty. With the formation of the Protectorate, excessive drinking figures both liberty from godly discipline and the freedom to practice an outlawed social ritual. Such drinking, as Lois Potter writes, offers "an example of the release from the prison of the body."[55] This assertion of liberty has been read as nostalgia for a carefree time; the *carpe diem* ethos of the Cavalier poets, influentially analyzed by Earl Miner as a poetic response to the "Cavalier winter," reasserts lost rites and traditions, including the merriment now suppressed in times of war.[56] But the invocation of health drinking in these poems is more complex than a model of nostalgia allows. Cavalier poets may indeed invoke former rituals and rites, such as maypole dances and Whitsun ales. But in their invocation of health drinking, the poets reshape a derided social practice into a political rite of passage. They take up a ritual associated with roistering as a native custom. Rather than looking nostalgically backward to a lost custom of health drinking, then, these poets much more ingeniously seek to claim as an honorable English ritual a social practice earlier mocked by a diverse range of writers. Furthermore, they seek to find freedom in a ritual deemed compulsory and tyrannical. In doing so they anticipate the yoking of addiction and inspiration familiar from Romantic poets on, while hearkening back to the devotional mode celebrated in sixteenth-century addiction discourses.

The drinking and liberty invoked by these poets is not individual—the right to drink when and how one wishes—but instead collective—the right to drink together. Loyal camaraderie depends upon participation in the drinking ritual. These Cavalier poets thus exercise liberty and express devotion through a form of compulsory sociality derided by an earlier generation of writers on healthing. The loss of self in a potentially violent, addictive relation—based in pledging and devotion—appears now as heroic, where it had been foolish or sinister. Edmund Waller's "For drinking of Healths," for example, evokes health drinking and martial community:

> Wine fills the veins, and healths are understood,
> To give our Friends a title to our blood:
> Who naming mee, doth warme his courage so,
> Shews for my sake what his bold hand would do.[57]

Health drinking ensures the good fellowship of a masculine community: these men own "title" to one another's "blood." But the mention of "courage" and "bold hand" also suggests battle, as health drinkers promise to defend one another, whether in the drinking house or in the theater of war. The bold hand on the battlefield becomes the drinker's hand, raising a glass; the blood flowing in war becomes the wine flowing among friends. In "Song in a Siege," Robert Heath writes, "Let's drink then as we us'd to fight, / As long as we can stand." The speaker taunts his audience, crying "Hee neither dares to die or fight, / Whom harmless fears from healths affright."[58] Similarly, Cotton writes that "Who dares not drink's a wretched *Wight*; / Nor can I think that man dares fight / All day, that dares not drink all night."[59] Both speakers attack those who abstain from drinking as cowards who avoid the battlefield. Furthermore, in uniting drinking and fighting as signs of daring, these poets bolster their spirits for the fight against godly opponents. Here drinking feats presage battle victory; abstinence or moderation leads to "fear" and "wretchedness."

Of course, to some such overdrinking compromised rather than bolstered male communities, a process studied earlier in the analysis of Falstaff. The Cavalier's argument on the virtue of excessive drinking indeed challenges the literature on moderation as a primary masculine virtue, in contrast to emasculating excess.[60] Yet the moderation celebrated in one male community might be undermined by another. As Bernard Capp puts it, "The world of drink, gaming and roistering condemned by the conduct books represented an alternative model of manhood, always attractive to some."[61] Cavalier poets celebrate drunken excess as an expression of martial valor, and indeed health drinking becomes a rite of passage or obligation for inclusion in the Cavalier community, as drinking marks one's allegiance against godly interlopers. If earlier writers expressed suspicion of those men who demand health pledging as a form of friendship, a generation later Cavalier poets condemn those men unwilling or unable to drink. Indeed, as Angela McShane reveals, over the course of the seventeenth century it became "increasingly dangerous" to refuse the "secular sacrament of the loyal-health."[62]

Brome's poetry is especially clear in articulating this code of compulsory healthing. In "The Club," tired drinkers deserve banishment ("He that has a *heart* that's drowsie, / Shall be surely banished hence"), and those remaining health drinkers must stand their ground:

> Let the glass still run its round,
> And each *good-fellow* keep his ground,
> And if there be any *flincher* found,
> We'l have his soul *new-coyn'd.*
> (*SP*, 98)

The "flincher" (i.e., the man who passes the cup or abstains) will be beaten into shape, "his soul *new-coyn'd*," while the rest of the men defend their territory.[63] Through such gestures toward compulsory conviviality, Brome, arguably much more overtly than Herrick or Waller before him, acknowledges the sociopolitical downside of insisting upon drunkenness as a strategy of resistance. The community, in its violent opposition to the godly, turns punches and swords on themselves when left to their own devices in tavern or country house. Passing or abstaining, one is called a "flincher" or "wight," worthy of beating. The bravado in these poems effectively collapses the godly enemy and the moderate friend, turning all but the inebriated health drinker into an enemy target. Health drinking in these poems welcomes an audience into an apparently convivial community, but the speakers openly announce the violence and compulsion behind their gestures of hospitality.[64]

Thus poetic speakers, as inebriated health drinkers, experience precisely addiction's paradox, claiming political agency even as they depict their own deprived wills. But the masculinist ethos of health drinking puts extra pressure on this addictive mode, as participants experience a double bind: men are compelled to drink as a sign of manhood and yet they are at times derided as emasculated or dehumanized when they drink for the wrong reasons, namely peer pressure and lack of will. Cavalier representations insist it is mandatory to return a pledge, and this compulsory drinking serves to unify the community even as it compromises the individual drinker. The contradiction evident in many depictions of the practice in Cavalier verse—between drinking as a means of inclusion in a masculine community and as a compulsory gesture undermining the drinker's agency—resonates with other addictive dynamics tracked in this project. The balancing of will and release, coercion and dedication, proves the hallmark of such addictive practices, but healthing takes such a balance to an uncomfortable conclusion, as drinkers experience violence whether abstaining or participating. Modern concerns about addiction's violent power in ravaging the user are thus literalized in the case of health drinking. Even its most eager proponents recognize the compulsion behind the practice, which is experienced from within—as the drinker habituates to the practice—and from without—as fellow drinkers bully their allies.

Conclusion

One might claim, against this argument on health drinking as an addictive practice, that these verses are merely witty, convivial, symposiastic, or indeed Anacreontic. Goading or bullying, one might argue, denotes simply a community of drinkers, not social tyranny. And certainly in mingling drink and writing in "To live merrily," for example, Herrick nods to classical models of inspiration: Plato's symposium, Anacreon's verse, and Plutarch's table talk.[65] But the moderate pleasure of the banquet, and the interplay of wine and the civil conversation attendant on such feasts, contrasts with the more aggressive ritual of health drinking. The dominant language of healthing verse, as seen above, is not reason but excess; not health but sickness; not civil conversation but fighting. As tempting as it might be to group health drinking with the civil conviviality illuminated by critics such as Michelle O'Callaghan, Timothy Raylor, and Adam Smyth, the two are distinct forms of sociability.[66] These scholars help establish how conviviality involves drinking and pleasure but does not promote excess. Moderation is at odds with the dissipation of health drinking. Indeed, "convivial civility," to follow Michael Jeanneret, appears in "refined manners, moderate appetites, and the creation of a small society of close friends who are both educated and Epicurean."[67] To note the representations of drunken excess, then, is not to dismiss the pleasures of feasting or mirth. Instead, one can recognize both the pleasure of sensual ingestion, the joy of drinking, and the warmth of community on the one hand, and the compulsion and challenge to autonomy of group binge drinking on the other.

To study health drinking is thus to illuminate as addictive—namely, as devotional, compulsory, transformative, and potentially diseased—a practice that might otherwise seem merely convivial, escapist, or customary. And this practice might easily seem so because, as the above survey reveals, within the space of a few decades the dominant discourse on healthing moves from satire of such drinking—as a foreign practice, in Camden; as boisterous fellowship in Jonson; and as false friendship in Gascoigne and Shakespeare—to suspicion of those who fail to drink, in Waller, Lovelace, Brome, and Cotton. But it is worth noting how these lyric poets celebrate the features of health drinking that a prior generation of dramatists, and contemporary sectarian preachers, condemn: it is excessive, it is compulsory, and it is damaging to one's health.

Indeed, precisely those qualities of healthing that drew most negative attention among dramatists and satirists now attract the Cavaliers to the practice: it allows them to express loyalty and devotion to a like-minded community, losing their autonomy in the process. The self-shattering of health drinking is precisely the point. For as sectarians take up and vastly intensify the critique of health drinking, Cavaliers respond by celebrating the derided practice as an innovative form of communion. The link between the Eucharist and healthing exposes how health drinking might be celebrated in a time of increasing faction. With parishes policing the communion table and excluding royalists from participation, health drinking served as an alternate form of the Eucharist: "The increasingly popular loyal-healthing ritual," McShane writes, "seemed to offer an alternative vehicle for social and political union; combining customary practices of hospitality and communal 'caritas.'"[68] Religious devotion links to marital unity, as the "resort to the fellowship of hard-drinking comrades-in-arms," Jerome de Groot notes, functions "as an antidote to disappointment and defeat."[69]

The cultural struggle over health drinking exposes the tensions at stake in addictive phenomenon more broadly: ravishment, compulsion, devotion, and liberty prove alternately celebrated and condemned. The devotion and commitment of the addicted health drinker helps secure community, be it in a time of political strife or romantic uncertainty; but this degree of transformative commitment, in ravaging the addict and compromising his or her will, also signals the dangerous compulsion and indeed annihilation that can accompany such excess. As writers from Shakespeare and Jonson through Waller and Brome depict, health drinking is a practice of compulsory conviviality: the practice at once compels participants into a form of devotional community and encourages excessive, even dangerous, drinking. Sectarian writers and early modern dramatists for the most part condemn or mock such compulsion. By contrast, Cavalier poets largely celebrate it: mandatory, excessive drinking helps cull the community to the "fit though few."

Even as Cavalier poets viewed their healthing as loyal political action helping to sustain the royalist cause, in an historical irony, upon ascending the throne King Charles II disagreed. He condemned and outlawed the compulsory conviviality of health drinking in the first year of his restoration. His 1660 proclamation against the practice claims:

> There are . . . [a] sort of men, of whom we have heard much, and are sufficiently ashamed, who spend their time in Taverns, Tipling-houses, and Debauches, giving no other evidence of their affection to us, but in Drinking Our Health, and Inveighing against all others, who are not of their own dissolute temper; . . . We hope . . . that they will cordially renounce all the Licentiousness, Prophaneness, and Impiety, with which they have been corrupted and endeavored to corrupt others, and that they will, hereafter, become examples of Sobriety and Virtue, and make it appear, that what is past, was rather the Vice of the Time, then of the Persons and so the fitter to be forgotten together.[70]

However unsuccessful this proclamation might have been—not only did health drinking continue, as the writings of John Wilmot, the Earl of Rochester, Thomas D'Urfey, Samuel Pepys, and a host of others attest, but it became nearly obligatory, as the research of McShane has helped illuminate—nevertheless its language is prescient.[71]

The "vice of the time" has, to a large degree, been overlooked or underinterpreted by scholars. But it is not therefore "the fitter to be forgotten." Instead, the canon of healthing literature constructed here recounts, in reverse miniature, a broader story of addiction. A social phenomenon largely condemned by one community allows transformative freedom for another. This recuperation of an addictive phenomenon exposes the shifts in historical vantage point on the excess, devotion, and compulsion at stake in healthing specifically, and addiction more generally. What one generation deems a corrupt practice not only of binge drinking but also of groupthink offers comfort to a subsequent generation, as its poets turn, in the era's severest political storms, to the transformative, communal, and shattering practice of addiction.

The drama of healthing thus encapsulates the tensions and paradoxes attending addiction: the release of self in devotional relation invites, at best, admiration for its achievement of communion, even as it also appears, in some articulations, to be a worrying and diseased overthrow of the will. As much as the long history of addiction tends toward the latter view of disease, this project has worked to restore the former view on devotion, striving in the process to reassess the achievement of being in relation.

Epilogue

Why Addiction?

> Parenting, even more than coupling, requires the core addictive features of compulsive behavior despite negative consequences.
>
> —Maia Szalavitz, *Unbroken Brain*

At one point in *The Argonauts,* Maggie Nelson recounts her experience as a beginning graduate student in a seminar featuring the guest speaker Jane Gallop. Presenting new work before her audience, Gallop offered a slide show, including a number of photos of herself naked with her baby boy. She was trying, Nelson writes, "to talk about photography from the standpoint of the photographed subject," a subjective position she coupled with that of being a mother. Nelson recounts her own feelings about this talk: she was surprised, and pleased; she liked it and found parts of it "irresistibly interesting." But when the talk ended she heard the seminar's respondent ridicule Gallop for "mediocrity, naiveté, and soft-mindedness." As Nelson puts it, "The tacit undercurrent of her argument, as I felt it, was that Gallop's maternity had rotted her mind."[1]

I rehearse this episode because it points to an opposition I had not recognized in my own thinking, one I had to overcome in order to complete this book: an opposition between what might be defined broadly as family life on the one hand, and intellectual life on the other. Reading Nelson's account elicited a reminder of my own transformation, at once intellectual and domestic: even if I'd never imagined maternity would rot my mind, I certainly didn't think it would help it. Without even knowing it, I had been upholding the divisions invoked by the seminar's respondent, and organized my life accordingly. The process of writing this book has helped to teach me otherwise, allowing me to do away with what had been a limiting structural opposition, what I might call the wall in the middle of my living room, imagined (erroneously) to house one of my home's support beams. Like it or not, my house got remodeled and I came to experience life without such a dividing wall.

It took becoming a parent to find the space and light to write this book in its current form. This was because, in writing a research monograph about early modern addiction, I came eerily to hear how many of the terms of addiction that I chronicled as pathological and diseased in my book resonated with my condition as the parent of a newborn: I lived in a state of compulsion, pushed from the outside and pulled from the inside. I existed in an innovative relationship to another, a relationship at once decided for me, and embraced by myself. In this imperative and reflexive state, I chose to love in the face of behavior I would previously have deemed unacceptable and dismaying: I offered devotion to a needy, irritable, unsympathetic person who cried through the night, ate every two hours, and had little if any concern for my own wellbeing.

What had seemed entirely clear—the desirability and valor of self-possession, the horror of dependency and compulsive behavior, the nightmare of being subject to or controlled by another—was now muddied. Experiencing commitment not as jail time but as attachment, loving in the face of difficulties and disappointments as well as joy, shifted my sense of what it means to be in relation to another. Furthermore, I was encountering two modes of compelled relation at the same time—the one experientially, as a new parent, and the other intellectually, as a researcher of addiction. Reassessing relationships of compulsion and possession from my vantage point both as a parent and a writer, I suddenly became much more attuned to the moralization not only in the sources for my book, but also in my own account of addiction in early drafts. Perhaps, I thought, it is worth reconsidering the condemnation of one set of commitments in the face of the overt idealization of the other. After all, one experience—motherhood—is ostensibly valorized. And book writing—a similarly overwhelming commitment—is at least tolerated. Yet these experiences are marked by many of the behaviors associated, in the modern era, with a parallel, yet often condemned, phenomenon—addiction.

My experience as a writing parent finally opened my ears to what the early modern archive had been telling me all along but which I'd been unwilling to hear. For during the process of researching the book, I felt frustrated that early modern writers deployed versions of the word "addict" to account for relationships that, to my mind, were clearly not addictions. Or at least, they were not addictions from the vantage point of the term's modern definition. Yet I now began to look, and listen, more carefully to these unwelcome archival invocations and to investigate how devotion to God or study or to one's beloved might constitute addiction. And I found that the apparently opposite outcomes of early modern addictions—leading to the celebration of one's faith, vocation, or partnership; or to sorrow, rejection, manipulation or illness—had obscured for me the shared process of being in relation, the process of being addicted in the first place.

This book is the result of listening to the range of early modern materials and hearing its evidence. Writing through this new insight on addiction as devotion led to a second surprise. Trying to account for the depth of the addictive process, particularly in its challenge to models of the self as sovereign, I found myself turning to a set of theorists who focus not on addiction, but on sexuality. Eve Sedgwick's essay on addiction in relation to the history of sexuality helped prompt this project from the very start. But in trying to understand modes of being outside of sovereign selfhood, I found the work of Lauren Berlant, Jennifer Nedelsky, Leo Bersani, and Tim Dean transformative to such an extent that my debt to feminist and queer theory forms a spine through this project. In the work of these social critics I found a space to think about questions of desire and subjectivity. My reading of Nelson's *The Argonauts* only deepened my sense of why the transformations I chart in this project might find theorization from those critics most attuned to forms of kinship and to the challenge that loving attachment presents for models of subjectivity based in possession and sovereignty. The queer theory I invoke throughout the project offered, for me, the most thoughtful, clear-eyed articulation of how to understand forms of attachment that simultaneously shape and unravel identity, clarify and overcome the individual, without devolving into moralism or medicalization.

If addiction, as Maia Szalavitz puts it, is a learning disorder, I might put it slightly differently. Addiction is a mode of learning. Szalavitz explores this learning as compulsive and largely undesirable, as evidenced by her experience of and research into drug addiction. My own work, while acknowledging the struggles of drug and alcohol addiction and the need for a sympathetic response, nevertheless seeks to expand our understanding of what addiction makes possible. I hope I have done so by uncovering early modern writers' alternate vantage point on this very familiar topic. For as the early moderns teach us, addiction—a relation based in devotion, compulsion, and commitment—holds the potential to open space and light into an otherwise divided scene of being.

Notes

PREFACE

1. "addict, *adj.*" *Oxford English Dictionary Online*. Oxford University Press, May 2017. http://www.oed.com.libproxy1.usc.edu/view/Entry/2175?rskey=NHAbsD&result=2&isAdvanced=true (accessed May 09, 2017): "addict, *adj.* < classical Latin *addīctus* assigned by decree, made over, bound, devoted, past participle of *addīcere* to assign, to make over by sale or auction, to award, to appoint, to ascribe, to hand over, surrender, to enslave, to devote, to sentence, condemn < *ad* ad- *prefix* + *dīcere* to speak, say (see dictum *n.*).

2. Ibid.

3. William Prynne, *Histrio-mastix. The players scourge, or, actors tragedy, divided into two parts* (London, 1633), 617, 325.

4. See Karen Britland, "Circe's Cup: Wine and Women in Early Modern Drama," in *A Pleasing Sinne: Drink and Conviviality in Seventeenth-Century England*, ed. Adam Smyth (London: D. S. Brewer, 2004), 109–26, and especially Tanya Pollard, who in *Drugs and Theatre in Early Modern England* (Oxford: Oxford University Press, 2005) illuminates the dense connections of narcotics and theater, from staged scenes of drug use to the theater's imagined effects on its audiences.

5. "addiction, *n.*". *OED Online*. Oxford University Press, May 2017. http://www.oed.com.libproxy1.usc.edu/view/Entry/2179 (accessed May 09, 2017). Contrast definitions 1a ("The state or condition of being dedicated or devoted *to* a thing") and 2 ("Predilection, inclination; an instance of this, a 'penchant.' *Obs*") with 1b ("Immoderate or compulsive consumption of a drug or other substance"), where Johnson notes John Philips's "addiction to tobacco."

6. Ibid., s.v. "devote, *v.*" See especially definition 2, "To give up, addict, apply zealously or exclusively."

7. Ibid., s.v. "devotion, *n.*," definition II, 5.

8. Ibid., s.v. "devoted, *adj.*," definition 2a.

9. I am grateful to Bruce R. Smith for his engagement with this portion of my argument. My first book, on the political, cultural, and literary impact of the 1534 Tudor statute on "treason by words," also investigated the thorny category of loyal or disloyal speech. *Treason by Words: Literature, Law, and Rebellion in Shakespeare's England* (Ithaca, NY: Cornell University Press, 2006).

10. Jeffrey Masten, *Queer Philologies: Sex, Language, and Affect in Shakespeare's Time* (Philadelphia: University of Pennsylvania Press, 2016), 15.

11. Jeffrey Masten, "Toward a Queer Address: The Taste of Letters and Early Modern Male Friendship," *GLQ: A Journal of Gay and Lesbian Studies* 10 no. 3 (2004): 374. Phil Withington writes, relatedly, how "early modern terms that are ostensibly unremarkable and familiar have

been found to carry meanings and uses—and to illuminate assumptions and practices—that are surprising and distinctly unfamiliar to the modern observer." "Company and Sociability in Early Modern England," *Social History* 32 no. 3 (2007): 296. Withington references his own work on the word "company," as well as the work of Craig Muldrew, *The Economy of Obligation: The Culture of Credit and Social Relations in Early Modern England* (Basingstoke: Palgrave, 1998); Naomi Tadmore, *Family and Friends in Eighteenth-Century England: Household, Kinship and Patronage* (Cambridge: Cambridge University Press, 2001); Andy Wood, "The Place of Custom in Plebeian Political Culture: England, 1550–1800," *Social History* 22 (1997): 46–60. See also Withington's work on the word "society" in *Society in Early Modern England: The Vernacular Origin of Some Powerful Ideas* (Cambridge: Polity Press, 2010).

12. Roland Greene, *Five Words: Critical Semantics in the Age of Shakespeare and Cervantes* (Chicago: University of Chicago Press, 2013), 8.

13. Ibid., 3–4.

14. There are direct references to addictive states, such as Olivia to melancholy and Falstaff to sack, but as Pollard reveals in *Drugs and Theatre*, theater stages drugging in a variety of forms, from the use of actual drugs such as poppy and mandragora to more symbolic engagements.

15. I use the masculine pronoun here, where I would otherwise use various inclusive pronouns, because the early modern actor on the public stage was presumed to be male.

16. Michael Goldman, *The Actor's Freedom: Towards a Theory of Freedom* (New York: Penguin Group, 1975), 9.

17. William B. Worthen, *The Idea of the Actor* (Princeton, NJ: Princeton University Press, 2014), 3.

18. See Konstantin Stanislavsky, *An Actor Prepares*, trans. Elizabeth Reynolds Hapgood (New York: Taylor and Francis, 1989) and *An Actor's Work*, trans. Jean Benedetti (New York: Routledge, 2008); Rhonda Blair, in *The Actor, Image, and Acting: Acting and Cognitive Neuroscience* (New York: Routledge, 2007), offers an analysis of this process of accessing inspiration through affect.

19. Joseph R. Roach, *The Player's Passion: Studies in the Science of Acting* (Newark: University of Delaware Press, 1985), 50.

20. Worthen, *Idea of the Actor*, 4.

21. Mary Nyquist, in *Arbitrary Rule: Slavery, Tyranny, and the Power of Life and Death* (Chicago: University of Chicago Press, 2013), writes of "voluntary psychological servitude," evident in the rule of custom, calling it "tyranny's willing counterpart" (62). Addiction also resonates with what Kathryn Schwarz, in *What You Will: Gender, Contract, and the Shakespearean Social Space* (Philadelphia: University of Pennsylvania Press, 2011), calls "volitional acquiescence" (21).

22. Jennifer Nedelsky, "Law, Boundaries, and the Bounded Self," *Representations* 30 (Spring 1990): 182.

23. Lauren Berlant, *Cruel Optimism* (Durham, NC: Duke University Press, 2011), 96.

24. Thus, as compelling as a study of addictive gaming, swearing, tobacco use, and/or hunting might be, the connection of the term "addict" with these activities is less frequent than for faith, love, and drinking.

25. I take the term "self-shattering" from Leo Bersani, "Is the Rectum a Grave?," *October* 43 (Winter 1987): 218, 222, and "Sociality and Sexuality," *Critical Inquiry* 26 no. 4 (Summer 2000): 646–47.

26. Berlant, *Cruel Optimism*, 3.

INTRODUCTION

1. For an exception to the division of addiction studies and early modern studies, see Jonathan Herring, Ciaran Regan, Darin Weinberg, and Phil Withington, *Intoxication and Society: Problematic Pleasures of Drugs and Alcohol* (Basingstoke: Palgrave, 2013). Although individual essays focus on modernity or early modernity, the collection as a whole surveys intoxication from the medieval to modern periods. Peter Mancall's study of early American alcohol trading and consumption in *Deadly Medicine: Indians and Alcohol in Early America* (Ithaca, NY: Cornell University Press, 1997) also, unusually, illuminates the sociopolitical forces driving alcohol sale and trading while also paying close attention to issues of excess and alcoholism. Articles on early modern addiction include Deborah Willis, "*Doctor Faustus* and the Early Modern Language of Addiction," in *Placing the Plays of Christopher Marlowe,* ed. Sara Munson Deats and Robert A. Logan (Farnham, Surrey: Ashgate, 2008), 136–148; Dennis Kezar, "Shakespeare's Addictions," *Critical Inquiry* 30 no. 1 (Autumn 2003): 31–62; Jessica Warner, "'Resolv'd to Drink No More': Addiction as a Preindustrial Concept," *Journal of Studies on Alcohol* 55 (1994): 689, and "'Before There Was Alcoholism': Lessons from the Medieval Experience with Alcohol," *Contemporary Drug Problems* 19 (1992): 409–29. The Christian physician Christopher C. H. Cook's book-length study of addiction, *Alcohol, Addiction, and Christian Ethics* (Cambridge: Cambridge University Press, 2006), begins with biblical and classical forms of addiction (Augustine, Aquinas) and then turns to nineteenth-century temperance literature in its search for an ethical response to the understandings of, and policy responses to, addiction.

2. Eve Kosofsky Sedgwick, "Epidemics of the Will," in *Tendencies* (New York: Routledge, 1994), 131.

3. Nancy Campbell, *Discovering Addiction: The Science and Politics of Substance Abuse Research* (Ann Arbor: University of Michigan Press, 2007); Griffith Edwards, ed., *Addiction: Evolution of a Specialist Field* (Oxford: Blackwell Science, Addiction Press, 2002). For a challenge to the disease model, see Herbert Fingarette, *Heavy Drinking: The Myth of Alcoholism as a Disease* (Berkeley: University of California Press, 1988); and the discussions provoked by Jeffrey A. Schaler in *Addiction Is a Choice* (Chicago: Open Court, 2000); Howard Kushner, "Taking Biology Seriously: The Next Task for Historians of Addiction?," *Bulletin of the History of Medicine* 80, no. 1 (Spring 2006): 115–43; Thomas Szasz, "The Discovery of Drug Addiction," in *Classic Contributions in the Addictions*, ed. Howard Shaffer and Milton Earl Burglass (New York, NY: Brunner/Mazel, 1981), 35–49. The American Medical Association first diagnosed alcoholism as a disease in 1956. The World Health Organization previously defined addiction as "an overpowering desire, need, or compulsion to continue taking a drug, a willingness to obtain it by any means, a tendency to increased dosage, and a psychological and occasionally physical dependence on the drug," but has more recently deployed the term "dependency" rather than "addiction." Cited in H. Isbell and W. M. White, "Clinical Characteristics of Addictions," *American Journal of Medicine* 14 (1953): 558–65.

4. Jeffrey Poland and George Graham, eds., "Introduction: The Makings of a Responsible Addict," *Addiction and Responsibility*, (Cambridge, MA: MIT Press, 2011), 6.

5. Lubomira Radoilska, *Addiction and Weakness of Will* (Oxford: Oxford University Press, 2013), 62. Philosophers offer sustained engagement with the question of "free will" and addiction. See also Cook, *Alcohol, Addiction, and Christian Ethics*; Mariana Valverde, *Diseases of the Will: Alcohol and the Dilemmas of Freedom* (Cambridge: Cambridge University Press, 1998); R. Jay Wallace, "Addiction as Defect of the Will: Some Philosophical Reflections," in *Normativity*

and the Will: Selected Papers on Moral Psychology and Practical Reason (Oxford: Oxford University Press, 2006), 165–89.

6. Radoilska, *Addiction*, 136.

7. Robert West and Jamie Brown, *The Theory of Addiction*, 2nd edition (Oxford: Wiley Blackwell, 2013), 1.

8. See Brian Harrison, *Drink and the Victorians* (London: Faber & Faber, 1971), 92.

9. Thomas Trotter, *An Essay, Medical, Philosophical, and Chemical, on Drunkenness, and Its Effects on the Human Body* (1804), 4th edition (London: Longman, Hurst, Kees, and Orme, 1810), 18. Trotter also deals with the issue of drunkenness in his 1786 treatise, *A View of the Nervous Temperament; Being a Practical Enquiry into the Increasing Prevalence, Prevention, and Treatment of Those Diseases Commonly Called Nervous, Bilious, Stomach, and Liver Complaints; Indigestion; Low Spirits, Gout, Etc.*, 2nd edition (London, 1786). On the disease concept, see also the study of E. Morton Jellinek, *The Disease Concept of Alcoholism* (New Haven, CT: Hillhouse Press, 1960).

10. Trotter, *Essay, Medical, Philosophical, and Chemical, on Drunkenness*, 46–47.

11. Benjamin Rush, *An Inquiry into the Effects of Ardent Spirits upon the Human Body and Mind* (1785), 8th edition (Boston: James Loring, 1823), 8. On alcoholism in nineteenth-century America, see Sarah W. Tracy, *Alcoholism in America: From Reconstruction to Prohibition* (Baltimore: Johns Hopkins University Press, 2005).

12. Benjamin Rush, *Medical Inquiries and Observations upon the Diseases of the Mind* (1812), 5th edition (Philadelphia: Grigg and Eliot, 1835), 264.

13. Harry G. Levine, "The Discovery of Addiction: Changing Concepts of Habitual Drunkenness in America," *Journal of Studies on Alcohol* 39 no. 1 (1978): 144.

14. Roy M. MacLeod, "The Edge of Hope: Social Policy and Chronic Alcoholism, 1870–1900," *Journal of the History of Medicine and Allied Sciences* 22 (1967): 223.

15. Ibid., 224.

16. Levine, "Discovery of Addiction," 144.

17. Ibid., 148.

18. Roy Porter, introduction to Thomas Trotter, *An Essay, Medical, Philosophical, and Chemical, on Drunkenness, and Its Effects on the Human Body*, ed. Roy Porter (New York: Routledge, 1988), xii.

19. MacLeod, "Edge of Hope," 217.

20. Of course, it is important to acknowledge that physicians are far from unified in naming the cause of addiction. While the dominant medical narrative on addiction stresses the shift in understanding from vice to disease, as traced above, there has also been resistance to medicalization. Caught between biological determinism and individual free will as two ways of thinking about the issue, physicians often refer to both poles in treating addiction.

21. In the century before Trotter, as Roy Porter writes, "practitioners not only saw drunkenness as a cause of disease, but regarded it as, in its own right, an addictive disorder." "The Drinking Man's Disease: The 'Pre-History' of Alcoholism in Georgian Britain," *British Journal of Addiction* 80 no. 4 (1985): 385. See, too, Jessica Warner, *Craze: Gin and Debauchery in an Age of Reason* (London: Profile Books, 2003); and Jonathan White, "'The Slow but Sure Poison:' The Representation of Gin and its Drinkers, 1736–1751," *Journal of British Studies* 42 (2003): 35–64.

22. Jonathan Herring, Ciaran Regan, Darin Weinberg, and Phil Withington, "Starting the Conversation," in Herring et al., *Intoxication and Society*, 3–4.

23. David Courtwright, *Forces of Habit: Drugs and the Making of the Modern World* (Cambridge, MA: Harvard University Press, 2001); Dave Boothroyd, *Culture on Drugs: Narco-Cultural*

Studies of High Modernity (Manchester: Manchester University Press, 2006); Bruce Alexander, *The Globalisation of Addiction: A Study in Poverty of the Spirit* (Oxford: Oxford University Press, 2008).

24. Peter W. Kalivas and Nora D. Voldow, "The Neural Basis of Addiction: A Pathology of Motivation and Choice," *American Journal of Psychiatry* 162 no. 8 (August 2005): 1403. See also Steven E. Hyman, Robert C. Malenka, and Eric J. Nestler, "Neural Mechanisms of Addiction: The Role of Reward-Related Learning and Memory," *Annual Review of Neuroscience* 29 (2006): 565. I am grateful to Michael Quick and Anuj Aggerwal for their suggestions for reading and for an invaluable series of conversations about the neuroscience of addiction.

25. See the study of Nasir H. Naqvi, David Rudrauf, Hanna Damasio, and Antoine Bechara, "Damage to the Insula Disrupts Addiction to Cigarette Smoking," *Science*, new series, 315, no. 5811 (Jan. 26, 2007), 531–34, which investigates the effect of brain lesions on drug addiction. The researchers found that smokers with brain damage involving the insula—a region in the brain linked to emotion and feelings, as well as to conscious urges—quit smoking easily, immediately, and without relapse. Viewing these neurobiological studies from a historical vantage point, we might say that modern research into addiction over the course of the last two centuries has become increasingly internalized, as perception of the phenomenon has shifted from medical psychology to brain chemistry. It is no longer a social disease of a culture, or even an individual, but instead a pathologized condition of a portion of an organ. For a response to such medicalization, see Timothy A. Hickman, who brings a historian's skill to analyzing and contextualizing the new "brain science" of addiction in "Target America: Visual Culture, Neuroimaging, and the 'Hijacked Brain' Theory of Addiction," *Past and Present* 222 no. S9 (2014): 207–26.

26. See Sue Vice, Matthew Campbell, and Tim Armstrong, eds. *Beyond the Pleasure Dome: Writing and Addiction from the Romantics* (Sheffield: Sheffield Academy Press, 1994).

27. Janet Ferrell Brodie and Marc Redfield, eds. *High Anxieties: Cultural Studies in Addiction* (Berkeley: University of California Press, 2002), 4.

28. Ibid., 4.

29. Stacey Margolis, "Addiction and the Ends of Desire," in Brodie and Redfield, *High Anxieties,* 19.

30. Anna Alexander and Mark S. Roberts, eds., "Introduction," *High Culture: Reflections on Addiction and Modernity* (State University of New York Press, 2003), 3; Jacques Derrida, "The Rhetoric of Drugs," in Alexander and Roberts, *High Culture,* 24; Bela Szabados and Kenneth G. Probert, eds., *Writing Addiction: Towards a Poetics of Desire and Its Others* (Regina, Saskatchewan, Canada: University of Regina, 2004). Shifts in the drug trade made certain substances more available, and the introduction of new methods of ingestion (namely the smoking of opium) radically altered the addictive landscape.

31. "addiction, *n.*" OED Online. May 2017. Oxford University Press. http://www.oed.com.libproxy1.usc.edu/view/Entry/2179?redirectedFrom=addiction (accessed May 10, 2017), definition 2. William Shakespeare, *Othello*, ed. E. A. J. Honigmann, with a new introduction by Ayanna Thompson, revised edition. Arden Shakespeare, 3rd series (London: Bloomsbury Arden Shakespeare, 2016).

32. George Joye, *The prophete Isaye, translated into englysshe* (1531), ch. 5, n.p.

33. Ibid.

34. Joye, *The Psalter of Dauid in Englyshe* (London: Thomas Godfray, 1534), "The argument into the xvii Psalm," n.p.; "the argument into the lxxxxij Psalm,", Liir; and "the argument until the lxxxxiiij Psalm,".n.p.

35. Ibid., "The fyfth Octonary," *The Psalter of Dauid in Englyshe,* n.p.

36. Joye, *The Unitie and Scisme of the Olde Chirche* (Antwerp, 1543), 6v.

37. John Bale, *The seconde part of the Image of both churches after the most wonderfull and heauenlye Reuelacyon of Saynt Iohan the Evangelist* (Antwerp, c.1545), 7v, 6v.

38. Polidore Vergil, *An abridgement of the notable woorke of Polidore Vergile.* Gathered by Thomas Langley (London, 1546), 34, 139.

39. Philip Nicolls, *The copie of a letter sente to one maister Chrispyne chanon of Exceter for that he denied ye scripture to be the touche stone or trial of al other doctrines* (London, [1548?]), C2v.

40. John Jewel, "An Homilee of good workes. And first of Fasting," in *The second tome of homilees of such matters as were promised, and intituled in the former part of homilees. Set out by the aucthoritie of the Queenes Maiestie: and to be read in every parishe church agreeably* (London, 1571), 174.

41. Humphrey Gifford, *A posie of gilloflowers* (London, 1580), 64–5.

42. Barnabe Googe, "Capricornus, the tenth Booke," in *The zodiake of life written by the godly and zealous poet Marcellus Pallingenius stellatus, wherein are conteyned twelve bookes disclosing the haynous crymes [and] wicked vices of our corrupt nature: and plainlye declaring the pleasaunt and perfit pathway unto eternall lyfe* (London, 1565), NNiiir. As Chapter 1 discusses in more detail, translations of Latin and continental texts include the terms "addict" and "addiction" either as a direct translation of the Latin original or as a perceived cognate for the French words for dedication and attachment. Thus, in the 1561 translation of Philip Melanchthon's *A famous and godly history* from Latin into English, Henry Bennet Callesian deploys the term "addict" as a direct translation of the original Latin, writing how the believers "addict them selves to the truthe." Philipp Melanchthon, *A famous and godly history contaynyng the lyues a[nd] actes of three renowmed reformers of the Christia[n] Church, Martine Luther, Iohn Ecolampadius, and Huldericke Zuinglius. . . . Newly Englished by Henry Bennet Callesian* (London, 1561), n.p.

43. Thomas Rogers, *Of the imitation of Christ, three, both for wisedome, and godlines, most excellent bookes; made 170 yeeres since by one Thomas of Kempis, and for the worthines thereof oft since translated out of Latine into sundrie languages* (London, 1580), 70, 191.

44. Donald Lupton, *The glory of their times. Or, The lives of ye primitiue fathers* (London, 1640), 369.

45. Raphael Holinshed, "The Historie of Scotlande," in *The firste volume of the chronicles of England, Scotlande, and Irelande* (London, 1577), 162.

46. Ibid., 230.

47. Guglielmo Gratarolo, *A direction for the health of Magistrates and Studentes. . . . Written in Latin by Guilielmus Gratarolus, and Englished, by T. N.* (1574), n.p.

48. John Huarte, *The examination of mens wits* (1594), B1r. See also Levinus Lemnius, *The sanctuarie of salvation, helmet of health, and mirrour of modestie and good maners* (London, 1592), and *The haven of pleasure containing a freemans felicitie, and a true direction how to live well*, by I. T. (London, 1597).

49. Lancelot Andrewes, *A Sermon Preached Before His Majestie at Whitehall the fifth of November last 1617,* (London, 1618), 2; see also *XCVI sermons by the Right Honorable and Reverend Father in God, Lancelot Andrewes, late Lord Bishop of Winchester* (London, 1629), 984.

50. Roger Edgeworth, "The 8th treatise or sermon," in *Sermons very fruitfull, godly, and learned, preached and sette foorth by Maister Roger Edgeworth* (London, 1557), Bbbbii.

51. William Baldwin, "The Spouse to the Younglynges, xvi," in *The canticles or balades of Salomon, phraselyke declared in Englysh metres* (London: William Baldwin, 1549), n.p. Baldwin's poem draws on the Song of Solomon and references the Church as the Spouse, speaking to its members.

52. Gregory Martin, "The Book of Psalmes: Psalm CXVIII," in *The holie Bible faithfully.* translated into English, out of the authentical Latin. By the English College of Doway (Douai, 1609–10), 228, 229.

53. Thomas Taylor, *The parable of the sower and of the seed* (London, 1621), 207.

54. Desiderius Erasmus, "The paraphrase of Erasmus upon the Epistle of S. Paule to Titus," in *The seconde tome or volume of the Paraphrase of Erasmus upon the Newe Testament conteynyng the epistles of S. Paul, and other the Apostles* (London, 1549), EEEEiiir.

55. Thomas Bilson, *The true difference betweene Christian subjection and unchristian rebellion wherein the princes lawfull power to commaund for trueth, and indeprivable right to beare the sword are defended against the Popes censures and the Iesuits sophismes* (London, 1585), 459; William Charke, *A treatise against the Defense of the censure, given upon the bookes of W. Charke and Meredith Hanmer, by an unknowne popish traytor in maintenance of the seditious challenge of Edmond Campion* (London, 1586), 320.

56. John Foxe, *Actes and monuments of matters most speciall and memorable, happenyng in the Church with an vniuersall history of the same . . . Newly revised and recognised, partly also augmented, and now the fourth time agayne published*, vol. 1 (London, 1583), 20, 101.

57. Ibid., 566.

58. John Jewel, "An Homilee agaynst gluttony and drunkenness," in *The second tome of homilees of such matters as were promised, and intituled in the former part of homilees. Set out by the aucthoritie of the Queenes Maiestie: and to be read in euery parishe church agreeably* (London, 1571), 203.

59. Henry VIII, *A glasse of the truthe* (London, 1532), B3r.

60. Keith Edwin Wrightson, "The Puritan Reformation of Manners, with Special Reference to the Counties of Lancashire and Essex, 1640–1660" (Ph.D. diss., Cambridge University, 1973), 13, 15. On the Puritan Reformation of Manners, see also Keith Wrightson, "Postscript: Terling Revisited," in *Poverty and Piety in an English Village: Terling, 1525–1700*, ed. Keith Wrightson and David Levine, 2nd edition (Oxford University Press, 1995), 186–220; Martin Ingram, "Reformation of Manners in Early Modern England," in *The Experience of Authority in Early Modern England*, ed. Paul Griffiths, Adam Fox, and Steve Hindle (London: Macmillan Press, 1996), 47–88; Marjorie Keniston McIntosh, *Controlling Misbehavior in England, 1370–1600* (Cambridge: Cambridge University Press, 1998); Peter Lake, "Defining Puritanism—Again?," in *Puritanism: Transatlantic Perspectives on a Seventeenth-Century Anglo-American Faith*, ed. Francis J. Bremer (Boston: Massachusetts Historical Society, 1993), 3–29, which challenges the notion that a reformation of manners was limited to the efforts of the godly; David E. Underdown, *Fire from Heaven: The Life of an English Town in the Seventeenth Century* (London: Harper Collins, 1992). For a study of Cromwell's role in this reforming effort, see Bernard Capp, *England's Culture Wars: Puritan Reformation and Its Enemies in the Interregnum, 1649–1660* (Oxford: Oxford University Press, 2012).

61. Ingram, "Reformation of Manners," 68. The work of Margaret Spufford provides a corrective to overemphasis on innovation: "To think or to imply that such social control was new shows a certain shortness of historical perspective on the part of the historians concerned." "Puritanism and Social Control?," in *Order and Disorder in Early Modern England*, ed. Anthony Fletcher and John Stevenson (Cambridge: Cambridge University Press, 1985), 57.

62. Wrightson, "Puritan Reformation of Manners," 18.

63. A. Lynn Martin, *Alcohol, Violence, and Disorder in Traditional Europe* (Kirksville, MO: Truman State University Press, 2009), 23.

64. "Early medieval writers and their patristic predecessors," Hugh Magennis notes in his

analysis of Old English sermons, "are particularly exercised about the sin of drunkenness." *Anglo-Saxon Appetites: Food and Drink and Their Consumption in Old English and Related Literature* (Dublin: Four Courts Press, 1999), 103.

65. Jonathan Wilcox, ed. *Aelfric's Prefaces*, Durham Medieval Texts no. 9 (Durham: University of Durham, 1994), 112–13, trans. at 129.

66. Richard Morris, ed., *Old English Homilies of the Twelfth Century: From the Unique Ms. B.14.52 in the Library of Trinity College, Cambridge*, 2nd series (London: Early English Text Society, 1873), 36.

67. Unlike the Ten Commandments, the seven deadly sins are a list coming not from the Bible but from Pope Gregory I, who revised a previous list of sins to arrive at the seven: lust, gluttony, avarice, sloth, wrath, envy, and pride. These sins overlapped and were conceived in relationship to one another, so that one sin might lead to others. See Elaine Treharne, "Gluttons for Punishment: The Drunk and Disorderly in Early English Homilies," 24th Annual Brixworth Lecture, 2nd series, no. 6 (Leicester: University of Leicester, 2007); Morton W. Bloomfield, *The Seven Deadly Sins: An Introduction to the History of a Religious Concept, with Special Reference to Medieval English Literature* (East Lansing: Michigan State College Press, 1952); Siegfried Wenzel, "The Seven Deadly Sins: Some Problems of Research," *Speculum* 43, no. 1 (January 1968): 1–22; John Livingston Lowes, "Chaucer and the Seven Deadly Sins, *PMLA* 30, no. 2 (1915): 237–371.

68. Geoffrey Chaucer, *The Canterbury Tales*, ed. F. N. Robinson (Oxford: Oxford University Press, 2008), 316.

69. As Gluttony boasts "Mans florchynge Flesch, / Fayre, frele, and fresch, / I rape to rewle in a rese / To kloye in my kynde," in *The Castle of Perseverance*, ed. David N. Klausner (Kalamazoo, MI: Medieval Institute Publications, 2010), scene x. If accounts of drunkenness might be formulaic, they are also complex. Langland, even as he rehearses all of the common concerns about drunkenness in terms of health, finances, and blasphemy, refuses to offer pious condemnations of drunkenness. His catalogue of the alehouse dwellers in *Pierce Plowman*—ranging from the shoe seller and the rat-catcher to the cobbler pawning his cloak and the horse dealer his hood to help pay for their drinking—undercuts easy moralizing. Similarly the sermonizing content of "The Pardoner's Tale"—as critics and students have long noted—is undermined by Chaucer's satirical portrait of the Pardoner himself, compromising the didactic effect of the warnings against drunkenness. Indeed, Chaucer's Pardoner anticipates a shift in writings about drunkenness, from the homiletic warnings within the frame of the seven deadly sins to the attacks on drunkenness through the lens of social satire. See John M. Bowers, "'Dronkenesse is Ful of Stryvyng:' Alcoholism and Ritual Violence in Chaucer's *Pardoner's Tale*," *ELH* 57 (1990): 757–84; Lowes, "Chaucer and the Seven Deadly Sins"; John Watkins, "The Allegorical Theatre: Moralities, Interludes, and Protestant Drama," in *The Cambridge History of Medieval English Literature*, ed. David Wallace (Cambridge: Cambridge University Press, 1999), 767–93.

70. "Although the scheme lived on in some catechisms," Siegfried Wenzel writes of the deadly sins, "after the sixteenth century it no longer played an important role in analyses of human behavior or church teaching" ("Seven Deadly Sins," 22). Certainly, writers continue to turn to language of the deadly sins, as in Thomas More's unfinished treatise *The Four Last Things* (1522), in which Gluttony exhibits "the dropsy, the colic, the stone, the strangury, the gout, the cramp, the palsy, the pox, the pestilence and the apoplexy." *The Four Last Things*, ed. D. O'Connor (London: Burns Oates and Washbourne, 1935), 78. More frequently, though, this schema appears through satire, as in Thomas Nashe's *Pierce Penniless* (London, 1592), Thomas Lodge's *Wits Misery, and the Worlds Madnesse* (London, 1596), and George Gascoigne, *A Delicate Diet for daintie mouthde Droonkardes* (London, 1576).

71. Skelton's portrait of Maude Ruggy showcases a woman suffering from alcoholic diseases, including gout, palsie, and dropsy, who drags herself into the alehouse, props herself against the wall, and drinks whatever she can; "The Tunning of Elynour Rumming," in *John Skelton: Selected Poems*, ed. Gerald Hammond (New York: Routledge, 2003). See Peter C. Herman, "Leaky Ladies and Droopy Dames: The Grotesque Realism of Skelton's The Tunnynge of Elynour Rummynge," in *Rethinking the Henrician Era: Essays on Early Tudor Texts and Contexts*, ed. Peter C. Herman (Chicago: University of Illinois Press, 1993), 145–67.

72. John Downame, *Foure treatises tending to disswade all Christians from foure no lesse hainous than common sinnes; namely the abuses of swearing, drunkenesse, whoredome, and briberie* (London, 1609), 83. Samuel Ward, in *Woe to Drunkards: A sermon by Samuel Ward, preacher of Ipswich* (London: A. Math for John Marriott, 1622), offers another such attack on drunkenness: "A drinker goes, as a foole to the stocks and an ox to the slaughterhouse, having no power to withstand the temptation, but in hee goes with him to the tipling house, not considering that the Chambers are the Chambers of death; and the guests, the guests of death; and there hee continues as one bewitched or conjured in a spell out of which he returnes not til he hath emptied his purse of money, his head of reason, and his heart of all his former seeming grace" (13). The tavern holds a necromantic power over customers, luring them with a rival spirit away from pious living. The anonymous author(s) of *A looking glasse for drunkards: Or, The hunting of drunkennesse Wherein drunkards are vnmasked to the view of the world. Very conuenient and vsefull for all people to ruminate on in this drunken age* (London, M. Flesher for F. C[oules], 1627) similarly counsels readers to "take heed of haunting taverns, inns, and alehouses which are the occasions of drunkenesse, for whosoever will avoyd the sinne must avoid the occasions which lead thereunto it: it may be truly said, the way to the alehouse is the drunkards path" (ch. 3), while Robert Harris, *Drunkard's Cup* (1619), writes, "Come to a mans house, and where is hee? His wife knowes not; ask the servants, they know not; when will he be home? They cannot shew you; yes they can, but they blush to speak; forsooth the matter is this: there's his house, but his dwelling is at the Alehouse, and when all his money is spent" (18).

73. William Perkins, *A godly and learned exposition or commentarie upon the three first chapters of the Revelation* (1595), 217; Prynne, *Histrio-mastix*, act 7, scene 3.

74. In the period between 1608 and 1641, eighteen tracts appeared on the topic of drunkenness. These tracts were reprinted multiple times, with writings by Henry Smith, Robert Harris, Samuel Ward, and Richard Younge reappearing every few years. These pamphlets either preach against drunkenness along with other sins, or they single out drunkenness for special consideration, dedicating an entire sermon or pamphlet to the topic.

75. Downame, *Foure treatises*, 83. Downame's most prominent patron was Thomas Egerton, Lord Ellesmere and Lord Chancellor of England. See P. S. Seaver, "Downame, John," in *Oxford Dictionary of National Biography* (Oxford: Oxford University Press, 2004). For another argument against habitual, daily drunkenness, offered the same year as Downame's, see Thomas Thompson, *A Diet for a Drunkard, delivered in two sermons at St. Nicholas Church in Bristoll, 1608* (London: Richard Backworth, 1612). See also Junius Florilegus [pseud.], *The odious, despicable, and dreadfull condition of a drunkard* (London: R. Cotes, 1649).

76. Downame, *Foure treatises*, 101.

77. Galen, *On Diseases and Symptoms*, trans. Ian Johnston (Cambridge University Press, 2006). See Nicolas Culpeper, *The English Physitian* (1652); John Walkington, *The Optic Glass of Humours* (London, 1631); William Vaughan, *Natural and Artificial Directions for health, derived from the best Philosophers, as well moderne, as ancient* (London: Richard Bradocke, 1600). On the challenge, and variety, of regulation—with a masterful discussion of what counts as

"enough"—see Jennifer Richards, "Health, Intoxication, and Civil Conversation in Renaissance England," *Past and Present* 222 no. S9 (2014): 177ff. "Moderation," she writes, "is almost never qualified" (177). Even excess could be deemed, as Joshua Scodel argues, in *Excess and the Mean in Early Modern English Literature* (Princeton, NJ: Princeton University Press, 2002), rational choice.

78. Louise Hill Curth and Tanya M. Cassidy, "'Health, Strength, and Happiness': Medical Constructions of Wine and Beer in Early Modern England," in Smyth, *Pleasing Sinne*, 152.

79. Ibid., 144; Richards, "Health, Intoxication, and Civil Conversation," 177–78. Here we might distinguish drunkenness from intoxication or as Richards puts it, "light drunkenness" (185), which can prove an aid to civil conversation.

80. Holinshed, *Firste volume of the chronicles of England, Scotlande, and Irelande*, 96.

81. John Hoskins, fellow of New-College in Oxford, Minister and Doctor of Law, *Sermons preached at Pauls Crosse* (1615), 30–31. See also Downame, who writes of the daily drinker, "Their braine will bear it [drink] without any great alteration" (*Foure treatises*, 83); and Ward, who writes in *Woe to Drunkards* (1622), "Wine takes away the heart, and spoyles the braine" (10).

82. Downame, *Foure treatises*, 96.

83. "Excessive and intemperate drinking," write the anonymous author of *The odious, despicable, and dreadfull condition of a drunkard*, "hath brought upon him [the drinker] a world of diseases and infirmities" including "crudities, rhemes, gowtes, dropsies, aches, imposthumes, apoplexies, inflammations, plurisies, consumptions" (23).

84. *Looking glasse for drunkards*, Chap. 2, n.p. Thomas Beard, too, catalogues how "excessive drinking breedeth creudities, rheumes, impostumes, gouts, consumptions, apoplexies, and such like," in *The Theatre of Gods Judgements . . . now secondly printed* (London, 1612), 430–31.

85. Roy Porter, *Disease, Medicine and Society in England, 1550–1860*, 2nd edition (Cambridge: Cambridge University Press, 1995), 20. See also Curth and Cassidy, "'Heath, Strength, and Happiness,'"143–60; Michael C. Schoenfeldt, *Bodies and Selves in Early Modern England: Physiology and Inwardness in Spenser, Shakespeare, Herbert, and Milton* (Cambridge: Cambridge University Press, 1999); Gail Kern Paster, *Humoring the Body: Emotions and the Shakespearean Stage* (Chicago: University of Chicago Press, 2004). On early modern medicine, see Andrew Wear, *Knowledge and Practice in English Medicine, 1550–1680* (Cambridge: Cambridge University Press, 2000); Margaret Healy, *Fictions of Disease in Early Modern England: Bodies, Plague, and Politics* (Basingstoke: Palgrave, 2001); Jennifer C. Vaught, ed., *Rhetorics of Bodily Disease and Health in Medieval and Early Modern England* (Farnham, Surrey: Ashgate Press, 2010).

86. Florilegus [pseud.], *The odious, despicable, and dreadfull condition*, 5.

87. Mary Nyquist, *Arbitrary Rule: Slavery, Tyranny, and the Power of Life and Death* (Chicago: University of Chicago Press, 2013), 22. Nyquist's analysis helps draw attention to the pernicious legacy of ethico-spiritual slavery in providing a philosophical basis for chattel slavery, a tyranny defended on the grounds of the natural slavery of certain communities, nations, or peoples. See also David Brion Davis, *The Problem of Slavery in Western Culture* (Ithaca, NY: Cornell University Press, 1966), who surveys the early Christian notion of slavery as part of natural law or God's divine will, as a result of the fall from grace: "for some two thousand years men thought of sin as a kind of slavery. One day they would come to think of slavery as sin," 90.

88. Maynwaringe, *Vita sana & longa the preservation of health and prolongation of life proposed and proved* (London, 1669), 74–5. This Maynwaringe text is also reprinted in *Two broadsides against tobacco* (London, 1672); Samuel Ward and Samuel Clark's *A Warning-piece to all Drunkards and Health-Drinkers* (1682); and [John Hancock], *The Touchstone, or Trial of Tobacco* (1676).

89. Warner, "'Resolv'd to Drink No More,'" 689. Warner also explores early addiction through the example of the gin craze in *Craze.* On attitudes toward excessive drinking see also Porter, "The Drinking Man's Disease"; John Brewer and Roy Porter, eds. *Consumption and the World of Goods* (London: Routledge, 1993); and Roy Porter and Mikulás Teich, eds. *Drugs and Narcotics in History* (Cambridge: Cambridge University Press, 1995). Porter has helped illuminate the prehistory of addiction in early physicians' writings, studying how eighteenth-century medical writers, including Barnard Mandeville, George Cheyne, Thomas Wilson, and John Coakley Lettson, address habitual drunkenness as a disease before the research of Trotter and Rush.

90. Warner, "'Resolv'd to Drink No More,'" 685.

91. Cook, *Alcohol, Addiction, and Christian Ethics*, xii.

92. Mark Hailwood, *Alehouses and Good Fellowship in Early Modern England* (London: Boydell and Brewer, 2014), 225. Of course, such drinking culture suffered internal tensions, as Hailwood's analysis reveals.

93. Hugh Crompton, *Pierides, or the Muses Mount* (London: J. G. for Charles Webb, 1658), 109. On toasts and drinking contests see Gina Bloom, "Manly Drunkenness: Binge Drinking as Disciplined Play," in *Masculinity and the Metropolis of Vice, 1550–1650*, ed. Amanda Bailey and Roze Hentschell (New York: Palgrave Macmillan, 2010), 21–44; Angela McShane, "Material Culture and 'Political Drinking' in Seventeenth-Century England," *Past and Present* 222 no. S9 (2014): 249; Michelle O'Callaghan, "Tavern Societies, the Inns of Court, and the Culture of Conviviality in Early Seventeenth-Century London," in Smyth, *Pleasing Sinne*, 37–54, and *The English Wits: Literature and Sociability in Early Modern England* (Cambridge: Cambridge University Press, 2007); Alexandra Shepard, *Meanings of Manhood in Early Modern England* (Oxford University Press, 2006); Adam Smyth, "'It were far better be a *Toad*, or *Serpant*, then a Drunkard': Writing about Drunkenness," in Smyth, *Pleasing Sinne*, 193–210; Richard Valpy French, *The History of Toasting or Drinking of Healths in England* (London: National Temperance Publication Depot, 1881).

94. Thomas D'Urfey, *Wit and Mirth: Pills to Purge Melancholy*, vol. 6 (London, 1720), 184. On the term "merry" as a synonym for "drunk," see, for example, the following court case: Nehemiah Brettargh, as described by William Blundell of Little Crosby, went 'merry to bed' one night and was found dead the next morning, *Transactions of the Historical Society of Lancashire and Cheshire,* ed. E. M. Hance and T. N. Norton, vol. 36 (Liverpool: Adam Holden, 1887), from Lancashire Mortuary Letters 1666–1672, no. 37.

95. William Hornby, *The Scourge of Drunkennes* (London, 1618), C4v. Hornby, in rehearsing this familiar defense of drunken fellowship, goes on to counsel against it.

96. William Prynne, *Healthes Sicknesse. Or a compendious and brief discourse; proving the drinking and pledging of Healthes, to be Sinfull, and utterly Unlawfull unto Christians* (London: Augustine Mathewes 1628), A3r. The concept of "company," invoked by Prynne, proves multiple and various, as Phil Withington reveals in "Company and Sociability in Early Modern England": one's association with company—through dedication to a particular community—might offer both an exercise in self-definition and a mode of implication.

97. Hailwood, *Alehouses and Good Fellowship* writes of the "affectionate social bonds" at the heart of the early modern alehouse's appeal, 113; James Brown, "Alehouse Licensing and State Formation in Early Modern England," in Herring et al., *Intoxication and Society*, 110–32; Judith M. Bennett, "Conviviality and Charity in Medieval and Early Modern England," *Past and Present*, no. 134 (Feb. 1992): 19–41, and *Ale, Beer, and Brewsters in England: Women's Work in a Changing World, 1300–1600* (Oxford: Oxford University Press, 1996); Peter Clark, *The Alehouse: A Social*

History (London: Longman, 1983); Steven Earnshaw, *The Pub in Literature: England's Altered State* (Manchester: Manchester University Press, 2000); Beat Kümin, *Drinking Matters: Public Houses and Social Exchange in Early Modern Central Europe* (Basingstoke: Palgrave-Macmillan, 2007); Beat Kümin and B. Ann Tlusty, eds. *The World of the Tavern: Public Houses in Early Modern Europe* (Farnham: Ashgate, 2002); A. Lynn Martin, "Drinking and Alehouses in the Diary of an English Mercer's Apprentice, 1663–1674," in *Alcohol: A Social and Cultural History*, ed. Mack P. Holt (NY: Berg, 2006), 93–106; Withington, "Company and Sociability in Early Modern England."

98. Smyth, introduction to Smyth, *Pleasing Sinne*, xv. Smyth frames his collection around a set of questions that remain crucial for understanding early modern alcohol culture: "Was alcohol a source of health, or illness? A force for social bonding, or a catalyst for disorder and rebellion? A marker of social grace, or a sign of debasement?" Introduction to Smyth, *Pleasing Sinne*, xiii–xxv; xiv.

99. Codes of conduct for drinking men (such as drinking heavily without succumbing to drunkenness) and for women (such as drinking lightly and visiting the tavern only when accompanied by a man or by female friends) meant that complex gender relations were negotiated within drinking spaces. See Bennett, *Ale, Beer, and Brewsters in England*; Bloom, "Manly Drunkenness"; Bernard Capp, "Gender and the Culture of the English Alehouse in Late Stuart England," in *The Trouble with Ribs: Women, Men and Gender in Early Modern Europe*, ed. Anu Korhonen and Kate Lowe (Helsinki: Helsinki Collegium for Advance Studies, 2007), 103–27; A. Lynn Martin, *Alcohol, Sex, and Gender in Late Medieval and Early Modern Europe* (London: Palgrave Macmillian, 2001); Patricia Fumerton, "Not Home: Alehouses, Ballads, and the Vagrant Husband in Early Modern England," *Journal of Medieval and Early Modern Studies* 32 no. 3 (Fall 2002): 493–518; Benjamin Roberts, "Drinking like a Man: The Paradox of Excessive Drinking for Seventeenth-Century Dutch Youths," *Journal of Family History* 29 no. 3 (2004): 237–52; Alexandra Shepard, *Meanings of Manhood*, and "'Swil-bolls and tos-pots': Drink Culture and Male Bonding in England, c.1560–1640," in *Love, Friendship and Faith in Europe, 1300–1800*, ed. Laura Gowing, Michael Hunter, and Miri Rubin (Palgrave Macmillan, 2005), 110–30.

100. Charles Ludington, *The Politics of Wine in Britain: A New Cultural History* (Basingstoke: Palgrave, 2013); Cedric C. Brown, "Sons of Beer and Sons of Ben: Drink as a Social Marker in Seventeenth-Century England," in Smyth, *Pleasing Sinne*, 3–20; Keith Wrightson, "Alehouses, Order, and Reformation in Rural England, 1590–1660," in *Popular Culture and Class Conflict, 1590–1914: Explorations in the History of Labour and Leisure*, ed. Eileen Yeo and Stephen Yeo (Hassocks: Harvester Press, 1981).

101. Angela McShane [Jones], "Roaring Royalists and Ranting Brewers: the Politicization of Drink and Drunkenness in Political Broadside Ballads from 1640 to 1689," in Smyth, *Pleasing Sinne*, 69–88; Charles C. Ludington, "'Sometimes to your country true': The Politics of Wine in England, 1660–1714," in Smyth, *Pleasing Sinne*, 89–108. For an important challenge to the link of drunkenness to lower social orders in early modern England see Phil Withington, "Renaissance Drinking Cultures and Popular Print," in Herring et al., *Intoxication and Society*, 135–52.

102. In their studies of Shakespeare and the Cavalier poets respectively, C. L. Barber and Earl Miner illuminate the ancient rites, including customary drinking rituals, behind many early modern literary texts, in C. L. Barber, *Shakespeare's Festive Comedy* (Princeton, NJ: Princeton University Press, 1959); and Earl Miner, *The Cavalier Mode from Jonson to Cotton* (Princeton, NJ: Princeton University Press, 1971). Subsequent critics developed this study of ritual by drawing on the work of Mikhail Bakhtin, whose model of the carnivalesque illuminates the political work achieved through drunken excess. Shakespeare's Eastcheap or Jonson's Bartholomew Fair,

as Michael Bristol, Peter Stallybrass, and Peter Lake have shown, represent the convivial freedom exercised by the folk community against such official culture. See Mikhail Bakhtin, *Rabelais and His World*, trans. Helene Iswolsky (Cambridge, MA: MIT Press), 1968; Michael Bristol, *Carnival and Theatre: Plebeian Culture and the Structure of Authority in Renaissance England* (New York: Metheun, 1985); Peter Stallybrass and Allon White, *The Politics and Poetics of Transgression* (Ithaca, NY: Cornell University Press, 1986); Peter Lake, with Michael C. Questier, *The Anti-Christ's Lewd Hat: Protestants, Papists and Players in Post-Reformation England* (New Haven, CT: Yale University Press, 2002). Excessive drinking signals a rebellious challenge to authority, whether that authority takes the form of a puritan magistrate or a usurping king. This link between drunkenness and rebellion also appears in Leah Sinanoglou Marcus's *The Politics of Mirth: Jonson, Herrick, Milton, Marvell and Defense of Old Holiday Pastimes* (Chicago: Chicago University Press, 1986), which, along with a host of subsequent studies on Cavalier culture, illuminates how drinking served royalist ends, offering a political challenge to dour religious opponents. See also Ruth Connolly and Tom Cain, eds., "*Lords of Wine and Oile": Community and Conviviality in the Poetry of Robert Herrick* (Oxford: Oxford University Press, 2011); Robert Wilcher, *The Writing of Royalism, 1628–1660* (Cambridge University Press, 2001); Thomas Corns, *Uncloistered Virtue: English Political Literature, 1640–1660* (Oxford: Clarendon Press, 1992); Jerome De Groot, *Royalist Identities* (Basingstoke, Hampshire: Palgrave Macmillan, 2004).

103. In its study of excess with achievement, this project participates, as Richards puts it in her study of discretion, in the recent reassessment of what has long been viewed as the opposition between regulation and excess. As she argues, "The preoccupation with restraint and excess has left the conviviality of moderate intoxication, light-headedness, and its rhetorical practice—the witty adaptation of sayings—overlooked and undervalued" ("Health, Intoxication, and Civil Conversation," 172).

104. Henry Crosse, *Vertues common-wealth: or The high-way to honour Wherin is discovered* (London, 1603), T2r.

105. William Perkins, *A treatise of man's imaginations* (London, 1607), 136.

106. George Benson, *Sermon at Paules Cross* (London, 1609), 75; Thomas Cooper, *The Churches Deliverance* (London, 1609), 73. See also John Boys, *The autumne part from the twelfth Sundy* (London, 1613), who claims "drunkennesse a point of good fellowship," 163.

107. I do not seek to locate "addicts" in the archives. Instead, I concur with Withington, in "Introduction: Cultures of Intoxication," *Past and Present* 222 no. S9 (2014): 9–33, when he writes of the challenge of tracking drunkenness as a phenomenon: "The 'ecstasies' of most people remain far beyond the historical record and, by their very nature, are notoriously difficult to recollect and convey" (11).

108. Greene, *Five Words*, 7.

CHAPTER I

Note to epigraph: John Henry Jones, ed. *The English Faust Book* (Cambridge: Cambridge University Press, 1994), 92.

1. Christopher Marlowe, *Doctor Faustus*, ed. Michael Keefer (Toronto: Broadview Press, 2007), prologue, lines 15–16. Hereafter cited above, with the standard act, scene, and line divisions. All citations are to this edition unless otherwise noted. Keefer prints the A text (although he occasionally prefers and prints the B text reading of a speech) on the grounds that it is more authentic, whereas the B text shows signs of censorship and corruption. This chapter follows

Keefer in finding the A text more reliable and also more germane to the chapter's argument on Marlowe's engagement with Calvin.

2. Willis, "*Doctor Faustus* and the Early Modern Language of Addiction," 144.

3. The term "Calvinism" arguably places too much emphasis on a singular theologian, and therefore, as Nigel Voak notes in *Richard Hooker and Reformed Theology* (Oxford: Oxford University Press, 2003), "'Reformed' is now increasingly preferred as the more satisfactory, inclusive alternative" (xvii). This chapter deploys both terms, recognizing the broader reformed interests of Foxe and Perkins while also emphasizing the specific influence of Calvin on their writings. On Calvinism in England, see Philip Benedict, *Christ's Churches Purely Reformed: A social history of Calvinism* (New Haven, CT: Yale University Press, 2002); R. T. Kendall, *Calvin and English Calvinism to 1649* (Oxford: Oxford University Press, 1979); Peter Lake, *Anglicans and Puritans? Presbyterianism and English Conformist Thought from Whitgift to Hooker* (London: Unwin Hyman, 1988), and "Calvinism and the English Church, 1570–1635," *Past and Present* 114 (1987): 32–76; John T. McNeill, *The History and Character of Calvinism* (Oxford: Oxford University Press, 1954); Anthony Milton, *Catholic and Reformed: The Roman and Protestant Churches in English Protestant Thought, 1600–1640* (Cambridge: Cambridge University Press, 1995); David Harry Stam, "England's Calvin: A Study of the Publication of John Calvin's Works in Tudor England" (Ph.D. diss., Northwestern University, 1978); Nicholas Tyacke, *Anti-Calvinists: The Rise of English Arminianism c. 1590–1640* (Oxford: Oxford University Press, 1987). On the publication history of Calvin in England see Stam, "England's Calvin," 10n15 and Appendix 2, 243–46.

4. See Margaret Ann O'Brien, "Christian Belief in Doctor Faustus." *ELH* 37 (1970): 1–11; Michael Hattaway, "The Theology of Marlowe's *Doctor Faustus*," *Renaissance Drama*, new series, 3 (1970): 51–78; David Riggs, "Marlowe's Quarrel with God," in *Critical Essays on Christopher Marlowe*, ed. Emily C. Bartels (London: Prentice Hall, 1997), 39–58; Robert Ornstein, "Marlowe and God: The Tragic Theology of *Dr. Faustus*," *PMLA* 83 (1968): 1378–85; John Stachniewski, *The Persecutory Imagination: English Puritanism and the Literature of Religious Despair* (Oxford: Clarendon Press, 1991); A. D. Nutall, *The Alternative Trinity: Gnostic Heresy in Marlowe, Milton, and Blake* (Oxford: Clarendon Press, 1998). On the ferocity of Calvinist theories of election, see Keefer, "Introduction," in Marlowe, *Doctor Faustus*, 41; Kristen Poole, "Dr. Faustus and Reformation Theology," in *Early Modern English Drama*, ed. Garrett Sullivan, Patrick Cheney, and Andrew Hadfield (Oxford: Oxford University Press, 2006), 98; Pauline Honderich, "John Calvin and *Doctor Faustus*," *Modern Language Review* 68 no. 1 (1973): 1–13.

5. Alan Sinfield, *Faultlines: Cultural Materialism and the Politics of Dissident Reading* (Berkeley: University of California Press, 1992), 236. Debating the play's staging of election, scholars turn to the two texts of *Dr. Faustus*, the 1604 A text and the longer, revised (and frequently dismissed as inferior) 1616 B text, to assess how and why Marlowe (or others) might have revised his presentation of Calvinist theology. Leah Marcus, in concert with Keefer and others, deems the A text to be straight-line Calvinist Protestantism, whereas the B text proves Arminian in its acknowledgement of the believer's role in his own salvation. See Leah Marcus, "Textual Indeterminacy and Ideological Difference: The Case of 'Doctor Faustus,'" *Renaissance Drama*, new series, 20 (1989): 1–29.

6. While reformers such as Perkins and Foxe demonstrate their engagement with Calvin's thought, Calvin also, notably, involved himself in English clerical concerns, dedicating one of his publications of 1548 to Edward Somerset, the Lord Protector, and offering him a letter of advice on reform that same year. Calvin received further encouragement from Thomas Cranmer, who asked him to write King Edward VI with frequency in order to help the reformed cause in England. See McNeill, *History and Character of Calvinism*, 310–11.

7. Indeed, as Elizabeth Spiller writes, "*Doctor Faustus* is a play about books," staging the engagement and ravishment of the title character by books. "Marlowe's Libraries: A History of Reading," in *Christopher Marlowe in Context,* ed. Emily C. Bartels and Emma Smith (Cambridge: Cambridge University Press, 2013), 101. See also Paul Budra, "*Doctor Faustus:* Death of a Bibliophile," *Connotations* 1 no. 1 (1991): 1–11.

8. "Cicero to Appio Pulchro," in Abraham Flemming, trans., *A panoplie of epistles, or, a looking glasse for the vnlearned.* Gathered and translated out of Latine into English (London, 1576), 18. The Latin original reads: "iis studiis eaque doctrina, cui me a pueritia dedi," in "III.x: Cicero to Appio Pulchro, Laodicea, early in May 50 B.C.," in Cicero, *Letters to His Friends,* vol. 1, trans. W. Glynn Williams, Loeb Classical Library (London: Heinemann, 1927), 224. This Loeb volume translates the phrase as "that study and that learning to which I have devoted myself from boyhood" (225).

9. Cicero, *The familiar epistles of M. T. Cicero Englished and conferred with the French, Italian and other translations* (London: Printed by Edward Griffin, 1620): "Cicero to Marcus Coelius Aedile Curule, Epistle 13," 79; and "Cicero to Aulus Torquatus, Epistle 4," 266. The Latin originals read "studiosus studiorum etiam meorum" and "litterae, quibus semper studui," in Cicero, *Letters to His Friends,* 132 and 444.

10. "Trebonius to Cicero (June 8, Athens)," in Flemming, trans., *Panoplie of epistles,* 131.

11. "The life of Lucius Annaeus Seneca, described by Justus Lipsius," in Seneca, *The workes of Lucius Annaeus Seneca, both morrall and natural,* trans. Thomas Lodge (London, 1614), C3v. This is a translation of Justus Lipsius's edition of Seneca's works, *Annaei Senecae Philosophi Opera, Quae Exstant Omnia, A Iusto Lipsio emendata, et Scholiis illustrata* (Antwerp: Plantijn-Moretus, 1605).

12. "Lucius Annaeus Seneca, His Epistles to Lucilius: Epistle LXXII," in Seneca, *Workes of Lucius Annaeus Seneca,* 296. The full quote reads, "But as soone as I have made an end of this (say wee) I will wholly dedicate my selfe, and if I can end this troublesom matter, I will addict my selfe unto studie. Thou must not expect till thou have leasure to follow Philosophie. Thou must contemne all other things, to be alwayes with her." The original reads "Deinde ipsi nobis dilationem damus: 'cum hoc peregero, toto animo incumbam' et 'si hanc rem molestam composuero, studio me dabo.' Non cum vacaveris, philosophandum est; Omnia alia neglegenda, ut huic adsideamus," in "The Epistles of Seneca, Epistle LXXII," in Seneca, *Ad Lucilium Epistulae Morales,* trans. Richard M. Gummere, Loeb Classical Library (London: Heinemann, 1920), 96.

13. Seneca, Lucius Annaeus, "The Tranquilitie and Peace of the Mind," in Seneca, *Workes of Lucius Annaeus Seneca,* 644.

14. "Of a Wise-mans rest and retirement," in Seneca, *Workes of Lucius Annaeus Seneca,* 907.

15. Poole, "Dr. Faustus and Reformation Theology," 102.

16. "A Treatise of Anger, Written by Lucius Annaeus Seneca to His Friend Novaus," from Seneca, *Workes of Lucius Annaeus Seneca,* 515. The Latin reads, "Commota enim semel et excussa mens ei servit quo impellitur. Quarundam rerum initia in nostra potestate sunt, ulteriora nos vi sua rapiunt nee regressum relinquunt" in Seneca, *Moral Essays,* vol. 1, trans. John W. Basore, Loeb Classical Library (London: Heinemann, 1928), 124.

17. On the widespread influence of Calvinist theology in England, Benedict, in concert with Collinson, Tyacke, and Lake, argues that the English church of the 1580s and 1590s "drew its theological inspiration from continental theology and was fundamentally Reformed in outlook" (*Christ's Churches Purely Reformed,* 232). Over half of Elizabeth's initial ecclesiastical appointments were returning Marian exiles steeped in the continental Reformed faith, Calvin

having welcomed them to Geneva during the period of their exile, where they were allowed to organize their own church (Benedict, *Christ's Churches Purely Reformed*, 244; Stam, "England's Calvin," 11; McNeill, *History and Character of Calvinism*, 311). This historiographical emphasis on the relationship of English to Continental theologians—in contrast to the earlier insistence on the distinction of the English church, against both Rome and Geneva—has been especially valuable, Benedict argues, in breaking free of English historiographical "insularity" (*Christ's Churches Purely Reformed*, 232).

18. Jean Calvin, *A harmonie vpon the three Euangelists, Matthew, Mark and Luke with the commentarie of M. Iohn Caluine*. Faithfully translated out of Latine into English, by E. P. Whereunto is also added a commentarie vpon the Euangelist S. Iohn, by the same author (London, 1584), 364; The British Library 1005.c.14 has a printer error labeling page 364 as 348. In this edition of Calvin's *A harmonie vpon the three Euangelists*, forms of the word "addiction" appear fifty-three times, since the Latin verb forms of *addīcere* are translated into the English as "addiction" to express devotion to Christ, God, and scripture. Calvin's Latin reads "eos demum ad percipiendam Evangelii gratiam esse idoneos qui posthabitis omnibus aliis desideriis, ad eam potiendam sua studia & se totos addicunt" in Jean Calvin, *Harmonia ex Evangelistis tribus composite, Matthaeo, Marco, et Luca* (Geneva, 1582), 175.

19. Jean Calvin, *The Holy Gospel of Jesus Christ, according to John, with the Commentary of M. John Calvin*. Faithfully translated out of Latine into English by Christopher Fetherstone (London, 1584), 457 ("Nunquam ergo in hoc officio constanter perget, nisi in cuiuscorde sic regnabit amor Christi, ut sui oblitus, totumque se illi addicens, impedimenta Omnia superset." Calvin, *Harmonia ex Evangelistis*, 597). Three versions of Calvin's *A harmonie vpon the three Euangelists* appeared in England between 1580 and 1610. This long, composite publication consists of two separately numbered parts, which may have been intended for separate publication: the translation of *A harmonie* by Eusebius Paget and *A commentarie vpon the Euangelist S. Iohn* by Christopher Fetherstone. This joint publication first appeared 1584 with a variant edition that same year. In 1610 another edition appeared, printed by T. Adams.

20. Calvin, *Holy Gospel of Jesus Christ*, 46 ("tunc demum se illi addicere coeperunt, ut Messiam agnoscerent qualis iam illis praedicatus fuerat." Calvin, *Harmonia ex Evangelistis*, 412–13). He also writes in *The Holy Gospel of Jesus Christ* that "this interrogation importeth as much as if Christe did exhort him, to follow the Massias and to addict himselfe wholy unto him" (235).

21. Calvin, *Holy Gospel of Jesus Christ*, 240 ("eos in Dei ovile vere colligi, ut censeantur in eius grege, qui se uni Christo addicunt." Calvin, *Harmonia ex Evangelistis*, 500).

22. Calvin, *Harmonie vpon the three Euangelists*, 459 ("Nam quicunque se Christo simpliciter addicent, nec quic quam è suo capite assingere tentahunt Evangelio, nunquam eos certa lux deficiet." Calvin, *Harmonia ex Evangelistis*, 223).

23. On Marlowe's direct engagement with Calvin's *Harmonie*, see Adrian Streete, "Calvinist Conceptions of Hell in Marlowe's *Doctor Faustus*," *Notes and Queries* (December 2000): 430–32.

24. Benedict, *Christ's Churches Purely Reformed*, 245. On the rise of English Calvinism in the late Elizabethan period, Tyacke writes that "nowhere was that ascendancy more obvious than at Cambridge University" (*Anti-Calvinists*, 28).

25. Benedict, *Christ's Churches Purely Reformed*, 245. Benedict notes that six to eight of his books were produced each year, between 1578 and 1581.

26. *Bibliotheca Calviniana*, eds. Rodolphe Peter and Jean-François Gilmont (Geneva: Librairie Droz, 1991–2000), 2: 839–42; 2: 921–26.

27. Calvin, *Harmonie vpon the three Euangelists*, 47 ("Deus tam benigne pro sua

misericordia egerit cum populo ut eum redimeret: nempe ut redempti, se totos addicant & deuoueant colendo salutis suae authori." Calvin, *Harmonia ex Evangelistis*, 20).

28. Calvin, *Harmonie vpon the three Euangelists,* 148 ("ut posthabitis omnibus aliis curis, se totos Ecclesiae, cui destinati sunt, addicant ac deuoueant." Calvin, *Harmonia ex Evangelistis*, 68).

29. Calvin, *Holy Gospel of Jesus Christ,* 153 ("Fieri non posse quin se Christo addicant quicunque Dei sunt discipuli, & Deo esse indociles qui Christum [rei]iciunt." Calvin, *Harmonia ex Evangelistis*, 460).

30. Calvin, *Holy Gospel of Jesus Christ,* 142 ("quia enim Christi virtutem adulterant qui ventri & rebus terrenis sunt addicti, quid in se quaerendum sit & qua de causa quaerendus sit disputant." Calvin, *Harmonia ex Evangelistis*, 455).

31. Calvin, *Harmonie vpon the three Euangelists,* 204, and *Holy Gospel of Jesus Christ*, 142.

32. Calvin, *Holy Gospel of Jesus Christ,* 142 ("quia pro ingenii nostri crassitie semper rebus terrenis addicti sumus, ideo prius corrigit ingenitum illum nobis morbum, quam ostendat quid agendum fit." Calvin, *Harmonia ex Evangelistis*, 455).

33. Calvin, *Harmonie vpon the three Euangelists,* 203–4 ("Verum quidem est, fideles ipsos nunquam ita in solidum addictos esse dei obsequio, quin retrahantur subinde vitiosis carnis cupiditatibus." Calvin, *Harmonia ex Evangelistis*, 94).

34. Stam, "England's Calvin," 114, 115. With two editions of *Job* in 1574, subsequent editions appeared in 1579, 1580, and 1584 (STC 4446, 4446a, and 4447). Golding dedicated his translation to Robert Dudley, the Earl of Leicester. About the publication of these sermons in England, Stam writes, "In completely distinct type settings, each edition consisted of over 800 pages of closely-printed double columns. They represented large capital commitments in type, printing machinery, and paper, substantial labor costs for composition, printing, and collation. . . . Elizabethan printers undertook such monumental tasks . . . not simply from religious zeal but also because a healthy reading audience made them worthwhile in financial terms" (101).

35. Jean Calvin, *Sermons de M. Jean Calvin sur le livre de Job. Recueillis fidelement de sa bouche selon qu'il les preschoit* (Geneva: Jean I de Laon, 1563), 408, and *Sermons of Master John Calvin, upon the Booke of Job*, trans. Arthur Golding (London, 1574), 412. This is Calvin's "LXXX Sermon on the Book of Job." *Adonner* is a complex term and appears to have a genealogy resonant with that of "addiction." While signifying dedication, according to La Curne de Sainte-Palaye (Jean-Baptiste de), *Dictionnaire historique de l'ancien langage françois* ou *Glossaire de la langue françoise depuis son origine jusqu'au siècle de Louis XIV*, 10 vols. (Paris: H. Champion, 1875–1882) the verb *adoner* designates the giving of oneself, "De là s'*adonner* à une chose, pour s'y accorder." In the *Larousse* (and in concert with the *OED* definition of "addiction" seen in Chapter 1), s'adonner becomes a potentially damaging inclination or passion: "Se livrer à une activité par inclination et avec ardeur; en particulier, se laisser aller à un penchant néfaste: S'adonner à la boisson" in *Larousse,* consulted on May 15, 2017*:* http://www.larousse.fr/d*ictionnaires/francais/s_adonner.* Modern French deploys the term *addiction* or *dépendence* for this harmful dedication; the French *addiction* appears to derive from English, but as seen above, the French *adonner* becomes the English word "addiction" in Calvin's translations, which then crosses back into French in its modified form of *addiction* as a pathological compulsion. French synonyms for addiction also include *attaché*.

36. Calvin, *Sermons de M. Jean Calvin sur le livre de Job*, 408, and *Sermons of Master John Calvin, upon the Booke of Job*, 412.

37. Calvin, *The Psalms of David and others, with M. John Calvin's Commentaries*, trans. Arthur Golding (London, 1571), n.p. In the original French: "Dieu toutesfois par sa providence secrete me feit finalement tourner bride d'un autre costé. Et premièrement, comme ainsi soit

que je fusse si obstinément adonné aux superstitions de la Papauté, qu'il estoit bien mal-aisé qu'on me peust tirer de ce bourbier si profound, par un conversion subite il donta et rangea à dicilité mon Coeur, lequel, eu esgard à l'aage, estoit par trop edurcy en tells chose," in Calvin, *Commentaires de Jehan Calvin sur le livre des Pseaumes* (Paris: Librairie de Ch. Meyreuis et compagnie, 1859), viii; and in Latin, "Deus tamen arcano providentiae suae fraeno cursum meum aliò tandem reflexit. Ac primo quidem, quum superstitionibus Papatus magi[s] perfinaciter addictus essem, quam ut facile esset e tam profundo luto me extrahi, animum meum, qui pro aetate nimis obduruerat, subita conuersione ad docilitatem subegit," in Calvin, *Librum Psalmorum, Joannis Calvini Commentarius* (Geneva: Nicolaus Barbirius et Thomas Courteau, 1564), iii. On Calvin's conversion, see Egil Grislis, "Menno Simons on Conversion: Compared with Martin Luther and Jean Calvin," *Journal of Mennonite Studies* 11 (1993): 55–75, 57; and F. Bruce Gordon, *Calvin* (New Haven, CT: Yale University Press, 2009), 33.

38. Quoted in Françoise Wendel, *Calvin: The Origins and Development of His Religious Thought*, trans. Philip Mairet (New York: Harper & Row, 1963), 37.

39. Alexandra Walsham, "Skeletons in the Cupboard: Relics After the English Reformation," *Past and Present* 205 no. S5 (2010): 121.

40. This sort of dedication appears in devotional texts in England, counseling the believer toward complete surrender to God. Thomas Becon writes of "true disciples" in this way: "Thus: when a pure mynd is ioyned with the word . . . they suffer all things for the wordes sake, their study is wholly to love & obey God, & yeldeth fruict a thousand fold." *A new postil conteinyng most godly and learned sermons vpon all the Sonday Gospelles, that be redde in the church thorowout the yeare* (London, 1566), 123. Similarly Rudolf Gwalther, in his "Cxxiij Homelie" upon "The xviij chapter upon the Actes of the Apostles," writes how even grief should not compromise a believer's devotion to God: "No affection ought to pull away men truly addicted unto God, from him." *An hundred, threescore and fiftene homelyes or sermons, vppon the Actes of the Apostles, written by Saint Luke: made by Radulpe Gualthere Tigurine, and translated out of Latine into our tongue, for the commoditie of the Englishe reader* (London, 1572), 694.

41. Calvin, *Psalms of David*, n.p. In the original French it reads: "Ayant doncques receu quelque goust et cognoissance de la vraye piété, je fus incontinent enflambé d'un si grand désir de proufiter, qu'encores que je ne quittasse pas du tout les autres estudes, je m'y employoye toutesfois plus laschement" (Calvin, *Commentaires de Jehan Calvin*, viii); in Latin, it is: "Itaque aliquot verae pietatis gustu imbutus, tanto proficiendi studio exarsi, ut reliqua studia quanuis non abitcerem, frigidius tamen sectarer" (Calvin, *Librum Psalmorum*, iii).

42. Foxe, *Actes and monuments*, vol. 2 (first half), 1031. On the publishing history of Foxe, see Thomas S. Freeman and Elizabeth Evenden, *Religion and the Book in Early Modern England: The Making of John Foxe's "Book of Martyrs"* (Cambridge: Cambridge University Press, 2011); Christopher Highley and John N. King, eds., *John Foxe and His World* (Aldershot: Ashgate, 2002); John N. King, *Foxe's "Book of Martyrs" and Early Modern Print Culture* (Cambridge: Cambridge University Press, 2006); John R. Knott, *Discourses of Martyrdom in English Literature, 1563–1694* (Cambridge: Cambridge University Press, 1993).

43. Foxe, *Actes and monuments*, vol. 2 (first half) (London, 1583), 1075.

44. Foxe, *Actes and monuments*, vol. 1 only (London, 1583), 20. If for Foxe, as for Calvin and Perkins, idolatry represents one form of addiction, then an equally dangerous and rival form of addiction appears with excessive faith in one's own will. Foxe writes that those martyred under Queen Mary serve as examples to "inspire you so that ye be not addict to your owne selfe will or wyt." Foxe, *Acts and Monuments*, vol. 2 (first half), 1617.

45. The fact that so many editions saw print before Perkins's death in 1602 is all the more

surprising given that his first publication appeared in 1589. See Kristen Poole, *Supernatural Environments in Shakespeare's England: Spaces of Demonism, Divinity, and Drama* (Cambridge: Cambridge University Press, 2011), 259; and Kendall, *Calvin and English Calvinism*, 53n.

46. Poole, *Supernatural Environments*, 151; William Perkins, *A treatise tending unto a declaration whether a man be in the estate of damnation or in the estate of grace and if he be in the first, how he may in time come out of it: if in the second, how he maie discerne it, and perseuere in the same to the end* (London, 1590), 209. See also William Perkins, *The works of that famous and worthie Minister of Christ, in the University of Cambridge, M. W. Perkins* (Cambridge, 1605). As Kendall notes, Perkins proves indebted to Beza as much as Calvin (*Calvin and English Calvinism*, 51).

47. William Perkins, *The combat betweene Christ and the Diuell displayed: or A commentarie upon the temptations of Christ* (London, 1606), 31. Tyacke writes that "symptomatic" of the Calvinist ascendency at Cambridge was Perkins, "one of the most widely read English writers" (*Anti-Calvinists*, 28). As noted above, Cambridge library inventories attest to the widespread availability of Calvin's editions and translations. "By the end of the sixteenth century Perkins has replaced the combined names of Calvin and Beza as one of the most popular authors of religious works in England," Kendall notes (*Calvin and English Calvinism*, 52–53).

48. Perkins, *A godlie and learned exposition*, 97.

49. Ibid., 148.

50. William Perkins, *A commentarie or exposition, upon the five first chapters of the Epistle to the Galatians* (Cambridge, 1604), 619.

51. Lambert Daneau, *A dialogue of witches, in foretime named lot-tellers, and now commonly called sorcerers. . . .* Written in Latin by Lambertus Danaeus. And now translated into English (London, 1575), C1r, C5r.

52. Ibid., D7r.

53. Ibid., E5r.

54. Ibid., K4v.

55. Michael Goldman, in "Marlowe and the Histrionics of Ravishment," *Two Renaissance Mythmakers: Christopher Marlowe and Ben Jonson*, ed. Alvin Kernan (Baltimore, 1977), 22–40, writes that Faustus's books "have ravished him, but he is also dissatisfied with them" (24) for he wants something deeper in his studies than mere show.

56. Genevieve Guenther, "Why Devils Came When Faustus Called Them," *Modern Philology* 109 no. 1 (2011): 46–70, explores how Faustus's attraction to necromancy arises from his refusal of the Christian dialectic prizing the soul above the body; he instead celebrates the body and seeks magic as a means of preserving himself in his understanding of material existence: damnation in hell, while walking and disputing, seems preferable to salvation in heaven, with its erasure of the body.

57. Edward Snow, "Marlowe's *Doctor Faustus* and the Ends of Desire," in Kernan, *Two Renaissance Mythmakers*, 79.

58. Spiller writes, alternately, that Faustus's bibliophilia is set "alongside gluttony, lust and envy, making it the newest sin in the age of print" ("Marlowe's Libraries," 108).

59. Calvin, *Harmonie upon the three Euangelists*, 364.

60. Ibid., 457; Luke Wilson, *Theaters of Intention* (Stanford: Stanford University Press, 2000).

61. Calvin, *Harmonie upon the three Euangelists*, 148.

62. Here, by way of contrast, we might think of Milton's internalized view of contract in *Discipline and Divorce*, where, as Victoria Kahn writes, contract serves as "a symbol of ethical and political self-determination rather than a matter of negotiation or dependence upon

another." "'The Duty to Love:' Passion and Obligation in Early Modern Political Theory," *Representations* 68 (Autumn 1999): 94.

63. A. N. Okerlund, "The Intellectual Folly of Dr. Faustus," *Studies in Philology* 74 no. 3 (1977): 268.

64. Calvin, *Sermons de M. Jean Calvin sur le livre de Job*, 408, and Calvin, *Sermons of Master John Calvin, upon the Booke of Job*, 412.

65. John Henry Jones, ed., *The English Faust Book* (Cambridge: Cambridge University Press, 1994), 98, line 255.

66. Foxe, *Actes and monuments*, vol. 1 only, 20.

67. While this chapter emphasizes the scholarly addiction expressed in these lines, Faustus's expression of dedication and surrender has also been interpreted in erotic terms. See Kate Chedgzoy, "Marlowe's Men and Women: Gender and Sexuality" in *The Cambridge Companion to Christopher Marlowe,* ed. Patrick Cheney (Cambridge: Cambridge University Press, 2004), 245–61, and David Clark, "Marlowe and Queer Theory," in *Christopher Marlowe in Context,* ed. Bartels and Smith, 232–41.

68. Guenther, "Why Devils Came When Faustus Called Them."

69. See *Faustus* 1.1.50–52; 2.1.32, 45, 55, 110.

70. Richard A. Posner, *Law and Literature*, 3rd edition (Cambridge, MA: Harvard University Press, 2009), 152.

71. Posner, *Law and Literature*, 152.

72. Daniel Yeager writes, "Despite the lateness of the hour, Faustus could have avoided the contract—which demanded full performance on the part of Mephastophilis before Faustus's performance was due—by repudiating the contract." "Marlowe's *Faustus:* Contract as Metaphor?," *University of Chicago Law School Roundtable* 2 no. 2 (1995): 613.

73. Yeager, "Marlowe's *Faustus:* Contract as Metaphor?," 607.

74. Posner, *Law and Literature*, 153.

75. Posner expands on this point: "And being a man of honor and in his own way a hero, he makes no effort to break his contract when the time comes for him to deliver what he has sold. He substitutes the sanctity of contract for the sanctity of God, and thus cannot imagine a God of mercy but only one of justice" (*Law and Literature*, 150–51). Yeager also notes that, unlike the insecure Mephastophilis, "Faustus, contrastingly, believes the contract is inviolable" ("Marlowe's *Faustus:* Contract as Metaphor?," 607).

76. William M. Hamlin argues that this complex relationship between desire and doubt reflects a "series of cyclical trajectories wherein Faustus's habit of casting doubt is preempted by an experience of euphoric ravishment–ravishment that yields in turn to new casting of doubt." "Casting Doubt in Marlowe's 'Doctor Faustus,'" *Studies in English Literature, 1500–1900* 41 no. 2 (Spring 2001): 262.

77. Yeager, "Marlowe's *Faustus:* Contract as Metaphor?," 617.

78. William Tyndale, *The Newe Testament dylygently corrected and compared with the Greke by Willyam Tindale, and fynesshed in the yere of our Lorde God A.M.D. & xxxiiij. in the moneth of Nouember* (1534), Cccvi.

79. On the disappointing nature of Faustus's magic, see Richard Halpern, "Marlowe's Theater of Night: *Doctor Faustus* and Capital," *ELH* 71 no. 2 (2004): 455–95, and Stephen Orgel, "Tobacco and Boys: How Queer was Marlowe," in *The Authentic Shakespeare* (New York: Routledge, 2002), 211–30. Orgel writes, "For all its talk of the perils of boundless ambition, there is a continuous sense of disappointment in the play, a sense that Faustus isn't ambitious enough" (224).

80. Poole, "Dr. Faustus and Reformation Theology," 104.

81. As Calvin writes, "the more a mans vices are, so much the more fiercely doth he with loftie words extol free wil," in "The Commentarie of M. John Calvine upon the Gospel of John," from Calvin, *Harmonie upon the three Euangelists*, 207.

82. Calvin, *Harmonie upon the three Euangelists*, 457 ("Nunquam ergo in hoc officio constanter perget, nisi in cuiuscorde sic regnabit amor Christi, ut sui oblitus, totumque se illi addicens, impedimenta Omnia superset." Calvin, *Harmonia ex Evangelistis*, 597).

83. Calvin, *Harmonie upon the three Euangelists*, 203–4.

84. Ibid., 142 ("quia pro ingenii nostri crassitie semper rebus terrenis addicti sumus, ideo prius corrigit ingenitum illum nobis morbum, quam ostendat quid agendum fit." Calvin, *Harmonia ex Evangelistis*, 455).

85. Eric Rasmussen, introduction to *Doctor Faustus*, ed. David Bevington and Eric Rasmussen, The Revels Plays (Manchester: Manchester University Press, 1993), 21.

86. Poole, in "Dr. Faustus and Reformation Theology," elucidates how audience members might nevertheless find comfort in this staging of election, even as they struggle to understand the nature of salvation and damnation during a period of intense theological shifts.

87. Roma Gill writes that "The behaviour of Marlowe's Faustus seems to follow Perkins's Calvinist theology quite closely." Introduction to *Doctor Faustus*, New Mermaid edition (London: Ernest Benn, 1965), xviii.

CHAPTER 2

Note to epigraph: Giorgio Agamben, *Potentialities: Collected Essays in Philosophy*, trans. Daniel Heller-Roazen (Stanford, CA: Stanford University Press, 1999), 204.

1. Recent studies of service in Shakespeare include Elizabeth Rivlin, *The Aesthetics of Service in Early Modern England* (Evanston, IL: Northwestern University Press, 2012); David Schalkwyk, *Shakespeare, Love and Service* (Cambridge: Cambridge University Press, 2008); David Evett, *Discourses of Service in Shakespeare's England* (New York: Palgrave, 2005).

2. David Schalkwyk, "Is Love an Emotion?: Shakespeare's *Twelfth Night* and *Antony and Cleopatra*," *symplokē* 18 no. 1–2 (2010): 102–103. See also his "The Discourses of Friendship and the Structural Imagination of Shakespeare's Theater: Montaigne, *Twelfth Night*, De Gournay," *Renaissance Drama* 38 (2010): 141–71.

3. Schalkwyk, "Is Love an Emotion?," 110.

4. N. R. Helms, "Conceiving Ambiguity: Dynamic Mindreading in Shakespeare's *Twelfth Night*," *Philosophy and Literature* 36 no. 1 (2012): 125.

5. See Jason Scott-Warren, "When Theaters Were Bear-Gardens; or, What's at Stake in the Comedy of Humors," *Shakespeare Quarterly* 54 no. 1 (Spring 2003): 63–82; he explores the play as a comedy of humours, and charts the baiting and exposure of a range of characters, from Viola and Olivia to Malvolio and Andrew. The audience draws pleasure, as in the comedy of humors genre more generally, from the exposure of each character's humoral predispositions. Viola is able to "prove herself to be fully human through her heroic command of language," while Andrew and Malvolio are both "unmasked" for their "humanity that is merely apparent," for they are "liars, apes, and pretenders to eloquence and wit" (79).

6. Helms, "Conceiving Ambiguity," 131. For an alternate reading that emphasizes not transformation but continuity in the shift from melancholy to love, see Amy L. Smith and Elizabeth Hodgson, "'A Cypress, not a bosom, hides my heart': Olivia's Veiled Conversions," *Early Modern Literary Studies* 15 no. 1 (2009–2010).

7. William Shakespeare, *Twelfth Night, Or What You Will*, ed. Keir Elam, Arden Shakespeare, 3rd series (London: Bloomsbury, 2008). Hereafter cited above, with the standard act, scene, and line divisions. All citations are to this edition unless otherwise noted. On melancholy, see Drew Daniel, *The Melancholy Assemblage: Affect and Epistemology in the English Renaissance* (New York: Fordham University Press, 2013); Douglas Trevor, *The Poetics of Melancholy in Early Modern England* (Cambridge: Cambridge University Press, 2004); Paster, *Humoring the Body*; Lynn Enterline, *Tears of Narcissus: Melancholia and Masculinity in Early Modern Writing* (Stanford, CA: Stanford University Press, 1995); and Adam Kitzes, *The Politics of Melancholy from Spenser to Milton* (London: Routledge, 2006).

8. Elam, "Introduction," in ibid., 53.

9. Daniel, *Melancholy Assemblage*, 230. Daniel's analysis of melancholy as an assemblage helps illuminate this interconnection in *Twelfth Night* by insisting on the "extended, provisional, and modular set of relations between and across material elements, and relationships between and across individual subjects" (*Melancholy Assemblage*, 12).

10. John Banister, *An antidotarie chyrurgicall containing great varietie and choice of all sorts of medicines that commonly fal into the chyrurgions vse* (London, 1589), 105; Thomas Dekker, *The belman of London. Bringing to light the most notorious villanies that are now practised in the kingdome* (London, 1608), n.p.

11. *The third booke of Amadis de Gaule . . . Written in French by the Lord of Essars, Nicholas de Herberay. . . . Translated into English by A. M.* (London, 1618), 26. See also Samuel Purchas, who in *Purchas his pilgrimage* (London, 1613) recounts how, in a region of Africa, men are "alwayes encombred with melancholy, they addict themselves to no pleasures," 531.

12. Gervase Markham, *Cauelarice, or The English horseman contayning all the arte of horsemanship, as much as is necessary for any man to understand* (1607), 2. Another text invoking humors and addiction is that of Leo Africanus, *A geographical historie of Africa, written in Arabicke and Italian by Iohn Leo a More, borne in Granada, and brought up in Barbarie* (1600).

13. Michel de Montaigne, "Of three commerces or societies," *Essays*, trans. John Florio (London, 1613), 458.

14. Nyquist, *Arbitrary Rule*, 22.

15. Michel de Montaigne, "Sur trois types de relations," *Les Essais, Livre III* (1595), *Traduction en français modern du texte d'édition de 1595* (Paris: Guy de Penon, 2009), 47.

16. John Florio, *A Worlde of Wordes: or, Most copious, and exact Dictionarie in Italian and English* (London, 1598), 97, 9.

17. Ibid., 97, 100.

18. Ibid., 455.

19. Randle Cotgrave, *A Dictionary of the French and English Tongues* (London, 1611), n.p.

20. Guy Miège, *A New Dictionary French and English, with another English and French* (London, 1677), n.p.

21. Thomas Cooper, *Thesaurus Linguae Romanae et Britannicae* (London: Henry Denham, 1584), n.p.

22. Ibid. Examples include "Deuouere & constituere alicui hostiam. Cic. To vow. Deuouere se amicitiæ alicuius. Cæs. To addict & give himselfe to ones friendship for euer."

23. Ibid.

24. John Baret, *An Alveary or Triple Dictionary, in English, Latin, and French* (London: Henry Denham, 1574), n.p.

25. Thomas Thomas, *Dictionarium Linguae Latinae et Anglicanae* (Cambridge: Richardum Boyle, 1587). One might also, as explored in Chapter 1, be addicted to study, as Baret's examples

in his definition of "addict" indicate, ("Cic. Addict: and given, to the studie of learning. . . .") or as Thomas's definition of study reveals: "Stūdiōsus, studiosior, & studiosissimus, a, um. Studious, diligent, earnest, painfull, constant, addict, or firmly disposed: that fauoureth or loveth wel or much that hath a great desire, or giveth and setteth his minde to a thing: desirous of learning: also learned, Quint." Or, more generally J. B. [John Bullokar], *An English Expositor: teaching the interpretation of the hardest words in our language* (London: John Legatt, 1616) defines "addict" as "To apply, or give ones selfe much to any thing" (B3v).

26. Cotgrave, *Dictionary*. The labor of addiction appears in his definition of *s'Employer*, as meaning "To indeuor, labour, be earnest upon, to set himselfe about; to give, applie, or addict himselfe unto," as well as his definition of *Entendement* as "Vnderstanding, apprehension, conceit; iudgement; intelligence; naturall knowledge. *Mettre son entendement à*. To give his mind to, set his wit on; addict, or applie himselfe unto; to intend, or attend, with diligence."

27. Thomas, *Dictionarium Linguae*.

28. Bruce R. Smith, Introduction to *Twelfth Night: Texts and Contexts* (New York: Palgrave, 2001), 8.

29. Tim Dean, *Unlimited Intimacy: Reflections on the Subculture of Barebacking* (Chicago: University of Chicago Press, 2009); Bersani, "Is the Rectum a Grave?," 217–22; and "Sociality and Sexuality," 646–47.

30. Tim Dean, "Sex and the Aesthetics of Existence," *PMLA* 125 no. 2 (2010): 389. Bersani coins "homo-ness" to describe this process, offering, as Dean writes, "a deliberately paradoxical claim that '[h]omo-ness is an anti-identitarian identity'" (390). See Leo Bersani, *Homos* (Cambridge, MA: Harvard University Press, 1995), 101. This shattering of self-identification in order to open oneself up to the love of (or for) another resonates with early modern invocations of addicted love, despite historical distance and despite the fact that Bersani and Dean theorize expressions of physical intimacy, in contrast to the various—and often metaphysical—expressions of love ultimately at stake in *Twelfth Night*. As Bersani writes, in reference to St. Catherine of Genoa, "Self-annihilation is the precondition for union." Leo Bersani and Adam Phillips, *Intimacies* (Chicago: University of Chicago Press, 2008), 54.

31. Richard Strier writes that "Schoenfeldt mocks the idea that Shakespeare may have thought there was 'some ethical joy implicit in the virtue of giving' (93) but this is exactly what Hubler and Empson are able to show." *Unrepentant Renaissance: From Petrarch to Shakespeare to Milton* (Chicago: University of Chicago Press, 2011), 20.

32. Kathryn Schwarz, in *What You Will: Gender, Contract, and Shakespearean Social Space*, (Philadelphia: University of Pennsylvania Press, 2011), explores the ways in which consent to social strictures on femininity might prove a form of resistance for early modern women, thereby challenging binaristic insistence on passive conformity versus active rebellion.

33. On the benefits of self-regulation, see Schoenfeldt, *Bodies and Selves*. On, alternately, the value of passions and predispositions, see Strier, *Unrepentant Renaissance*, and "Against the Rule of Reason: Praise of Passion from Petrarch to Luther to Shakespeare to Herbert," in *Reading the Early Modern Passions: Essays in the Cultural History of Emotion*, ed. Gail Kern Paster, Katherine Rowe, and Mary Floyd-Wilson (Philadelphia: University of Pennsylvania Press, 2004).

34. See Heather Dubrow, *Echoes of Desire: English Petrarchism and its Counterdiscourses* (Ithaca: Cornell University Press, 1995); Danila Sokolov, *Renaissance Texts, Medieval Subjectivities: Rethinking Petrarchan Desire from Wyatt to Shakespeare* (Pittsburgh, PA: Duquesne University Press, 2017); William J. Kennedy, *The Site of Petrarchism: Early Modern National Sentiment in Italy, France, and England*, (Baltimore, MD: Johns Hopkins University Press, 2003); and on

the lyric "I" see Carla Freccero, "Tangents (of Desire)," *Journal for Early Modern Cultural Studies*, 16, no. 2 (Spring 2016): 91–105.

35. Guiseppe Mazzotta, *The Worlds of Petrarch* (Durham, NC: Duke University Press, 1993), 3. On issues of sovereignty in Petrarchism, see Melissa Sanchez, who in *Erotic Subjects: The Sexuality of Politics in Early Modern English Literature* (Oxford: Oxford University Press, 2011) argues, "Petrarchan courtship," with its martyred lover and unattainable mistress, "continued to shape" the ways the English understood "sovereignty and obedience in the seventeenth century" (25).

36. Ross Knecht, "'Invaded by the World': Passion, Passivity, and the Object of Desire in Petrarch's *Rime sparse*," *Comparative Literature* 63 no. 3 (2011): 237.

37. Timothy Reiss, *Mirages of the Selfe: Patterns of Personhood in Ancient and Early Modern Europe* (Stanford, CA: Stanford University Press, 2003), 303.

38. Ross Knecht, "The Grammar of Passion: Language and Affect in Early Modern Literature" (Ph.D. diss., New York University, 2011), 53. See also Melissa E. Sanchez, "'In My Selfe the Smart I Try': Female Promiscuity in 'Astrophil and Stella.'" *ELH* 80, no. 1 (Spring 2013): 1–27.

39. Philip Sidney, "Astrophil and Stella 61," *Sir Philip Sidney: The Major Works*, ed. Katherine Duncan-Jones (Oxford: Oxford University Press, 2008), 177. On Sidney's poem see Ross Knecht, "The Grammar of Passion: Language and Affect in Early Modern Literature" (Ph.D. diss., New York University, 2011), chap. 1.

40. This paradox resonates with a much broader literature on governance and sovereignty, counseling devoted service to the king. See Sanchez, *Erotic Subjects.*

41. David Schalkwyk, "Love and Service in Twelfth Night and the Sonnets," *Shakespeare Quarterly* 56 no. 1 (Spring 2005): 92.

42. Schalkwyk gestures toward such addictive love when he writes, "It is Viola's abject submission of her will to Orsino's desire . . . which opens the space for an intimacy that encompasses more than the mere social advancement of a favored servant or 'Pittiful thriuor'" ("Love and Service," 90). If he goes on to argue that such devoted service compromises love by diminishing the possibility for reciprocity, nonetheless his analysis helps draw out the crucial role of selfless devotion in facilitating the growth of love in the first place.

43. William Gouge, *Of Domesticall Duties Eight Treatises* (London, 1622), 593–94.

44. Schalkwyk, "Love and Service," 85.

45. Lorna Hutson, "On Not Being Deceived: Rhetoric and the Body in *Twelfth Night*," *Texas Studies in Literature and Language* 38 no. 2 (Summer 1996): 160.

46. Ibid., 162.

47. See Michael Neil, *"Servile Ministers": Othello, King Lear and the Sacralization of Service* (Vancouver: Ronsdale Press, 2004).

48. On pity's disruptive potential, see Heather James, "Dido's Ear: Tragedy and the Politics of Response," *Shakespeare Quarterly* 52 (2001): 360–82.

49. Maurice Hunt, "Malvolio, Viola, and the Question of Instrumentality: Defining Providence in *Twelfth Night*," *Studies in Philology* 90 no. 3 (Summer 1993): 289.

50. Olivia's experience of love does not encompass, of course, "one self king" dominating her in the way Orsino fantasizes. Indeed, Sir Toby announces that Olivia will not marry above her station, presumably in part because it would compromise her own estate. Nevertheless she does acknowledge the power of love to overcome and redefine her.

51. Elam, in Shakespeare, *Twelfth Night*, 164; Robert Burton, *Anatomy of Melancholy* (London, 1652), 202–12.

52. Elam, in Shakespeare, *Twelfth Night*, 53.

53. Daniel, *Melancholy Assemblage,* 15.

54. Knecht, "'Invaded by the World,'" 245.

55. Elam, in Shakespeare, *Twelfth Night,* 53.

56. Barnabe Riche, "Apollonius and Silla" in *Rich His Farewell to the Military Profession* (1583), Jiii.

57. Ibid.

58. As Schalkwyk writes, "The logic of the oath or promise as a speech act requires an identity stable enough to bear the responsibility—the projection into the future—that is the essence of such utterances" ("Is Love an Emotion," 123).

59. Jed Rubenfeld, "The Riddle of Rape-by-Deception and the Myth of Sexual Autonomy," *Yale Law Journal* 122 no. 6 (April 2013): 1372–1443.

60. Ibid., 1421.

61. Stephen Dickey, "Shakespeare's Mastiff Comedy," *Shakespeare Quarterly* 42 no. 3 (Fall 1991): 272–73.

62. Schalkwyk writes, "Love requires both attachment and commitment, even if the emotions that inform it may vary, and it is something not achieved as a settled emotional state but rather navigated, dialogically and performatively, through emotives or passionate utterances" ("Is Love an Emotion?," 123).

63. Valerie Traub, *Desire and Anxiety: Circulations of Sexuality in Renaissance Drama* (New York: Routledge, 1992), 143.

64. Laurie Shannon, "Nature's Bias: Renaissance Homonormativity and Elizabethan Comic Likeness," *Modern Philology* 98 no. 2 (November 2000): 187.

65. Montaigne, "Of Friendship," *Essays*, 94.

CHAPTER 3

Note to epigraph: Lauren Berlant, *Cruel Optimism* (Durham, NC: Duke University Press, 2011), 1.

1. William Shakespeare, *King Henry IV, Part 2*, ed. James C. Bulman, Arden Shakespeare, 3rd series (London: Bloomsbury Arden Shakespeare, 2016). Hereafter cited above, with the standard act, scene, and line divisions. All citations of the play are to this edition unless otherwise noted. While critics frequently reference Falstaff's sack speech, single studies of it are rare. See Alan D. Isler, "Falstaff's Heroic Sherris," *Shakespeare Quarterly* 22 no. 2 (Spring 1971): 186–88; and Jill L. Levenson, "Shakespeare's Falstaff: 'The Cause That Wit Is in Other Men,'" *University of Toronto Quarterly* 74 no. 2 (2005): 722–28.

2. On Falstaff's gluttony see Elena Levy-Navarro, *The Culture of Obesity in Early and Late Modernity* (Basingstoke: Palgrave, 2008); Michael Davies, in "Falstaff's Lateness: Calvinism and the Protestant Hero in *Henry IV*," *Review of English Studies* 56 no. 225 (2005): 351–78, who explores Falstaff as a reprobate within Calvinist discourses of salvation; Joshua B. Fisher, in "Digesting Falstaff: Food and Nation in Shakespeare's *Henry IV* plays," *Early English Studies* 2 (2009): 1–23, who explores the relationship between culinary consumption and national anxiety; Nina Taunton, in "Food, Time and Age: Falstaff's Dietaries and Tropes of Nourishment in *The Comedy of Errors*," *Shakespeare Jahrbuch* 145 (2009): 91–105, who teases out how Shakespeare stages anxieties about old age through dietary language—of gluttony in the case of Falstaff and abstinence in the case of Henry IV. For a medical argument about Falstaff's condition, see Henry Buchwald and Mary E. Knatterud, in "Morbid Obesity: Perceptions of Character and Comorbidities in Falstaff," *Obesity Surgery* (2000): 402–8, who classify Falstaff in relation to the fifteen

characteristics of obesity; by contrast, see Barbara Everett, in "The Fatness of Falstaff: Shakespeare and Character," *Proceedings of the British Academy* 76 (1990): 109–28, who explores how Falstaff's fatness figures Shakespeare's creative richness.

3. Thomas, *Dictionarium Linguae Latinae et Anglicanae.*

4. My reading emphasizes Falstaff's devotion to Hal within the context of their tavern fellowship. On the homoerotic nature of Hal and Falstaff's relationship, see Heather Findlay, "Renaissance Pederasty and Pedagogy: The 'Case' of Shakespeare's Falstaff," *Yale Journal of Criticism* 3 no. 1 (Fall 1989): 229–39; Jonathan Goldberg, *Sodometries: Renaissance Texts, Modern Sexualities* (Stanford, CA: Stanford University Press, 1992); Vin Nardizzi, "Grafted to Falstaff and Compounded with Catherine: Mingling Hal in the Second Tetralogy," in *Queer Renaissance Historiography: Backward Gaze*, ed. Vin Nardizzi, Stephen Guy-Bray, and Will Stockton (Surrey: Ashgate, 2009); Laurie Shannon, *Sovereign Amity: Figures of Friendship in Shakespearean Contexts* (Chicago: University of Chicago Press, 2002); Traub, *Desire and Anxiety.*

5. On Carnival in the *Henriad*, see David Ruiter, *Shakespeare's Festive History: Feasting, Festivity, Fasting and Lent in the Second Henriad* (Farnham, Surrey: Ashgate, 2003); François Laroque, *Shakespeare's "Battle of Carnival and Lent"* (New York: Macmillan, 1998) and *Shakespeare's Festive World: Elizabethan Seasonal Entertainment and the Professional Stage*, trans. Janet Lloyd (Cambridge: Cambridge University Press, 1991); Neil Rhodes, "Shakespearean Grotesque: The Falstaff Plays," from *Elizabethan Grotesque* (New York: Routledge, 1980), reprinted in *Thomas Nashe*, ed. Georgia Brown (Farnham, Surrey: Ashgate, 2011), 47–93; Charles Whitney, "Festivity and Topicality in the Coventry Scene of *I Henry IV*," *English Literature Renaissance* 24 no. 2 (Spring 1994): 410–48; Hugh Grady, "Falstaff: Subjectivity Between the Carnival and the Aesthetic," *Modern Language Review* 96 no. 3 (2001 July): 609–23.

6. Falstaff is one of Shakespeare's most popular characters: "Falstaff has become a kind of god in the mythology of modern man." Dover Wilson, *The Fortunes of Falstaff* (New York: Cambridge University Press, 1961), 128; "When we are wholly human . . . we become most like either Hamlet or Falstaff." Harold Bloom, *Shakespeare: The Invention of the Human* (New York: Penguin Books, 1998), 745.

7. Shakespeare's play demands, as Richard Strier argues, Falstaff's rejection: "It is impossible not to know that [Hal] is doing the right thing in rejecting Falstaff," and indeed Shakespeare "was pre-committed to the rejection of the comic character" of Falstaff. "Shakespeare Against Morality," in *Reading Renaissance Ethics*, ed. Marshall Grossman (New York: Routledge, 2007), 211. Yet as this chapter rehearses in closing, Strier offers a moving argument on the costs of this rejection.

8. Berlant, *Cruel Optimism*, 27. Here Berlant writes of the work of Dean and Bersani, to which she is indebted, that its "focus is on the optimism of attachment, and is often itself optimistic about the negations and extensions of personhood that forms of suspended intersubjectivity demand from the lover/reader."

9. Ibid.

10. Shannon explores how Hal necessarily rejects Falstaff as a condition of good rule: "Friendship discourses actually constitute kingliness—not as a model to be followed, as advice books to princes had argued, but as a practice for sovereigns to eschew" (*Sovereign Amity*, 166). On Falstaff's threat to sovereignty, see also Kristen Poole, *Radical Religion from Shakespeare to Milton: Figures of Nonconformity in Early Modern England* (Cambridge: Cambridge University Press, 2000); David Womersley, "Why Is Falstaff Fat?" *Review of English Studies: A Quarterly Journal of English Literature and the English Language* 47 no. 185 (1996): 1–22.

11. On the relationship of *1 Henry IV* to *2 Henry IV*—whether they are the two halves of

one longer play; two separately imagined, independent plays; or one initial play that proved so popular as to demand an (unexpected) sequel—see James C. Bulman's introduction to Shakespeare, *King Henry IV, Part 2*, in which he writes that "*Part Two* is arguably a sequel that need not rely on *Part One* to be understood" (9), and David Scott Kastan's introduction to William Shakespeare, *King Henry IV, Part 1*, ed. David Scott Kastan, Arden Shakespeare, 3rd series (London: Thomson Learning, 2002), 17–23, in which he writes that *2 Henry IV* "seems to me less a continuation than a commentary. *Part Two* does not so much bring the events of *Part One* to conclusion as reimagine the actions of the first play in a more sombre key" (22–23); and Laurie Shannon, who uses *Henry IV* to describe the two plays which are "one play in two bodies" (*Sovereign Amity*, 166).

12. William Fulbecke, *A Booke of Christian Ethicks or Moral Philosophie: Containing The true difference and opposition, of the two incompatible qualities, Vertue, and Voluptuousnesse* (London: Richard Jones, 1587), 23.

13. Florilegus, *Odious, despicable, and dreadfull condition*, 22–24.

14. Beard, *Theatre of Gods Judgements*, 430–31. Similarly, Dr. Everard Maynwaringe writes of disease and drunkenness in *Vita sana & longa the preservation of health and prolongation of life proposed and proved* (London, 1669) and in *Two broad-sides against tobacco*, 21–22, 24, 26.

15. *A looking glasse for drunkards* (1627) also describes how drunkenness creates "multitudes of diseases in the body of man, as apoplexies, falling sicknesses, palsies, dropsies, consumptions, giddinesse of the head, inflammation of the blood and liver, distemper of the brain, deprivation of the sense, and whatnot" (A4r). William Vaughan, in *Natural and Artificial Directions for Health* (1600), counsels that one must drink moderately "least the belly-God hale you at length captive into his prison house of gurmandise, where you shalbe afflicated with . . . many diseases" (74). The drinker, Downame writes, "is brought into grievous diseases, as dropsies, gouts, palsies, apoplexie, and such like" (*Foure treatises*, 96–97).

16. William Shakespeare, *King Henry IV, Part 1*, ed. David Scott Kastan, Arden Shakespeare, 3rd series (London: Thomson Learning, 2002). Hereafter cited above, with the standard act, scene, and line divisions. All citations are to this edition unless otherwise noted.

17. Historians of science and lexicographers behind the *OED* put this definition later, but one of this project's contributions lies in tracing an earlier use of the term addiction in relation to alcohol.

18. Downame, *Foure treatises*, 83.

19. Ibid., 79.

20. Junius Florilegus. *The odious, despicable, and dreadfull condition of a drunkard*, 6.

21. Ward, *Woe to Drunkards*, 10.

22. Richard Younge, [R. Junius, pseud.], *The Drunkard's Character: or, A True Drunkard with such sinnes as raigne in him* (London, 1638), 233–34.

23. E. G. [Edmund Gardiner], *The Triall of Tabacco. Wherein, his worth is most worthily expressed: as, in the name, nature, and qualitie of the sayd hearb; his speciall use in all Physicke* (London, 1610), 16v–17r.

24. Downame, *Foure treatises*, 93.

25. Trotter, *Essay, Medical, Philosophical, and Chemical, on Drunkenness*.

26. Florilegus. *Odious, despicable, and dreadfull condition of a drunkard*, 6.

27. Downame, *Foure treatises*, 93. On the relationship between tyranny and slavery, and particularly on the exploitation of so-called voluntary psychological servitude to brutal political ends, see Nyquist, *Arbitrary Rule*.

28. Thompson, *Diet for a Drunkard*, 14.

29. Everett, "Fatness of Falstaff," 127.

30. "Shakespeare goes to great lengths," Ian Frederick Moulton writes, "to show us just how serious Falstaff's corruption is." "Fat Knight, or What You Will: Inimitable Falstaff," in *A Companion to Shakespeare's Works*, ed. Richard Dutton and Jean Howard (Oxford: Blackwell, 2003), 229. Hugh Grady makes a version of this point when he writes, "Any attentive reader or viewer of *1 Henry IV* can work up a long list of ethically dubious actions by Falstaff. . . . How, then, does he come away with readers' and viewers' good feelings?" *Shakespeare, Machiavelli, and Montaigne: Power and Subjectivity from Richard II to Hamlet* (Oxford: Oxford University Press, 2002), 153. His answer concerns Falstaff's relationship to emergent modernity, since Falstaff represents the pleasure repressed or denied by capitalism and the Protestant state.

31. Falstaff's much-vaunted verbal skills also show some signs of degeneration by the end of *2 Henry IV*. This is in part, I hypothesize, why the producers of *The Hollow Crown* (dir. Richard Eyre for BBC, 2013) relocate the sack speech to the beginning of *1 Henry IV*, where the speech might appear comical; by the end of their brilliant but bleak *2 Henry IV*, it would no longer appear even appropriate, Falstaff having lost a good deal of his vigor, wit, and playfulness.

32. Benjamin Bertram, "Falstaff's Body, the Body Politic, and the Body of Trade," *Exemplaria* 21 no. 3 (2009): 297. See Karen Raber, "Shakespeare's Fluid Mechanics," in *Culinary Shakespeare*, ed. Amy Tigner and David Goldstein (Pittsburgh: Duquesne University Press, 2016), 75–96, on the economic consequences of importation of foreign wines; and Barbara Sebek, " 'Wine and sugar of the best and fairest': Canary, the Canaries and the Global in Windsor," *Culinary Shakespeare*, ed. Tigner and Goldstein, 41–56, who illuminates the tensions in the Anglo-Spanish wine trade in connection with Falstaff's representation.

33. David Goldstein, *Eating and Ethics in Shakespeare's England* (Cambridge: Cambridge University Press, 2013).

34. Falstaff does, however, hope to dine at one point in the play, when he asks Master Gower, who seems to have some insight into Hal's location and condition, to dinner: "Master Gower, shall I entreat you with me to dinner?" (*2 Henry IV*, 2.1.166), and "Will you sup with me, Master Gower?" (2.1.171). Unable to recruit this dinner companion, and indeed ridiculed by the Lord Chief Justice for his request, Falstaff appears more and more isolated. Shannon has a more hopeful reading of Falstaff in the countryside, emphasizing a "bucolic" Gloucestershire scene with Falstaff, Shallow, and friends, in which the word "merry" is repeated more than anywhere else in the plays. But, she writes, "Falstaff instantly abandons his place in the bucolic scene to go to the new king" (*Sovereign Amity*, 182).

35. Montaigne, "Of Friendship," in *Essays*, 93. In addition to Laurie Shannon's, Jeffrey Masten's reading of this passage has influenced my own; Masten, *Textual Intercourse: Collaboration, Authorship, and Sexualities in Renaissance drama* (Cambridge: Cambridge University Press, 1997), 35; and "Toward a Queer Address," 377.

36. Ibid.

37. Shannon, *Sovereign Amity*, 167.

38. These lines might be directed at Poins. But as Kastan points out in his edition, the language is more typical of Falstaff's relationship with Hal.

39. The scene might be played for laughs, as in a production at the University of Sidney in which Falstaff caught the bottle Hal had thrown at him and drank out of it; described in Derek Peat, "Falstaff Gets the Sack," *Shakespeare Quarterly* (Fall 2002): 379–85.

40. Montaigne, *Essays*, 93.

41. Ibid., 93–94.

42. See Philip Stubbes, *The Anatomie of Abuses: contayning a discoverie, or briefe summarie of*

such Notable Vices and Imperfections, as now raigne in many Christian Countreyes of the Worlde. (London: Richard Jones, 1583): "The Drunkard in his drunkenness," as Stubbes argues, "killeth his freend, revileth his lover, discloseth secrets and regardeth no man: . . . so that I will not feare to call drunkerds beasts, and no men, and much wurse then beasts, for beasts never exceed in such kind of excesse, or superfluitie" (Jivr–v).

43. To make another lexicographical point, in his encomium Falstaff argues that sack causes "inflammation," fighting the weakness and weariness attendant on a malady like melancholy. His use of this term "inflammation" is innovative—Shakespeare's play is one of the first texts to deploy it to describe a mental state, such as a passion, or an excited condition. Richard Hooker enlists the term in the same year, in his *Lawes of Ecclesiastical Polity*, to signal the ardor of keen devotion. Even as Shakespeare and Hooker might use the term innovatively, inflammation more generally signals a pathological condition in medical literature: "a morbid process affecting some organ or part of the body, characterized by excessive heat, swelling, pain, and redness" (*OED* definition 3, 2nd edition, 1989). An earlier version was first published in the *New English Dictionary*, 1900. Precisely at the moment he defends sack, then, Shakespeare deploys a term that also evokes drinking's deleterious effects.

44. Falstaff collapses two distinct communities of drinkers into one here: first, in his view of witty drinkers he evokes classical precedents such as Anacreontic verse; and second, his link of sack to fighting recalls the familiar and raucous groups of drunken soldiers returning to England from the Dutch wars, formerly armed with weapons, now with sack. Drunken excess offered one means of forming community for those men otherwise isolated, impoverished, and traumatized after returning from war. On classical drinking, see Scodel, *Excess and the Mean*; Stella Achilleos, "The *Anacreontea* and a Tradition of Refined Male Sociability," in Smyth, *Pleasing Sinne*, 21–36; O'Callaghan, *English Wits*, and "Tavern Societies"; Alison Findlay, "Theatres of Truth: Drinking and Drama in Early Modern England," in *A Babel of Bottles: Drinkers and Drinking Places in Literature*, ed. James Nicholls and Susan Owen (Sheffield: Sheffield University Press, 2000), 21–39. On male drinking culture see B. Ann Tlusty, in "The Public House and Military Culture in Early Modern Germany," in Kümin and Tlusty, *World of the Tavern*, 137–53, who highlights the "exaggerated norms of masculine behavior" (150) evident among soldiers drinking in taverns; and Capp, in "Gender and the Culture of the English Alehouse," who illuminates the masculine culture of boisterous drinking and fighting. See also B. Ann Tlusty, *Bacchus and Civic Order: The Culture of Drink in Early Modern Germany* (Charlottesville: University of Virginia Press, 2001); Martin, *Alcohol, Sex, and Gender*; Shepard, *Meanings of Manhood*. Both Adam Smyth's study of drunkenness and my own study of toasts draw attention to the pleasures of heavy drinking for certain male communities. See Smyth, "'It were far better to be a *Toad* or a *Serpant*, then a Drunkard'"; Rebecca Lemon, "Compulsory Conviviality in Early Modern England," *ELR* 43 no. 3 (September 2013): 381–414. On loutish drinking culture, see Anna Bryson, *From Courtesy to Civility: Changing Codes of Conduct in Early Modern England* (Oxford: Clarendon Press, 1998); Shepard, *Meanings of Manhood*, 100–102; Bloom, "Manly Drunkenness,"23.

45. If Falstaff has long been read as a surrogate father to Hal, the sack speech takes this relational process a step further in imaginatively invoking a small city of drinkers, produced out of, or at least mirroring, Falstaff and Hal's relationship. For readings of Falstaff as an alternative father, see Coppélia Kahn, *Man's Estate: Masculine Identity in Shakespeare* (Berkeley: University of California Press, 1981); Meghan C. Andrews, "Gender, Genre, and Elizabeth's Princely Surrogates in *Henry IV* and *Henry V*," *Studies in English Literature, 1500–1900* 54 no. 2 (Spring 2014): 375–99; Richard P Wheeler, "Deaths in the Family: The Loss of a Son and the Rise of

Shakespearean Comedy," *Shakespeare Quarterly* 51 no. 2 (Summer 2000): 127–53. Alternatively see Traub, who, in *Desire and Anxiety*, explores Falstaff as the abjected mother that Hal discards.

46. Montaigne, as discussed above, describes what Falstaff experiences: a seized will, a sense of losing himself in relation. It is worth noting, however, that Montaigne's account of reciprocal friendship concludes with a mirror image: the speaker, having admitted to his own incapacity, then describes how friendship affects his partner, "which likewise having seized all his will, brought it to loose and plunge it selfe in mine, with a mutuall greedinesse, and with a semblable concurrance." Montaigne, *Essays,* 93.

47. Hal, like his father, represents a "lonely, other-directed, and singularly burdened *monarchical* being" (Shannon, *Sovereign Amity*, 180).

48. As he says to Poins after his jest with Francis the drawer, "I am now of all humours that have showed themselves humours since the old days of Goodman Adam to the pupil age of this present twelve o'clock at midnight" (*1 Henry IV*, 2.4.90–94). He boasts of his own range, but as Kastan's gloss also suggests, he melds himself to his audiences: "I will fit myself to all humours: I will game with a gamester, drink with a drunkard, be civil with a citizen, fight with a swaggerer, and drab with a whore-master" (1.3.136–38), as John Cooke's character Spendall puts it in *Greene's Tu Quoque*, in *A Select Collection of Old Plays*, ed. Robert Dodsley, Isaac Reed, and Octavius Gilchrist, vol. 7 (London, 1825), 2–98.

49. On King Henry IV's use of such economic and transactional language see Lemon, *Treason by Words,* 52–78.

50. David Schalkwyk writes, a "loyal servant," in "service," is defined as "a commingling of affect and structure, devotion and self-interest, abandon and control" (*Shakespeare, Love and Service*, 168, 172).

51. Shannon, *Sovereign Amity*, 168.

52. William Shakespeare, *King Henry V*, ed. T. W. Craik, Arden Shakespeare, 3rd series (London: Bloomsbury Arden Shakespeare, 1995).

53. Strier, "Shakespeare Against Morality," 212.

54. Marjorie Garber, *Shakespeare After All* (New York: Knopf Doubleday, 2008), 325. On Falstaff as Vice, see Robert Weimann, *Shakespeare and the Popular Tradition in the Theater*, ed. Robert Schwartz (Baltimore: Johns Hopkins University Press, 1978), 128–31. On the range of Falstaff's roles, see Arthur Kinney, "Shakespeare's Falstaff as Parody," *Connotations* 12 no. 2–3 (2002–3): 105–25.

55. "Most simply, Falstaff is not how meaning is renewed, but rather how meaning gets started." Harold Bloom, ed. *Henry IV, Part 2* (New York: Chelsea House Publishers, 1987), 6.

56. Joseph R. Roach, *The Player's Passion: Studies in the Science of Acting*. (Newark: University of Delaware Press, 1985), 42.

57. Ibid., 41–42; with thanks to Barbara Mello for drawing my attention to this point.

CHAPTER 4

Note to epigraph: Sedgwick, "Epidemics of the Will," 131.

1. William Shakespeare, *Othello*, ed. E. A. J. Honigmann, with a new introduction by Ayanna Thompson. Arden Shakespeare. Rev. 3rd series (London: Bloomsbury Arden Shakespeare, 2016). All citations are to this edition unless otherwise noted.

2. Louis C. Charland puts it this way: "Addicts display agency. However, agency of this sort

is not the same as capacity. For our purposes 'capacity' will be understood to mean decision-making capacity. . . . Decision-making capacity is a fundamental concept of Western medical law and ethics. It derives from the doctrine of informed consent, which requires that consent be informed, free, and 'capable' in order to be deemed valid." "Decision-Making Capacity and Responsibility in Addiction," in *Addiction and Responsibility*, ed. Jeffrey Poland and George Graham (Cambridge, MA: MIT Press, 2011), 147. See also Steven J. Morse, "Addiction and Criminal Responsibility," in Poland and Graham, *Addiction and Responsibility*, 159–200.

3. Valverde, *Diseases of the Will.*

4. Africanus, *A geographical historie of Africa*. The *Historie*'s relationship to *Othello* has been substantially analyzed by critics, who illuminate its status as a bestseller, prejudicial ethnography, and sophisticated historiography. Emily C. Bartels analyzes Africanus as "securing his Christian, European self at the expense of his 'Other' identity as a Moor," while also illuminating the *Historie*'s "insistence on the primacy and contingency of Africa's discrete communities." "Making More of the Moor: Aaron, Othello, and Renaissance Refashionings of Race," *Shakespeare Quarterly* 41 no. 4 (1990): 435; and *Speaking of the Moor: From Alcazar to Othello* (Philadelphia: University of Pennsylvania Press, 2008), 150. See also her "Othello and Africa: Postcolonialism Reconsidered," *William and Mary Quarterly*, 3rd series, 54 no. 1 (1997): 45–64. Jonathan Burton analyzes "a form of textual mimicry," whereby the *Historie* undermines the "fantastical and anti-Islamic tendencies of earlier works on Africa." " 'A most wily bird': Leo Africanus, *Othello* and the trafficking in difference," in *Post-Colonial Shakespeare*, ed. Ania Loomba and Martin Orkin (New York: Routledge, 1998): 44. The *Historie*'s relation to *Othello* remains tantalizing because, as Ayanna Thompson writes, Shakespeare "tells a tale that echoes the fascinating reality of Johannes Leo Africanus's life," tracing a "well-born, educated and experienced African who works his way into the upper echelons of white, European power" (introduction to Shakespeare, *Othello*, 15).

5. Africanus, *A geographical historie of Africa*, book 2, n.p.; "Summary of Religions," 77, 377. He also claims that people from Zanzibar are "addicted to sorcery and witchcraft," people from Biafar and Medra in Ethiopia are "addicted to inchantments, witchcrafts, and all kind of abominable sorceries," and dwellers in the Arabian gulf are "addicted unto Magick and inchantments, and doe bring to passe matters incredible" ("Certain Answers," 27; 41, 47).

6. On Othello's jealousy, see Mary Floyd-Wilson, who writes that "ironically, and counter intuitively for modern readers, Othello is undone by his attachment to constancy rather than by barbarous mutability." *English Ethnicity and Race in Early Modern Drama* (Cambridge: Cambridge University Press, 2003), 154; and Rebecca Olson, who writes, "In *Othello*, however, Shakespeare seems to go out of his way to establish that the Moor Othello is *not* predisposed to jealousy." " 'Too Gentle': Jealousy and Class in *Othello*," *Journal for Early Modern Cultural Studies* 15 no. 1 (Winter 2015): 6.

7. Ian Smith, "We Are Othello: Speaking of Race in Early Modern Studies," *Shakespeare Quarterly* 67 no. 1 (2016): 109. See also Kyle Grady, "Othello, Colin Powell, and Post-Racial Anachronisms," *Shakespeare Quarterly* 67 no. 1 (2016): 68–83.

8. James Siemon, "Making Ambition Virtue? *Othello*, Small Wars, and Marital Profession," in *Othello: The State of Play*, ed. Lena Cowen Orlin (London: Bloomsbury Arden Shakespeare, 2014), 190.

9. For Othello, Lena Cowen Orlin writes that the "acquisition of the trappings of a household reverberates with his discovery of the anxiety of possession." "Desdemona's Disposition," in *Shakespearean Tragedy and Gender*, ed. Shirley Nelson Garner and Madelon Sprengnether (Bloomington: Indiana University Press, 1996): 185.

10. Jeremy Horder, "Pleading Involuntary Lack of Capacity," *Cambridge Law Journal* 52 no. 2 (July 1993): 298. For a comprehensive analysis of modern mental incapacity doctrines see Arlie Loughnan, "Mental Incapacity Doctrines in Criminal Law," *New Criminal Law Review* 15 no. 1 (Winter 2012): 1–31.

11. The legal ruling on Dred Scott, a slave, as a human being but not an individual (meaning, not a person or citizen with legal rights) stands as one notorious example of such differentiation. See James T. McHugh, "What Is the Difference Between a 'Person' and a 'Human Being' Within the Law," special issue on public law, *Review of Politics* 54 no. 3, (Summer 1992), 445–61.

12. See Rebecca Williams, "The Current Law of Intoxication: Rules and Problems," in *Intoxication and Society: Problematic Pleasures of Drugs and Alcohol*, ed. Jonathan Herring, Ciaran Regan, Darin Weinberg, and Phil Withington (Basingstoke: Palgrave Macmillan, 2013), 267–82. See also Alan Bogg and Jonathan Herring, who analyze the law's failure to take addiction into account, except in the case of murder in "Addiction and Responsibility," in ibid., 247–67.

13. Edmund Plowden, *The commentaries, or Reports of Edmund Plowden: containing divers cases upon matters of law, argued and adjudged in the several reigns of King Edward VI, Queen Mary, King and Queen Philip and Mary, and Queen Elizabeth [1548–1579]*, Part 1 (1551; repr., London: H. Watts and W. Jones, 1792), 19. There is probably medieval precedent for this ruling; the ruling cites not only classical precedent for this view in Aristotle but also civilian views in "Bartholinus and others," as Plowden writes in one of his footnotes for this case. But my own research and that of legal historians continues to place this case as the first incident of English rulings on drunken capacity. On this legislation, see David McCord, "The English and American History of Voluntary Intoxication to Negate *Mens Rea*," *Journal of Legal History* 11 no. 3 (1990): 372, and R. U. Singh, "History of the Defence of Drunkenness in English Criminal Law," *Law Quarterly Review*, no. 196 (October 1933): 528.

14. *The Reports of Sir Edward Coke* (1603; J. Butterworth and Son, 1826), 573. This quote is taken from part 4, section 125a-b.

15. Sir Edward Coke, *Institutes of the Laws of England*, book 1 (London, 1628), 247a. How this proposition operated in practice remains unclear. It is possible that aggravation referred simply to a judicial discretion to take intoxication into account when sentencing a defendant.

16. Sir William Blackstone, *Commentaries on the Laws of England*, vol. 4 (Oxford: Clarendon Press, 1765–69), 25–26.

17. Blackstone (*Commentaries*, 26) citing the law of Pittacus of Mytilene (640–568 BC). Aristotle, *Nichomachean Ethics*, trans. W. D. Ross, in *Complete Works of Aristotle, Volume 2: The Revised Oxford Translation*, ed. Jonathan Barnes (Princeton: Princeton University Press, 2014), 1758.

18. Francis Bacon, *The Elements of the Common Lawes of England* (London: John Moore, 1630), 34.

19. Richard Hooker, *Laws of Ecclesiastical Polity: Preface, Book I, Book VIII*, ed. Arthur Stephen McGrade (Cambridge: Cambridge University Press, 1989), 86. Hooker provides the example on incest because he is speaking of men who transgress their own natures. Here he proves stricter than his biblical source: when Lot sleeps with his daughters in his drunkenness, he is depicted as unknowing and guiltless.

20. R. U. Singh writes, "It is very difficult to say if and to what extent, in actual practice of law, drunkenness was ever taken to operate as aggravation of an offence and meant infliction of a heavier sentence than would be due if the offence had been committed by a sober man. There is no English case on the point" ("Defence of Drunkenness," 531).

21. Court of Chancery: Six Clerks Office: Pleadings, series 1, Elizabeth I: C2/Eliz/B7/1. In the case of Thomas Hoskyn of Hungerford vs. Nicholas Munteyne of Bradford, the accusation concerned "making petitioner's servant drunk at Chippenham and obtaining from him kerseys and money" (Court of Chancery: Six Clerks Office: Early Proceedings, C1/22/193 c.1493–1500). In a case between Julian Mermean vs. John Bayly, the charge is "making petitioner's servant drunk and seizing a pack of her goods at Wells" (C1/15/85, c.1455–56). One could also, as a 1538–44 case indicates, be prosecuted as a drunken curate (C1/1056/25).

22. Incidents of drunkenness occasionally appear in session records; drunkenness was prosecuted through assize and ecclesiastical courts before the passage of the 1606 act. In the Caenarfonshire Quarter Sessions records, the charge first appears in a 1571 letter, mentioning articles laid against Rhys Thomas ap Gwilym as "a common drunkard, dizer and carder, keeper of disorder and misrule in his house" (XQS/1571/11; February 7, 1570–71). In records from Chester between 1559 and 1760, the first charge of "common drunkard and haunter of Ale houses" appears in 1613 (F.1, D.14, S. Middlewich, 20 April, 11 Jac. 1613), in *Quarter Session Records . . . for the County Palatine of Chester, 1559–1760*, ed. J. H. E. Bennett and J. C. Dewhurst (Cheshire: Record Society of Lancashire and Cheshire, 1940); in Worcestershire, the first indictment is in 1607, of "George Harris of Ombersley" as a "common Drunkard" (XIV.57, 99); two years later, "Edward Field of Kings Norton Gentleman" was indicted for "drunkenness" (LXXIII.37, 130), in *Worcestershire County Records. Division 1. Documents relating to Quarter Sessions. Calendar of the Quarter Sessions Papers*, vol. 1, *1591–1643*, ed. J. W. Willis Bund (Worcester: Ebenr. Baylis and Son, 1900).

23. Clark, *Alehouse*; Judith Hunter, "English Inns, Taverns, Alehouses and Brandy Shops: The Legislative Framework, 1495–1797," in Kümin and Tlusty, *World of the Tavern*, 65–82; Brown, "Alehouse Licensing and State Formation."

24. F. G. Emmison, *Elizabethan Life: Disorder. Mainly from Essex Sessions and Assize Records* (Chelmsford: Essex County Council, 1970), 212.

25. The 1603–4 canons required church courts to deal with persistent drunkenness and as a result the offense began to appear for the first time in the court's act books and presentments. The Canons of 1604, c.109 asks churchwardens to supply the names of those who "offend their bretheren, either by adultery, whoredome, incest, or drunkenness, or by swearing, ribaldry, usury, and any other uncleanness and wickedness of life." Gerald Bray, ed., *The Anglican Canons 1529–1947*, Ecclesiastical Law Society (London: Boydell and Brewer, 1998), 408–9. See R. H. Helmholz and J. H. Baker, eds., *The Oxford History of the Laws of England* (Oxford: Oxford University Press, 2004), 268; and F. G. Emmison, *Elizabethan Life: Morals & the Church Courts* (Chelmsford: Essex County Council, 1973).

26. "Act for the Repressing of the odious and loathsom Sin of Drunkenness" (4. Jacobus 5), in *A Statute against Drunkennesse* (London, 1644).

27. Few traces exist of this 1566 bill, other than references in *The House of Commons Journals*, noting that it was read on November 30 and committed on December 6.

28. SPD 12/282/43:

> Bill against excessive and common drunkenness
> Fforasmorche as the saide vice of drunkenness is not onlie of yt selfe a vile and detestable thinge in the abuse of god's good benefites whereby that is wasted by a few in excesse, which being moderately used would nourish and satisfie many and the saide drunken persone (the most part of them being of the woorst and inferior sorte of people) not only consume thence substance on occasion that often tymes dryves them to unlawfull shifte, and become more like brute beaste then reasonable

> creatures. But also out of the same vice dou spring divers other enow mysires within the comon weelthe, as loose and wanton belief + swearing and blasphemynge the name of almighte God, Quarrelless fyghtinge O bloodshede, manslaughter, yea and some tymes willfull and professed murdere, and divers other grevouse crymes and enormytes, and for that the saide sine dothe greatlie abounde in some places of thie Realeme, to the greate displeasure of almyghtie God, the rather as it is proposed becawse by no lawe alreadie established, the temporall magistrate hathe authoritie to correcte the saide offence for avoiding of the fowle and common sine of drunkenness Be it therefore ordeyned and Inacted by the Queenes moaste excellent majestie the Lorde Spiritual and temporall and the commons in this present parliament assembled, and by the authoritie of the Same that everie person and persons whiche . . . shalle founde to be a comon drunkarde by oftsoones falling into the saide vice, shall be required of presented indyed and fined as in case of a common Barretor and receive suche punyshement as a comon Barretor by the Lawes and statutes of this Realme.

The first version was read on February 17, 1585, and then again on October 31, 1601.

29. *The Journals of all the Parliaments of Queen Elizabeth, both of the House of Lords and House of Commons,* compiled by Sir Simonds D'Ewes of Stow-Hall in the County of Suffolk, revised and published by Paul Bowes, of the Middle-Temple, London (London: John Starkey, 1682), 629. The bill was heavily debated in both houses of parliament. The 1584–85 version was read in the House of Commons on October 31, then read again on November 3, when it went to committee. The bill was revised, and then a new version was read to the House of Commons by Dr. James the next day, November 4 or 5. This revised bill now circulated under its new title, "bill against drunkards and common haunters of Alehouses and Tavernes." This bill, read for a second time and committed on November 7, and was then shown by William Wray on November 28 and engrossed. The bill was then read in the House of Lords, for the first time on December 3 and for the second time on December 4. See T. E. Hartley, ed., *Proceedings in the Parliaments of Elizabeth I*, vol. 3, *1593–1601*. (London: Leicester University Press, 1995), 328–29.

30. *Journals of all the Parliaments of Queen Elizabeth*, 603, 623, 626.

31. *The Statutes at Large, from the First Year of King James the First to the Tenth Year of the Reign of King William the Third. . . . Volume the Third* (London, 1786), 68.

32. *Journals of all the Parliaments of Queen Elizabeth*, from the debate about the bill on swearing, with Gascock's offensive speech on justices of the peace recorded, 660–61.

33. *Proceedings and Debates of the House of Commons in 1620 and 1621.* Collected by a Member of that House. In Two Volumes, vol. 1 (Oxford: Clarendon Press, 1766), 532: Thurs 1 March 1620. "Drunkenness. Mr. Cary. To have the committee define a Drunkard. . . . Sir Edw. Wardor: to have drunkenness defined. Sir George Moore: to have the words, 'a common haunter of tipling houses,' added, instead of the word 'Drunkard.' "

34. Timothy Cunningham, comp., *The Journals of the House of Commons*, vol. 1, *1547–1628* (London: L. Hansard and Son, 1803), 1007.

35. P. W. Hasler, ed., *History of Parliament: The House of Commons, 1558–1603* (London: Secker and Warburg, 1981), 346.

36. Downame, *Foure treatises*, 79.

37. Ibid., 88.

38. William Shakespeare, *Richard II*, ed. Charles R. Forker, Arden Shakespeare, 3rd series

(London: Bloomsbury Arden Shakespeare, 2002), 2.3.165. This phrase describes Richard's advisors, Bushy, Bagot, and Greene.

39. Downame, *Foure treatises*, 88.

40. Ibid.

41. While concurring with legislators before him that a drunken person contracts his madness voluntarily, Hale amended this judgment with two conditions, one of which being "that although the *simplex* phrenzy occasioned *immediately* by drunkenness excuse not in criminals, yet if by one or more such practices, an *habitual* or *fixed* phrenzy be caused, though this madness was contracted by the vice and will of the party, yet this habitual and fixed phrenzy thereby caused puts the man into the same condition in relation to crimes, as if the same was contracted involuntarily at first." Sir Matthew Hale, *Historia placitorum coronae: The History of the Pleas of the Crown*, vol. 1 (Philadelphia: R. H. Small, 1847), 32; Blackstone amplified this point when he wrote, more generally, that "a vicious will without a vicious act is no civil crime, so, on the other hand, an unwarrantable act without a vicious will is no crime at all" (*Commentaries on the Laws of England*, 21). See also Parliament of Victoria Law Reform Committee, *Criminal Liability for Self-Induced Intoxication* (Melbourne: Government Printer, 1999), 31.

42. Singh, "Defense of Drunkenness," 541.

43. Horder, "Pleading Involuntary Lack of Capacity," 299.

44. These phrases come from Reniger v. Feogossa, Blackstone, Coke, and Bacon, which are cited above.

45. Jennifer Nedelsky, "Law, Boundaries, and the Bounded Self," *Representations* 30 (Spring 1990): 167.

46. On Cassio's military experience, see Andrew Sisson, "*Othello* and the Unweaponed City," *Shakespeare Quarterly* 66 no. 2 (Summer 2015): 137–66, in which he illuminates Cassio as "the citizen-soldier who, unlike the professional, moves seamlessly between peacetime and wartime occupations" (143).

47. Trotter, *Essay, Medical, Philosophical, and Chemical, on Drunkenness.*

48. This argument would be familiar to Shakespeare's audience, since it was circulating freely in contemporary medical and godly pamphlets on drinking. As surveyed in Chapter 3, excessive drinking was linked to a range of diseases, including "distillations, coughs, runnings of the nose, Apoplexies, Palsies, etc." Leonard Lessius, *Hygiasticon: or, the right course of preserving Life and Health unto extreme old age* (Cambridge: Roger Daniel, 1634), 151.

49. John Hoskins, *Sermons preached at Pauls Crosse* (London, 1615), 30–31; Samuel Ward, *Woe to Drunkards*, 10.

50. In "The Merry Hostess," for example, the speaker cries, "Here's a health to all English men who like ale," in contrast to "the bonny Scots" who enjoy merely her "stale ale." See also 272: "Match me this Wedding. Or, a health that was drunke in Sider and Perrie. And good strong beere to, which did make the lads mery. To a new court tune," and 536: "The merry Hostesse: Or, / a pretty new ditty, compos'd by an hostess that lives in the City: / to wrong such an Hostess were a great pitty, / by reason she caused this pretty new Ditty. / to the tune of *Buff Coat has no fellow*." *Roxburghe Ballads*, vol. 1, British Library Rox I.1–279.

51. On these substances as sleep aides, see Angelus Sala, *Opiologia: or, A treatise concerning the nature, properties, true preparation and safe vse and administration of opium*, trans. Thomas Bretnor (London: Nicholas Okes, 1618); Thomas Elyot, *Of the knowledeg whiche maketh a wise man* (London, 1533); Thomas Twyne, *The schoolemaster, or teacher of table philosophie* (London: Richarde Iones, 1576).

52. Indeed, Iago describes himself as a devil as soon as Cassio leaves (2.3.345).

53. See Cathy Shrank, "Beastly Metamorphoses: Losing Control in Early Modern Literary Culture," *Intoxication and Society*, 193–209.

54. Thomas, *Dictionarium Linguae Latinae et Anglicanae*, s.v. *Addīco*.

55. Thompson, introduction to Shakespeare, *Othello*, 37.

56. Paul Cefalu, "The Burdens of Mind Reading in Shakespeare's *Othello:* A Cognitive and Psychoanalytic Approach to Iago's Theory of Mind," *Shakespeare Quarterly* 64 no. 3 (Fall 2013): 277. On theory of mind in *Othello*, see also Laurie Maguire, "*Othello,* Theatre Boundaries, and Audience Cognition," in *Othello: The State of Play*, ed. Lena Cowen Orlin, Arden Shakespeare (London: Bloomsbury Arden Shakespeare, 2014), 17–44.

57. Alessandro Serpieri explores this negative invocation of bonds in the play, analyzing how images of the "bond" in the play are offered "in their most negative acceptations." "*Othello* and Venice: Discrimination and Projection," in *Visions of Venice in Shakespeare*, ed. Laura Tosi and Shaul Bassi (London: Routledge, 2011), 187.

58. Cefalu illuminates contentedness as "calmness," "silence," and "attunement to the unspoken rhythms Othello now enjoys with Desdemona" ("Burdens of Mind Reading," 271).

59. As Christopher Tilmouth writes, "Inwardness is situated at the boundary between the person and those to whom he relates, within the dialogic domain of intersubjectivity." "Passion and Intersubjectivity in Early Modern Literature," in *Passions and Subjectivity in Early Modern Culture*, ed. Brian Cummings and Freya Sierhuis (New York: Routledge, 2013), 16. On Othello's vulnerable condition in love, see Tzachi Zamir, *Double Vision: Moral Philosophy and Shakespearean Drama* (Princeton, NJ: Princeton University Press, 2007), 159ff.

60. "It is customary to view Othello as a tragedy that begins as a comedy," writes Laurie Maguire, "*Othello,* Theatre Boundaries, and Audience Cognition," 24.

61. Paul Yachnin, "Wonder-Effects: Othello's Handkerchief," in *Staged Properties in Early Modern English Drama*, ed. Jonathan Gil Harris and Natasha Korda (Cambridge: Cambridge University Press, 2002), 316. The handkerchief boasts an impressive critical history. See Ian Smith, "Othello's Black Handkerchief," *Shakespeare Quarterly* 64 no. 1 (Spring 2013): 1–25; Andrew Sofer, "Felt Absences: The Stage Properties of '*Othello's* Handkerchief," *Comparative Drama* 31 no. 3 (Fall 1997): 367–93; Lynda E. Boose, "Othello's Handkerchief: 'The Recognizance and Pledge of Love,'" *English Literary Renaissance* 5 (1975): 360–74; Harry Berger Jr. "Impertinent Trifling: Desdemona's Handkerchief," *Shakespeare Quarterly* 47 no. 3 (Fall 1996): 235–50.

62. On early modern friendship between men, see Alan Bray, "Homosexuality and the Signs of Male Friendship in Elizabethan England," *History Workshop Journal* 29 (1990): 1–19, reprinted in *Queering the Renaissance*, ed. Jonathan Goldberg (Durham: Duke University Press, 1994), 40–61; Alan Bray, *The Friend* (Chicago: University of Chicago Press, 2003); Mario DiGangi, *The Homoerotics of Early Modern Drama* (Cambridge: Cambridge University Press, 1997); Bruce R. Smith, *Homosexual Desire in Shakespeare's England: A Cultural Poetics* (Chicago: University of Chicago Press, 1991); Alan Stewart, *Close Readers: Humanism and Sodomy in Early Modern England* (Princeton, NJ: Princeton University Press, 1997); and Masten, *Textual Intercourse.* See also Valerie Traub's account of female friendship and erotic partnership in *The Renaissance of Lesbianism in Early Modern England* (Cambridge: Cambridge University Press, 2002).

63. On Iago as a drug peddler, offering words as drugs, see Bella Mirabella, "'A Wording Poet:' Othello among the Mountebanks," *Medieval and Renaissance Drama in England* 24 (2011): 150–75: his language is like "a drug, capturing the minds and hearts of those listening, while engendering anxiety and fear" (160–61). David Schalkwyk also illuminates Iago's linguistic power: he is "a figure completely adept at playing every language game *as if* he were acknowledging the other," but who, in the end, betrays an "unimaginable and unspeakable capacity to

remain completely detached from the usual engagements, commitment, and responses of human intercourse" (*Shakespeare, Love and Service*, 261).

64. Africanus, *A geographical historie of Africa,* book 2, n.p.

65. Ibid., book 1, 41.

66. Berlant, *Cruel Optimism*, 1.

67. Sisson, "Othello and the Unweaponed City," 157.

68. David Schalkwyk writes of Othello's vow, "Othello invokes 'heaven,' 'reverence,' and the 'sacred' in the name of a resolute constancy (traditionally associated with married love) that is the very opposite of 'sacred' or 'heavenly:' it is 'tyrannous'" (*Shakespeare, Love and Service*, 253).

69. Smith, "We Are Othello," 112.

70. Edmund Plowden, *The commentaries,* 19.

71. Sir Edward Coke, *The Reports of Sir Edward Coke*, 573.

72. Aristotle, *Nichomachean Ethics*, 1758.

73. Julia Reinhard Lupton, in *Citizen Saints: Shakespeare and Political Theology* (Chicago: University of Chicago Press, 2005), calls Othello's suicide a "death into citizenship," an execution of state service as a Christian fighting the Turks (121). Arthur L. Little Jr. movingly analyzes the play's link of race and religion within its "gut-wrenching, difficult reading of the confounded black body." "Re-Historicizing Race, White Melancholia, and the Shakespearean Property," *Shakespeare Quarterly* 67 no. 1 (2016): 98.

74. Lauren Berlant, "The Commons: Infrastructures for Troubling Times," *Society and Space* 34 no. 3 (2016): 394.

75. Ibid.

CHAPTER 5

Note to epigraphs: Charles Cotton, "To Mr. Alexander Brome, Epode," in *Poems on Several Occasions* (London, 1689), 523; [Charles Morton], *The Great Evil of Health-Drinking* (London: Jonathan Robinson, 1684), 10. The Huntington Library copy (HL 145582) includes William Prynne as the possible author.

1. Scodel, *Excess and the Mean*, 209.

2. McShane, "Material Culture and 'Political Drinking,'" 249.

3. See O'Callaghan, *English Wits* and "Tavern Societies," 45; and Marika Keblusek, "Wine for Comfort: Drinking and the Royalist Exile Experience, 1642–1660," in Smyth, *Pleasing Sinne*, 58.

4. Shepard, *Meanings of Manhood*, 101, 27, 100–102. See also Bryson, *From Courtesy to Civility*, 93.

5. "It was not uncommon for divines to use healthing as a religious metaphor, as in 1616, when Thomas Adams described Christ's sacrifice as 'drinking to us in his own bloud, a saving Health to all Nations'" (McShane, "Material Culture and Political Drinking," 257, citing Thomas Adams, *A Divine Herball* [1616]).

6. This legislation did not curb the practice: the misdemeanor most frequently prosecuted in 1651 was the drinking of healths, as noted by Paul H. Hardacre, *The Royalists During the Puritan Revolution* (The Hague: Martinus Nijhoff, 1956), 74–75. To give a brief sense of the scope of healthing, the *Calendar of State Papers* lists 467 incidents related to health drinking in the Stuart period. I am grateful to David Cressy for bringing my attention to some of these records.

7. Shakespeare, *Hamlet*, 1.4.11–12.

8. Shakespeare, *Othello*, 2.3.27–34.

9. McShane's "Material Cultural and Political Drinking," in analyzing the violence attendant on refusing a health, reveals how men and women would rather risk blows than the damnation potentially greeting a false oath (261). Her analysis precisely illuminates health drinking as both drinking and oath-taking.

10. I use the term "Cavalier poet" because of its critical legacy, while also recognizing that term is anachronistic, given that the term "Cavalier" appeared after 1642 as an insult for Charles I's supporters before being adopted by the "Cavaliers" themselves. The term "Cavalier poet" is also potentially misleading in that the political affiliation of some of the ostensibly Cavalier poets is not always or simply royalist. See Wilcher, *Writing of Royalism*; De Groot, *Royalist Identities*; James Loxley, *Royalism and Poetry in the English Civil Wars: The Drawn Sword* (Houndmills, Basingstoke: Macmillan Press, 1997).

11. To follow Thomas Corns's argument in *Uncloistered Virtue*, the representation of excessive drinking occurs in poems rife with "practices in which bonds of patronage and duty are consolidated" (111).

12. My conclusions on poetic representations of health drinking come out of my analysis of healthing poems in the following volumes, in addition to those cited directly in the chapter: *Choyce Drollery* (1656); *Antidote Against Melancholy: made up into pills* (London, 1661); *Merry Drollerie or, a collection of jovial poems, merry songs, witty drolleries* (London: J. W., for P. H., 1661); Hugh Crompton, *Bring a fardle of fancies, or a medley of musick, stewed in four ounces of the oyl of epigrams* (London: E. C., 1657), and *Pierides, or the Muses Mount* (London: J. G. for Charles Web, 1658); John Taylor, *Drinke and Welcome* (London: Anne Griffin, 1637); D'Urfey, *Wit and Mirth*; and David M. Vieth, ed., *The Complete Poems of John Wilmot, Earl of Rochester*. Yale Nota Bene Series (New Haven, CT: Yale University Press, 1962).

13. Tom Cain and Ruth Connolly, eds., *The Complete Poetry of Robert Herrick*, vol. 1 (Oxford: Oxford University Press, 2013), 96–97, lines 38–43. Hereafter abbreviated *CP* and cited parenthetically by page and line number. Cain and Connolly have established an earlier date of composition for a number of Herrick's poems, including "Welcome to Sack" and "Farewell to Sack," which circulated in the 1620s and 1630s. See also Paul Davis, "Maximum Parsimony," *Times Literary Supplement*, January 16, 2015, 3–5; and Tom Cain, who in "Robert Herrick," in *Oxford Dictionary of National Biography* (Oxford: Oxford University Press, 2004) notes that "*Hesperides* was and remains the only effort by an important English poet to publish his entire *oeuvre* in one organized collection."

14. Cedric C. Brown, "Drink as a Social Marker in Seventeenth-Century England," in Smyth, *Pleasing Sinne*, 5. On the social bonds forged through healthing, see Connolly and Cain, *"Lords of Wine and Oile."* On merry England, see Ronald Hutton, who in *The Rise and Fall of Merry England: The Ritual Year 1400–1700* (Oxford: Oxford University Press, 2005), 13–14, 58, explores the medieval origins of healthing in wassailing; and Storm Jameson, *The Decline Of Merry England* (London: Cassell and Company, 1930).

15. Bacchus features prominently in health drinking verse of the period. See also "The Ejection," in which the speaker drinks healths to "Bacchus and his fraternal crew" (17); and "The Puff," praising "the wholesome heat of Bacchus" (15) in Crompton, *Pierides.*

16. Jonson, "Leges Convivales," in David Bevington, Martin Butler, and Ian Donaldson, eds., *The Cambridge Edition of the Works of Ben Jonson* (Cambridge: Cambridge University Press, 2012), volume 5, 418–19. Hereafter abbreviated *WBJ* and cited parenthetically by volume, page, and line number.

17. "*LXVII: An Epistle answering to one that asked to be Sealed of the Tribe of Ben*," in *Ben*

Jonson, eds. C. H. Herford Percy and Evelyn Simpson, volume VIII (Oxford: Clarendon Press, 1947), 218, line 10.

18. Scodel, *Excess and the Mean*, 209.

19. Katharine Eisaman Maus, "Why Read Herrick?," in Connolly and Cain, "*Lords of Wine and Oile*," 36.

20. Scodel, *Excess and the Mean*, 227.

21. William Camden, *Annales, the true and royal history of the famous empresse Elizabeth, Queene of England, France, and Ireland*, trans. Abraham Darcie (London, 1625), 5. This practice of health drinking might also have been introduced by Dutch exiles who came to England during this period. Further, critiques of healthing may have served as attacks on the Dutch during a period increasingly marked by tensions in trade. See Andrew Fleck, who in "Marking Difference and National Identity in Dekker's *The Shoemaker's Holiday*," *SEL* 46 no. 2 (2006): 349–70, writes, "the genuinely Dutch figure of the skipper succumbs to English stereotypes of the gluttonous, drunken Dutchman" (365), and A. J. Hoenselaars, *Images of Englishmen and Foreigners in the Drama of Shakespeare and His Contemporaries* (Rutherford, N.J.: Fairleigh Dickinson University Press, 1992).

22. The history of health drinking in England is vexed. Caesar may have brought Roman health drinking practices into England with the invasion. See Hutton, *Merry England*, 13–14, 58; French, *History of Toasting*, esp. 18, 28, 40–41. On the heavy-drinking Englishman, see Hugh M. Thomas, who in *The English and the Normans: Ethnic Hostility, Assimilation, and Identity 1066–c.1220* (Oxford: Oxford University Press, 2005), 301–2, confirms the stereotype of English as drinkers in the works of Gerald of Wales, John of Salisbury, and Geoffrey of Vinsauf (who deems the country *Potatrix Anglia*).

23. Simon Schama, *The Embarrassment of Riches: An Interpretation of Dutch Culture in the Golden Age* (New York: Vintage Books, 1987), 190. I am grateful to Marjorie Rubright for this reference. Benjamin Roberts, a historian of the early modern Netherlands, writes that health drinking was "a ritualized aspect of Dutch cultural and social life" ("Drinking like a Man," 239). Drinking rituals existed for marriages, births, deaths, and various holidays, making healthing an established part of social life, even enjoying, as Tom Nichols writes, a "quasi-baptismal role . . . in secular society" in "Double Vision: The Ambivalent Imagery of Drunkenness in Early Modern Europe," *Past and Present* 222 no. S9 (2014): 163. He continues, "Through the frequent toasts raised by the officers in the Haarlem militia, the individual . . . was bound into the wider social body, becoming properly part of the citizenry" (164).

24. Thomas Nashe, The Unfortunate Traveller *and Other Works*, ed. J. B. Steane (London: Penguin Classics, 1972), 104.

25. Dabridgecourt Belchier, *Hans Beer-Pot: acting in the Low Countries by an Honest Company of Health-Drinkers* (1618): a young man claims he won't toast his father's health "to drinke away mine owne." His companions pressure him, but he insists, saying "tossepot Knight . . . I tell thee, ile no more."

26. Thomas Heywood, *Philocothonista or the Drunkard, Opened, Dissected, and Anatomized* (London: Robert Raworth, 1635).

27. August 8, 1586, "Quarrel between Count Hohenlohe and Capt. Edw. Norreys," from *Calendar of State Papers, Foreign: Elizabeth* (Holland 9/76), vol. 21, part 2, *June 1586–March 1587*, ed. Sophie Crawford Lomas and Allen B Hinds (London, 1927). Count Hollock is Count Philip Hohenlohe Langenburg; Edward Norris is younger brother to Sir John Norris. On this exchange, see Alexander Young, *History of the Netherlands* (Estes and Lauriate, 1884); Kevin Pask,

The Emergence of the English Author: Scripting the Life of the Poet in Early Modern England (Cambridge: Cambridge University Press, 1996).

28. Harleian MS 6395, Sir Nicholas L'Estrange, no. 499, Sir Fr. Needham, cited in William J. Thoms, ed., *Anecdotes and Traditions* (London: Camden Society, 1839), 68.

29. On April 1, 1606, at Essex House, for example, "a great quarrel between three gentlemen" broke out, on the occasion of drinking the earl of Southampton's health. See "James I: Volume 20, April 1606," in *Calendar of State Papers, Domestic: James 1, 1603–1610*, ed. Mary Anne Everett Green (London: HMSO, 1857), 308–14. The tangle of Essex and the Low Country wars appears in the work of John Taylor, *Drinke and Welcome* (London: Anne Griffin, 1637), C2v.

30. John Cotgrave, *Wits Interpreter* (1655), 333. On *Wits Interpreter*, see Smyth, "'It were far better be a *Toad*, or *Serpant*, then a Drunkard.'" His essay introduced me to this miscellany.

31. Bloom, "Manly Drunkenness."

32. Richard Brathwaite [Blasius Multibibus, pseud.], *A Solemne Joviall Disputation, theoreticke and praticke; briefly shadowing the law of drinking* (London: At the signe of the Red Eyes, E. Griffin, 1617), 21. On the literature on laws of drinking and convivial tavern societies, see O'Callaghan, *English Wits*; and Timothy Raylor, *Cavaliers, Clubs, and Literary Culture: Sir John Mennes, James Smith, and the Order of the Fancy* (Newark: University of Delaware Press, 1994).

33. Gascoigne, *Delicate Diet for daintie mouthde Droonkardes*, 18. A participant in the military culture referenced by Camden, Norris, and Nashe, Gascoigne sailed to the Low Countries as a soldier of fortune in 1572 and recounted his experiences in *Fruites of Warres* (1572) and *The Siege of Antwerp* (1576).

34. William Hornby, *The Scourge of Drunkenness* (London, 1618), B4v.

35. Prynne, *Healthes Sicknesse*, 6.

36. Charles Morton or William Prynne [Anon.], *The Great Evil of Health-Drinking: or a discourse wherein the original, evil, and mischief of drinking of healths, are discovered and detected; and the practice opposed* (London: Jonathan Robinson, 1684), 11.

37. *Looking glasse for drunkards*, ch. 3, n.p.

38. Richard Young [R. Junius, pseud.], *The Drunkard's Character*, 138.

39. Downame, *Foure treatises*, 93.

40. *Looking glasse for drunkards*, ch. 3, n.p. See also Timothy Gunton, *An extemporary answer to a cluster of drunkards, met together at Schiedam: made by Timothy Gunton, who was compelled thereto, upon his refusall to drink the kings health* (London, 1648), n.p.

41. Downame, *Foure treatises*, 80.

42. Samuel Ward, *Woe to Drunkards*, 26–27. See also Samuel Ward and Samuel Clark, *A Warning-piece to all Drunkards.* For other stories of felled health drinkers, see Henry Jessey, *The Lords loud call to England: being a true relation of some late, various, and wonderful judgments, or handy-works of God . . . As also of the odious sin of drinking healths* (London: 1660); and John Vicars, *The Looking-Glass for Malignants, enlarged. Or, the second part of Gods hand against God-haters* (London: I. Rothwell, 1645; first edition 1643), C2r.

43. Jonson, a student of Camden's at Westminster School, traveled to the Low Countries as a volunteer with the regiments of Sir Francis Vere and turned his experiences into satirical comedy in his subsequent plays. The character of Tucca in *Poetaster* encourages health drinking (3.4), and in *Cynthia's Revels* Cupid and Mercury engage in a conversation about the health drinking of Anaides, who "never kneels but to pledge Healths, nor prays but for a Pipe of Pudding-tabacco" (2.2). See David Bevington, Martin Butler, and Ian Donaldson, eds., *The Cambridge Edition of the Works of Ben Jonson* (Cambridge: Cambridge University Press, 2012).

44. Ben Jonson, *Every Man Out of His Humor*, ed. Helen Ostovich, The Revels Plays

(Manchester: Manchester University Press, 2001), 5.3.50–53. On Buffone as figure for the real-life buffoon Charles Chester, see Matthew Steggle, "Charles Chester and Ben Jonson," *SEL* 39 no. 2 (Spring 1999): 313–26.

45. Like Jonson, Thomas Dekker and Thomas Middleton link health drinking and swaggering to comic effect in *The Honest Whore, Part 1*. The gallants Fluello, Castruchio, and Pioratto taunt the patient Candido. Rather than shunning their worthless custom, as any reasonable person might, Candido rewards them with a beaker full of wine for pledging. His hospitality serves as further proof of his imperturbable character: even swaggering health drinkers don't move him to ire, despite their attempts to bully him into drinking. Fluello demands, "'Sfoot you shall pledge mee all" (1.5.162), threatening to steal a silver-and-gold beaker if Candido does not drink; Castruchio warns, "Pledge him, heele do't else" (1.5.166). But Candido calmly repeats to the men, "Oh, you must pardon me, I use it not" (1.5.159), "You know me syr, / I am not of that sin." (1.5.168–69). This scene hinges on the humorous encounter between two exaggerated character types, the impossibly patient Candido, and the hard-drinking city gallant. And it is the practice of health drinking that helps to throw their opposition into stark contrast. Thomas Dekker and Thomas Middleton, *Honest Whore, Part 1*, in *The Dramatic Works of Thomas Dekker*, ed. Fredson Bowers (Cambridge: Cambridge University Press, 1955), 2:1–130.

46. Brian Gibbons, *Jacobean City Comedy: A Study of Satiric Plays by Jonson, Marston and Middleton* (Cambridge, MA: Harvard University Press, 1968), 155.

47. Ben Jonson, *Epicene* in *The Cambridge Edition of the Works of Ben Jonson*, volume 3. See also *Epicene, or The Silent Woman*, ed. Richard Dutton, The Revels Plays (Manchester: Manchester University Press, 2003).

48. Ben Jonson, *Bartholomew Fair*, in *The Cambridge Edition of the Works of Ben Jonson*, volume 4. See also *The Selected Plays of Ben Jonson*, vol. 2, ed. Martin Butler (Cambridge: Cambridge University Press, 1989), 147–298; and George Chapman, Ben Jonson, and John Marston, *Eastward Ho*, ed. R. W. Van Fossen, The Revels Plays (Manchester: Manchester University Press, 2006), which features an extended drinking ritual as male tavern patrons encircle and drink to their female companion. Sir Petronel Flash presides over the toasts and in doing so steals Winifred, the wife of the jealous old usurer Security, right from under her husband's nose. In the climax of the scene, all the drinkers, including Flash and Security, circle around the disguised Winifred. They offer her toasts on their knees in a public display that would incense Security were he to realize the game featured his own wife. Instead the cuckolded Security celebrates the occasion, believing the woman to be Mistress Bramble: "I have one corner of my brain, I hope, fit to bear one carouse more. –Here, lady, to you that are encompassed there" (3.3.131–32) in *The Cambridge Edition of the Works of Ben Jonson*, volume 2.

49. Thus even comic invocations or defenses of healthing—ranging from Buffone to Toby—betray an edge of satire, as the drinker is gulled, ridiculed, or exposed. Health drinking follows Jonson's (and Cicero's) definition of comedy more generally. It is, as Cordatus puts it in *Every Man out of His Humour*, "pleasant and ridiculous, and accommodated to the correction of manners" (3.1.527–29).

50. Even as their characters might appreciate the transformative potential of healthing, these plays notably relegate the practice to a loutish subculture. The health drinkers in Dekker and Jonson almost universally play tricks on one another—cuckolding one man, ridiculing another, deceiving a third. Health drinking allows swaggerers and gallants, prodigal sons like Flash and miserable husbands like Otter and Security, to revel in temporary freedom. Shakespeare, too, develops this link between trickery and health drinking. Petruccio, for example, drinks to Kate's health upon their marriage in *Taming of the Shrew*, as if, as Gremio recounts, "He had

been aboard, carousing to his mates / After a storm" (3.2.161–62). Having "quaffed off the muscatel" he then "threw the sops all in the sexton's face" (3.2.162–63). Here healthing serves as part of Petruccio's "taming" ritual, in line with riding to the wedding ceremony on a nag—he mimics antisocial behaviors in an effort to humiliate Kate. See also the notable example of health drinking in Thomas Heywood's *The Iron Age, Part 1*, which posits the origins of the Trojan War in a health-drinking contest, when Menelaus entertains the visiting Paris and toasts him repeatedly.

51. Hamlet does not abstain from drinking, of course, and he promises to teach Horatio to "drink ere you depart" (1.2.174).

52. "To Mr. Alexander Brome, Epode," in Charles Cotton, *Poems on Several Occasions* (London, 1689), 513.

53. C. H. Wilkinson, ed., *The Poems of Richard Lovelace* (Oxford, 1930), 78, lines 13–20.

54. "Song I. The Royalist. Written in 1646," Alexander Brome, *Songs and Other Poems* (London: Henry Brome, 1664), 55. Hereafter abbreviated *SP* and cited parenthetically by page number. See also Roman R. Dubinski, ed., *Alexander Brome: Poems*, 2 vols. (Toronto: University of Toronto, 1982).

55. Lois Potter, *Secret Rites and Secret Writing: Royalist Literature, 1641–1660* (Cambridge: Cambridge University Press, 1989), 141. Further, Scodel has explored how Lovelace's excess is, given the circumstances, a reasonable response: the poet "justifies his non-Horatian excess with a Horatian appeal to decorum: extreme circumstances demand extreme remedies" (*Excess and the Mean*, 228).

56. Miner, *Cavalier Mode.*

57. Edmund Waller, *Poems upon several occasions* (London: Thomas Walkley, 1645), 77. See also Thomas Park, ed., *The Poetical Works of Edmund Waller* (London: J. Sharpe, 1806), 55–56.

58. Robert Heath, *Clarastella; together with Poems Occasional, Elegies, Epigrams, Satyrs* (London: Humphrey Moseley, 1650), 22–23.

59. "Ode," Cotton, *Poems on Several Occasions,* 445. In another "Ode," Cotton condemns those who "dare not pledge our *Loyal Bowls*" (ibid., 448).

60. Todd W. Reeser explores how "the ideal man of the Renaissance" exercised moderation in relation to "courage, diet, and prodigality" in *Moderating Masculinity in Early Modern Culture* (Chapel Hill: University of North Carolina Press, 2006), 14. Jean E. Howard and Phyllis Rackin, in *Engendering a Nation: A Feminist Account of Shakespeare's English Histories* (New York: Routledge, 1997), analyze how, in *Richard II*, Bolingbroke condemns his "young wanton and effeminate boy" who haunts taverns and spends his time with a "dissolute crew" (5.310–12). The tavern in the *Henriad* is thus marked, they argue, "as a feminized, theatrical space" (164–74; 165). On drinking and effeminacy, see also Britland, "Circe's Cup," 109–26. As powerful as the association between moderation and masculinity might have been, however, critics have equally illuminated how there was no singular notion of masculinity in the period. Certainly "for different social groups in early modern England," as Bruce R. Smith writes, "there were different masculinities." *Shakespeare and Masculinity* (Oxford: Oxford University Press, 2000), 57. He also notes that the "humanist man of moderation" was one of the "ideal types in Shakespeare's plays" (157). Yet as noted in chapter 3, "exaggerated norms of masculine behavior" were evident among soldiers drinking in taverns; see B. Ann Tlusty "The Public House and Military Culture" (Kümin and Tlusty, *World of the Tavern*, 150).

61. Bernard Capp, review of *Meanings of Manhood in Early Modern England*, by Alexandra Shepard, *Reviews in History* (January 2004), review no. 380, http://www.history.ac.uk/reviews/review/380. Capp elaborates that "one task facing future scholars will be to tease out . . . alternative meanings of manhood for those who embraced them." Gina Bloom is even more pointed

in her analysis of heavy drinking: for "working men of lower or middle status and youths . . . disorderly behaviors like heavy drinking could constitute a bid for an antipatriarchal, countercode of masculine conduct" ("Manly Drunkenness," 23). See also Martin, *Alcohol, Sex, and Gender*; and Shepard, in *Meanings of Manhood*, who offers a discussion of the disruptive excess of male drinking, as well as its association with good fellowship.

62. McShane, "Material Culture and 'Political Drinking,'" 250. See also her "Roaring Royalists and Ranting Brewers," in Smyth, *Pleasing Sinne*, 69–88. On health drinking in ballads see, for example, "The Royalist's Resolve" in *Roxburghe Ballads*, which claims "Come, drawer, some wine, /Or we'll pull down the sign, / For we are all jovial compounders; / We'll make the house ring /With healths to our King, / And confusion light on his confounders." Versions of this ballad also appear under the titles "The Compounders Song," in T. W. [Thomas Weaver], *Songs and Poems of Love and Drollery* (1654), 13–15, and "On the Goldsmiths Committee," in *Rump: Or an Exact Collection of the Choycest Poems and Songs relating to the Late Times* (London: Henry Brome, 1662), 235–37.

63. See also "Song XXV. The Prisoners" in Brome, *Songs and Other Poems*, 94.

64. As Robert Wilcher argues, Brome's drinking compromises his political efficacy: "There is clearly no hope of effective action to overturn the Protectorate from the rollicking Cavalier who authors these lines: 'we only converse with pots and with glasses, / Let the Rulers alone with their trade'" (*Writing of Royalism*, 333).

65. The seventeenth century saw the first significant flourishing of Anacreontic poetry in English, following the Henri Estienne's 1554 French edition of Anacreon, *Anacreontis Teji odae: Ab Henrico Stephano luce et Latinitate nunc primum donatae* (Paris: Henricum Stephanum, 1554), and initial efforts by Robert Greene, in *Orpharion* (London, 1589), and A. W., in Francis Davisons's *Poetical Rhapsody* (London, 1602). Herrick and Abraham Cowley paraphrased and/or translated the Greek verse, while Thomas Stanley offered the first full translation of Anacreon into English in 1651. On Herrick's relation to this tradition, see Corns, *Uncloistered Virtue*, esp. 96–100; Anne Baynes Coiro, *Robert Herrick's Hesperides and the Epigram Book Tradition* (Baltimore, MD: Johns Hopkins University Press, 1988); and Achilleos, "The *Anacreontea* and Refined Male Sociability." On Stanley, see Stella P. Revard, "Thomas Stanley and 'A Register of Friends,'" in *Literary Circles and Cultural Communities in Renaissance England*, ed. Ted-Larry Pebworth (London: University of Missouri Press, 2000), 148–72.

66. O'Callaghan, *English Wits*; Raylor, *Cavaliers*; Smyth, *Pleasing Sinne.* See also Withington, "Company and Sociability."

67. Michael Jeanneret, *A Feast of Words: Banquets and Table Talk in the Renaissance*, trans. Jeremy Whiteley and Emma Hughes (Cambridge: Polity Press, 1991), 179. For a literary and historical account of banquets, with analysis of a range of drama, including works by Middleton, Heywood, and Fletcher, see Chris Meads, *Banquets Set Forth: Banqueting in English Renaissance Drama* (Manchester: Manchester University Press, 2002).

68. McShane, "Material Culture and Political Drinking," 259. See also Nichols, "Double Vision," 147. Of Titian's *Adrians* painting he writes that "wine is pictured as sensually enervating: as in the Catholic Mass, wine functions in a transformative manner raising the intoxicant to a higher state of consciousness" (149–50).

69. De Groot, *Royalist Identities*, 30.

70. *A Proclamation against Vicious, Debauch'd, and Prophane Persons, by the King. At our court at Whitehal, the 30th day of May, in the twelfth year of our reign* (London: Christopher Barker, 1660). Health drinking continued as a social practice at least up through the eighteenth century.

71. By 1670, as McShane's research reveals, health drinking seemed truly compulsory, with even earlier detractors such as Richard Baxter conceding it prudent to participate in loyal-healthing. Such concessions, she writes, expose broad "acceptance of the need to demonstrate loyalty through the ritual of imbibing alcohol" ("Material Culture and Political Drinking," 273).

EPILOGUE

Note to epigraph: Maia Szalavitz, *Unbroken Brain: A Revolutionary New Way of Understanding Addiction* (New York: St. Martin's Press, 2016), 164.

1. Maggie Nelson, *The Argonauts* (Minneapolis, MN: Greywolf Press, 2015), 39–41.

Works Cited

NATIONAL ARCHIVES, LONDON

C1 Court of Chancery: Six Clerks Office: Early Proceedings, Richard II to Philip and Mary
C2 Court of Chancery: Six Clerks Office: Pleadings, series 1, Elizabeth I to Charles I
XQS Gwynedd Archives, Caernarfon Record Office, Quarter Sessions records
SP 12 State Papers, Domestic: Elizabeth I
SP 14 State Papers, Domestic: James I

PRINTED BOOKS

Achilleos, Stella. "The *Anacreontea* and a Tradition of Refined Male Sociability." In *A Pleasing Sinne: Drink and Conviviality in Seventeenth-Century England*, ed. Adam Smyth, 21–36. London: D. S. Brewer, 2004.

Adams, Thomas. *A Divine Herball*. London, 1616.

Africanus, Leo. *A geographical historie of Africa, written in Arabicke and Italian by Iohn Leo a More, borne in Granada, and brought vp in Barbarie. . . . Translated and collected by Iohn Pory, lately of Goneuill and Caius College in Cambridge*. London, 1600.

Agamben, Giorgio. *Potentialities: Collected Essays in Philosophy*. Trans. Daniel Heller-Roazen. Stanford, CA: Stanford University Press, 1999.

Albala, Ken. *Eating Right in the Renaissance*. Berkeley: University of California Press, 2002.

Alexander, Anna, and Mark S. Roberts, eds. *High Culture: Reflections on Addiction and Modernity*. Albany: State University of New York Press, 2003.

Alexander, Bruce. *The Globalisation of Addiction: A Study in Poverty of the Spirit*. Oxford: Oxford University Press, 2008.

Amadis de Gaule. The third booke of Amadis de Gaule . . . Written in French by the Lord of Essars, Nicholas de Herberay. . . . Translated into English by A. M. London, 1618.

Anacreon. *Anacreontis Teji odae: Ab Henrico Stephano luce et Latinitate nunc primum donatae*. Paris: Henricum Stephanum, 1554.

Andrewes, Lancelot. *A Sermon Preached Before His Majestie at Whitehall the fifth of November last 1617*. London, 1618.

———. *XCVI sermons by the Right Honorable and Reverend Father in God, Lancelot Andrewes, late Lord Bishop of Winchester*. London, 1629.

Andrews, Meghan. "Gender, Genre, and Elizabeth's Princely Surrogates in *Henry IV* and *Henry V*." *Studies in English Literature, 1500–1900* 54 no. 2 (Spring 2014): 375–99.

Antidote Against Melancholy: made up into pills. London, 1661.

Aristotle. *Nichomachean Ethics*. Trans. W. D. Ross. *Complete Works of Aristotle, Volume 2: The Revised Oxford Translation*. Ed. Jonathan Barnes. Princeton: Princeton University Press, 2014.

Bacon, Francis. *The Elements of the Common Lawes of England*. London: John Moore, 1630.

Bakhtin, Mikhail. *Rabelais and His World*. Trans. Helene Iswolsky. Cambridge, MA: MIT Press, 1968.

Baldwin, William. *The canticles or balades of Salomon, phraselyke declared in Englysh metres*. London: William Baldwin, 1549.

Bale, John. *The seconde part of the Image of both churches after the most wonderfull and heauenlye Reuelacyon of Saynt Iohan the Evangelist*. Antwerp, c.1545.

Banister, John. *An antidotarie chyrurgicall containing great varietie and choice of all sorts of medicines that commonly fal into the chyrurgions vse*. London, 1589.

Barber, C. L. *Shakespeare's Festive Comedy*. Princeton, NJ: Princeton University Press, 1959.

Baret, John. *An Alveary or Triple Dictionary, in English, Latin, and French*. London: Henry Denham, 1574.

Bartels, Emily C. "Making More of the Moor: Aaron, Othello, and Renaissance Refashionings of Race." *Shakespeare Quarterly* 41 no. 4 (1990): 433–54.

———. "Othello and Africa: Postcolonialism Reconsidered." *William and Mary Quarterly* 3rd series, 54 no. 1 (1997): 45–64.

———. *Speaking of the Moor: From Alcazar to Othello*. Philadelphia: University of Pennsylvania Press, 2008.

Beard, Thomas. *The Theatre of Gods Judgements . . . now secondly printed*. London, 1612.

Becon, Thomas. *A new postil conteinyng most godly and learned sermons upon all the Sonday Gospelles, that be redde in the church thorowout the yeare*. London, 1566.

Belchier, Dabridgecourt. *Hans Beer-Pot: acting in the Low Countries by an Honest Company of Health-Drinkers*. London, 1618.

Benedict, Philip. *Christ's Churches Purely Reformed: A Social History of Calvinism*. New Haven, CT: Yale University Press, 2004.

Bennett, J. H. E., and J. C. Dewhurst, eds. *Quarter Session Records . . . for the County Palatine of Chester, 1559–1760*. Cheshire: Record Society of Lancashire and Cheshire, 1940.

Bennett, Judith M. *Ale, Beer, and Brewsters in England: Women's Work in a Changing World, 1300–1600*. Oxford: Oxford University Press, 1999.

———. "Conviviality and Charity in Medieval and Early Modern England." *Past and Present* no. 134 (February 1992): 19–41.

Benson, George. *Sermon at Paules Cross*. London, 1609.

Berger Jr., Harry. "Impertinent Trifling: Desdemona's Handkerchief." *Shakespeare Quarterly* 47 no. 3 (Fall 1996): 235–50.

Berlant, Lauren. *Cruel Optimism*. Durham, NC: Duke University Press, 2011.

———. "The Commons: Infrastructures for Troubling Times." *Society and Space* 34 no. 3 (2016): 393–410.

Bersani, Leo. "Is the Rectum a Grave?" *October* 43 (Winter 1987): 197–222.

———. "Sociality and Sexuality." *Critical Inquiry* 26 no. 4 (Summer 2000): 641–56.

———. *Homos*. Cambridge, MA: Harvard University Press, 1995.

Bersani, Leo, and Adam Phillips. *Intimacies*. Chicago: University of Chicago Press, 2008.

Bertram, Benjamin. "Falstaff's Body, the Body Politic, and the Body of Trade." *Exemplaria* 21 no. 2 (2009): 296–318.

Bevington, David, Martin Butler, and Ian Donaldson, eds. *The Cambridge Edition of the Works of Ben Jonson*. 7 vols. Cambridge: Cambridge University Press, 2012.

Bibliotheca Calviniana. Eds. Rodolphe Peter and Jean-François Gilmont. 3 vols. Geneva, Switzerland: Librairie Droz, 1991–2000.

Bilson, Thomas. *The true difference betweene Christian subjection and unchristian rebellion wherein the princes lawfull power to commaund for trueth, and indepriuable right to beare the sword are defended against the Popes censures and the Iesuits sophismes*. London, 1585.

Blackstone, Sir William. *Commentaries on the Laws of England*. 4 vols. Oxford: Clarendon Press, 1765–69.

Blair, Rhonda. *The Actor, Image, and Acting: Acting and Cognitive Neuroscience*. New York: Routledge, 2007.

Bloom, Gina. "Manly Drunkenness: Binge Drinking as Disciplined Play." In *Masculinity and the Metropolis of Vice, 1550–1650*, ed. Amanda Bailey and Roze Hentschell, 21–44. Basingstoke: Palgrave Macmillan, 2010.

Bloom, Harold, ed. *Henry IV, Part 2*. New York: Chelsea House Publishers, 1987.

———. *Shakespeare: The Invention of the Human*. New York: Penguin Books, 1998.

Bloomfield, Morton W. *The Seven Deadly Sins: An Introduction to the History of a Religious Concept, with Special Reference to Medieval English Literature*. East Lansing: Michigan State College Press, 1952.

Bogg, Alan, and Jonathan Herring. "Addiction and Responsibility." In *Intoxication and Society: Problematic Pleasures of Drugs and Alcohol*, ed. Jonathan Herring, Ciaran Regan, Darin Weinberg, and Phil Withington, 247–67. Basingstoke: Palgrave, 2013.

Boose, Lynda E. "Othello's Handkerchief: 'The Recognizance and Pledge of Love." *English Literary Renaissance* 5 (1975): 360–74.

Boothroyd, Dave. *Culture on Drugs: Narco-Cultural Studies of High Modernity*. Manchester: Manchester University Press, 2006.

Bowers, Fredson. *The Dramatic Works of Thomas Dekker*. 4 vols. Cambridge: Cambridge University Press, 1953–61.

Bowers, John M. "'Dronkenesse is Ful of Stryvyng:' Alcoholism and Ritual Violence in Chaucer's *Pardoner's Tale*." *ELH* 57 (1990): 757–84.

Boys, John. *The autumne part from the twelfth Sunday*. London, 1613.

Brathwaite, Richard [Blasius Multibibus, pseud.]. *A Solemne Joviall Disputation, theoreticke and praticke; briefly shadowing the law of drinking*. London: At the signe of the Red Eyes, E. Griffin, 1617.

Bray, Alan. *The Friend*. Chicago: University of Chicago Press, 2003.

———. "Homosexuality and the Signs of Male Friendship in Elizabethan England." *History Workshop Journal* 29 (1990): 1–19.

Bray, Gerald, ed. *The Anglican Canons, 1529–1947*. Ecclesiastical Law Society. London: Boydell and Brewer, 1998.

Brewer, John, and Roy Porter, eds. *Consumption and the World of Goods*. London: Routledge, 1993.

Bright, Timothy. *A treatise of melancholie*. London, 1586.

Bristol, Michael. *Carnival and Theater: Plebian Culture and the Structure of Authority in Renaissance England*. New York: Methuen, 1985.

Britland, Karen. "Circe's Cup: Wine and Women in Early Modern Drama." In *A Pleasing Sinne: Drink and Conviviality in Seventeenth-Century England*, ed. Adam Smyth, 109–26. London: D. S. Brewer, 2004.

Brodie, Janet Ferrell, and Marc Redfield, eds. *High Anxieties: Cultural Studies in Addiction*. Berkeley: University of California Press, 2002.

Brome, Alexander. *Songs and Other Poems*. London: Henry Brome, 1664.

Brown, Cedric C. "Drink as a Social Marker in Seventeenth-Century England." In *A Pleasing Sinne: Drink and Conviviality in Seventeenth-Century England*, ed. Adam Smyth, 3–20. London: D. S. Brewer, 2004.

Brown, Georgia, ed. *Thomas Nashe*. Farnham, Surrey: Ashgate, 2011.

Brown, James. "Alehouse Licensing and State Formation in Early Modern England." In *Intoxication and Society: Problematic Pleasures of Drugs and Alcohol*, ed. Jonathan Herring, Ciaran Regan, Darin Weinberg, and Phil Withington, 110–32. Basingstoke: Palgrave, 2013.

Bryson, Anna. *From Courtesy to Civility: Changing Codes of Conduct in Early Modern England.* Oxford: Clarendon Press, 1998.

Buchwald, Henry, and Mary E. Knatterud, "Morbid Obesity: Perceptions of Character and Comorbidities in Falstaff." *Obesity Surgery* 10 no. 5 (2000): 402–8.

Budra, Paul. "*Doctor Faustus:* Death of a Bibliophile." *Connotations* 1 no. 1 (1991): 1–11.

Bullokar, John. *An English Expositor: teaching the interpretation of the hardest words in our language*. London: John Legatt, 1616.

Bulman, James C. Introduction to *King Henry IV, Part II*. Ed. James C. Bulman, 1–148. Arden Shakespeare, 3rd series. London: Bloomsbury Arden Shakespeare, 2016.

Burton, Jonathan. "'A Most Wily Bird': Leo Africanus, *Othello* and the Trafficking in Difference." In *Post-Colonial Shakespeare*, ed. Ania Loomba and Martin Orkin, 43–63. New York: Routledge, 1998.

Burton, Robert. *Anatomy of Melancholy.* London, 1652.

Cain, Tom. "Robert Herrick." In *Oxford Dictionary of National Biography*. Oxford: Oxford University Press, 2004.

Cain, Tom, and Ruth Connolly, eds. *The Complete Poetry of Robert Herrick*. Vol. 1 Oxford: Oxford University Press, 2013.

Calendar of State Papers, Domestic: James 1, 1603–1610. Ed. Mary Anne Everett Green. Vol. 20. London: HMSO, 1857.

Calendar of State Papers, Foreign: Elizabeth I. Vol. 21, Part 2, *June 1586–March 1587*. Ed. Sophie Crawford Lomas and Allen B Hinds. London: HMSO, 1927.

Calvin, Jean. *Commentaires de Jehan Calvin sur le livre des Pseaumes*. Paris: Librairie de Ch. Meyreuis et compagnie, 1859.

———. *Harmonia ex Evangelistis tribus composite, Matthaeo, Marco, et Luca*. Geneva, 1582.

———. *A harmonie upon the three Evangelists, Matthew, Mark and Luke with the commentarie of M. Iohn Caluine. Faithfully translated out of Latine into English, by E. P. Whereunto is also added a commentarie upon the Evangelist S. Iohn, by the same author.* London, 1584.

———. *The Holy Gospel of Jesus Christ, according to John, with the Commentary of M. John Calvin. Faithfully translated out of Latine into English by Christopher Fetherstone.* London, 1584.

———. *Librum Psalmorum, Joannis Calvini Commentarius*. Geneva: Nicolaus Barbirius et Thomas Courteau, 1564.

———. *The Psalms of David and others, with M. John Calvin's Commentaries*. Trans. Arthur Golding. London, 1571.

———. *Sermons de M. Jean Calvin sur le livre de Job. Recueillis fidelement de sa bouche selon qu'il les preschoit.* Geneva: Jean I de Laon, 1563.

———. *Sermons of Master John Calvin, upon the Booke of Job*. Trans. Arthur Golding. London, 1574.

Camden, William. *Annales, the true and royal history of the famous empresse Elizabeth, Queene of England, France, and Ireland.* Trans. Abraham Darcie. London, 1625.

Campbell, Nancy. *Discovering Addiction: The Science and Politics of Substance Abuse Research.* Ann Arbor: University of Michigan Press, 2007.

Capp, Bernard. *England's Culture Wars: Puritan Reformation and Its Enemies in the Interregnum, 1649–1660.* Oxford: Oxford University Press, 2012.

———. "Gender and the Culture of the English Alehouse in Late Stuart England." In *The Trouble with Ribs: Women, Men and Gender in Early Modern Europe*, ed. Anu Korhonen and Kate Lowe, 103–27. Helsinki: Helsinki Collegium for Advance Studies, 2007.

———. Review of *Meanings of Manhood in Early Modern England*, by Alexandra Shepard. *Reviews in History* (January 2004) review no. 380. http://www.history.ac.uk/reviews/review/380.

Cefalu, Paul. "The Burdens of Mind Reading in Shakespeare's *Othello:* A Cognitive and Psychoanalytic Approach to Iago's Theory of Mind." *Shakespeare Quarterly* 64 no. 3 (Fall 2013): 265–94.

Chapman, George, Ben Jonson, and John Marston, *Eastward Ho*. Ed. R. W. Van Fossen, The Revels Plays. Manchester: Manchester University Press, 2006.

Charke, William. *A treatise against the Defense of the censure, given upon the bookes of W. Charke and Meredith Hanmer, by an unknowne popish traytor in maintenance of the seditious challenge of Edmond Campion*. London, 1586.

Charland, Louis C. "Decision-Making Capacity and Responsibility in Addiction." In *Addiction and Responsibility*, ed. Jeffrey Poland and George Graham, 139–58. Cambridge, MA: MIT Press, 2011.

Chaucer, Geoffrey. *The Canterbury Tales*. Ed. F. N. Robinson. Oxford: Oxford University Press, 2008.

Chedgzoy, Kate. "Marlowe's Men and Women: Gender and Sexuality." In *The Cambridge Companion to Christopher Marlowe*, ed. Patrick Cheney, 245–61. Cambridge: Cambridge University Press, 2004.

Choyce Drollery. London, 1656.

Cicero. *The familiar epistles of M. T. Cicero Englished and conferred with the French, Italian and other translations*. London: Edward Griffin, 1620.

———. *Letters to His Friends*. Vol. 1. Trans. W. Glynn Williams. Loeb Classical Library. London: Heinemann, 1927.

Clark, David. "Marlowe and Queer Theory." In *Christopher Marlowe in Context*, ed. Emily C. Bartels and Emma Smith, 232–41. Cambridge: Cambridge University Press, 2013.

Clark, Peter. *The Alehouse: A Social History*. London: Longman, 1983.

Coiro, Anne Baynes. *Robert Herrick's Hesperides and the Epigram Book Tradition*. Baltimore, MD: Johns Hopkins University Press, 1988.

Coke, Sir Edward. *Institutes of the Laws of England*. Book 1. London, 1628.

———. *The Reports of Sir Edward Coke* (1603). London: J. Butterworth and Son, 1826.

Connolly, Ruth, and Tom Cain, eds. *"Lords of Wine and Oile": Community and Conviviality in the Poetry of Robert Herrick*. Oxford: Oxford University Press, 2011.

Cook, Christopher C. H. *Alcohol, Addiction, and Christian Ethics*. Cambridge: Cambridge University Press, 2006.

Cook, John. "*Greene's Tu-Quoque*" In *A Select Collection of Old Plays*, ed. Robert Dodsley, Isaac Reed, and Octavius Gilchrist. Vol. 7. London: Septimus Prowett, 1825.

Cooper, Thomas. *The Churches Deliverance*. London, 1609.

———. *Thesaurus linguae Romanae & Britannicae*. London, 1584.

Corns, Thomas. *Uncloistered Virtue: English Political Literature, 1640–1660*. Oxford: Clarendon Press, 1992.

Cotgrave, John. *Wits Interpreter*. London, 1655.

Cotgrave, Randle. *A Dictionary of the French and English Tongues*. London: A. Islip, 1611.

Cotton, Charles. *Poems on Several Occasions*. London, 1689.

Courtwright, David. *Forces of Habit: Drugs and the Making of the Modern World*. Cambridge, MA: Harvard University Press, 2001.

Crompton, Hugh. *Bring a fardle of fancies, or a medley of musick, stewed in four ounces of the oyl of epigrams*. London: E. C., 1657.

———. *Pierides, or the Muses Mount*. London, J.G. for Charles Webb, 1658.

Crosse, Henry. *Vertues common-wealth: or The high-way to honour*. London, 1603.

Culpeper, Nicolas. *The English Physitian*. London, 1652.

———. *The Essential Writings of Nicolas Culpeper*. New York: Kissinger Publishing, 2005.

Cunningham, Timothy, comp. *The Journals of the House of Commons*. Vol. 1, *1547–1628*. London: L. Hansard and Son, 1803.

Curth, Louise Hill, and Tanya M. Cassidy. "'Heath, Strength, and Happiness': Medical Constructions of Wine and Beer in Early Modern England." In *A Pleasing Sinne: Drink and Conviviality in Seventeenth-Century England*, ed. Adam Smyth, 143–60. London: D. S. Brewer, 2004.

Danaeus, Lambertus. *A dialogue of witches, in foretime named lot-tellers, and now commonly called sorcerers . . . Written in Latin by Lambertus Danaeus. And now translated into English*. London, 1575.

Daniel, Drew. *The Melancholy Assemblage: Affect and Epistemology in the English Renaissance*. New York: Fordham University Press, 2013.

Davies, Michael. "Falstaff's Lateness: Calvinism and the Protestant Hero in *Henry IV*." *Review of English Studies* 56 no. 225 (2005): 351–78.

Davis, David Brion. *The Problem of Slavery in Western Culture*. Ithaca, NY: Cornell University Press, 1966.

Davis, Paul. "Maximum Parsimony." *Times Literary Supplement*, January 16, 2015, 3–5.

Davison, Francis. *Poetical Rhapsody*. London, 1602.

Dean, Tim. "Sex and the Aesthetics of Existence." *PMLA* 125 no. 2 (2010): 387–92.

———. *Unlimited Intimacy: Reflections on the Subculture of Barebacking*. Chicago: University of Chicago Press, 2009.

De Groot, Jerome. *Royalist Identities*. Basingstoke, Hampshire: Palgrave Macmillan, 2004.

Dekker, Thomas. *The belman of London. Bringing to light the most notorious villanies that are now practised in the kingdome*. London: Nathaniell Butter, 1608.

Dekker, Thomas, and Thomas Middleton. *Honest Whore, Part 1*. In *The Dramatic Works of Thomas Dekker*. Ed. Fredson Bowers. Vol. 2. Cambridge: Cambridge University Press, 1955.

Derrida, Jacques. "The Rhetoric of Drugs." In *High Culture: Reflections on Addiction and Modernity*, ed. Anna Alexander and Mark S. Roberts, 19–44. Buffalo: State University of New York Press, 2003.

Dickey, Stephen. "Shakespeare's Mastiff Comedy." *Shakespeare Quarterly* 42 no. 3 (Fall 1991): 255–75.

DiGangi, Mario. *The Homoerotics of Early Modern Drama*. Cambridge: Cambridge University Press, 1997.

Downame, John. *Foure treatises tending to disswade all Christians from foure no lesse hainous than common sinnes, namely the abuses of swearing, drunkenesse, whoredome, and briberie*. London, 1609.

Dubinski, Roman R., ed. *Alexander Brome: Poems*. 2 vols. Toronto: University of Toronto, 1982.

Dubrow, Heather. *Echoes of Desire: English Petrarchism and its Counterdiscourses*. Ithaca: Cornell University Press, 1995.

D'Urfey, Thomas. *Wit and Mirth: Pills to Purge Melancholy*. 6 vols. London: W. Pearson, 1719–20.

Earnshaw, Steven. *The Pub in Literature: England's Altered State*. Manchester: Manchester University Press, 2000.

Edgeworth, Roger. *Sermons very fruitfull, godly, and learned, preached and sette foorth by Maister Roger Edgeworth*. London, 1557.

Edwards, Griffith, ed. *Addiction: Evolution of a Specialist Field*. Oxford: Blackwell Science, Addiction Press, 2002.

Elam, Keir. Introduction. *Twelfth Night, Or What You Will*. Ed. Keir Elam, 1–145. Arden Shakespeare, 3rd series London: Bloomsbury Arden Shakespeare, 2008.

Elyot, Thomas. *Of the knowledeg whiche maketh a wise man*. London, 1533.

Emmison, F. G. *Elizabethan Life: Disorder. Mainly from Essex Sessions and Assize Records*. Chelmsford: Essex County Council, 1970.

———. *Elizabethan Life: Morals & the Church Courts*. Chelmsford: Essex County Council, 1973.

Enterline, Lynn. *Tears of Narcissus: Melancholia and Masculinity in Early Modern Writing*. Stanford, CA: Stanford University Press, 1995.

Erasmus, Desiderius. "The paraphrase of Erasmus upon the Epistle of S. Paule to Titus." In *The seconde tome or volume of the Paraphrase of Erasmus upon the Newe Testament conteynyng the epistles of S. Paul, and other the Apostles*. London, 1549.

Everett, Barbara. "The Fatness of Falstaff: Shakespeare and Character." *Proceedings of the British Academy* 76 (1990): 109–28.

Evett, David. *Discourses of Service in Shakespeare's England*. New York: Palgrave, 2005.

Findlay, Alison. "Theatres of Truth: Drinking and Drama in Early Modern England." In *A Babel of Bottles: Drinkers and Drinking Places in Literature*, ed. James Nicholls and Susan Owen, 21–39. Sheffield: Sheffield University Press, 2000.

Findlay, Heather. "Renaissance Pederasty and Pedagogy: The 'Case' of Shakespeare's Falstaff." *Yale Journal of Criticism* 3 no. 1 (Fall 1989): 229–39.

Fingarette, Herbert. *Heavy Drinking: The Myth of Alcoholism as a Disease*. Berkeley: University of California Press, 1988.

Fisher, Joshua B. "Digesting Falstaff: Food and Nation in Shakespeare's *Henry IV* plays." *Early English Studies* 2 (2009): 1–23.

Fitzpatrick, Joan. *Food in Shakespeare: Early Modern Dietaries and the Plays*. Farnham, Surrey: Ashgate, 2007.

Fleck, Andrew. "Marking Difference and National Identity in Dekker's *The Shoemaker's Holiday*." *SEL* 46 no. 2 (2006): 349–70.

Flemming, Abraham, trans. *A panoplie of epistles, or, a looking glasse for the unlearned. Gathered and translated out of Latine into English*. London, 1576.

Florilegus, Junius [pseud.]. *The odious, despicable, and dreadfull condition of a drunkard*. London: R. Cotes, 1649.

Florio, John. *A Worlde of Wordes: or, Most copious, and exact Dictionarie in Italian and English*. London, 1598.

Floyd-Wilson, Mary. *English Ethnicity and Race in Early Modern Drama*. Cambridge: Cambridge University Press, 2003.

Foxe, John. *Actes and monuments of matters most speciall and memorable, happenyng in the Church with an vniuersall history of the same. . . . Newly revised and recognised, partly also augmented, and now the fourth time agayne published*. 2 vols. London, 1583.

Freccero, Carla. "Tangents (of Desire)." *Journal for Early Modern Cultural Studies*. 16, no. 2 (Spring 2016): 91–105.

Freeman, Thomas S., and Elizabeth Evenden. *Religion and the Book in Early Modern England: The Making of John Foxe's "Book of Martyrs."* Cambridge: Cambridge University Press, 2011.

French, Richard Valpy. *The History of Toasting or Drinking of Healths in England*. London: National Temperance Publication Depot, 1881.

Fulbecke, William. *A Booke of Christian Ethicks or Moral Philosophie: Containing The true difference and opposition, of the two incompatible qualities, Vertue, and Voluptuousnesse*. London: Richard Jones, 1587.

Fumerton, Patricia. "Not Home: Alehouses, Ballads, and the Vagrant Husband in Early Modern England." *Journal of Medieval and Early Modern Studies* 32 no. 3 (Fall 2002): 493–518.

Galen. *On Diseases and Symptoms*. Trans. Ian Johnston. Cambridge: Cambridge University Press, 2006.

Garber, Marjorie. *Shakespeare After All*. New York: Knopf Doubleday, 2008.

[Gardiner, Edmund]. E. G., *The Triall of Tabacco. Wherein, his worth is most worthily expressed: as, in the name, nature, and qualitie of the sayd hearb; his speciall use in all Physicke*. London, 1610.

Gascoigne, George. *A Delicate Diet for daintie mouthde Droonkardes*. London, 1576.

———. *Fruites of Warres*. London, 1572.

———. *The Siege of Antwerp*. London, 1576.

Gill, Roma, ed. Introduction to *Doctor Faustus*. New Mermaid edition. London: Ernest Benn, 1965.

Gibbons, Brian. *Jacobean City Comedy: A Study of Satiric Plays by Jonson, Marston and Middleton*. Cambridge, MA: Harvard University Press, 1968.

Gifford, Humphrey. *A posie of gilloflowers*. London, 1580.

Goldberg, Jonathan. *Sodometries: Renaissance Texts, Modern Sexualities*. Stanford: Stanford University Press, 1992.

———, ed. *Queering the Renaissance*. Durham, NC: Duke University Press, 1994.

Goldman, Michael. *The Actor's Freedom: Towards a Theory of Freedom*. New York: Penguin Group, 1975.

———. "Marlowe and the Histrionics of Ravishment." In *Two Renaissance Mythmakers: Christopher Marlowe and Ben Jonson*, ed. Alvin Kernan, 22–40. Baltimore: Johns Hopkins University Press, 1977.

Goldstein, David B. *Eating and Ethics in Shakespeare's England*. Cambridge: Cambridge University Press, 2013.

Googe, Barnabe. *The zodiake of life written by the godly and zealous poet Marcellus Pallingenius stellatus, wherein are conteyned twelve bookes disclosing the haynous crymes [and] wicked vices of our corrupt nature: and plainlye declaring the pleasaunt and perfit pathway unto eternall lyfe*. London, 1565.

Gordon, F. Bruce. *Calvin*. New Haven, CT: Yale University Press, 2009.

Gouge, William. *Of Domesticall Duties Eight Treatises*. London, 1622.

Grady, Hugh. "Falstaff: Subjectivity between the Carnival and the Aesthetic." Modern Language Review 96 no. 3 (2001 July): 609–23.

———. *Shakespeare, Machiavelli, and Montaigne: Power and Subjectivity from Richard II to Hamlet*. Oxford: Oxford University Press, 2002.

Grady, Kyle. "Othello, Colin Powell, and Post-Racial Anachronisms." *Shakespeare Quarterly* 67 no. 1 (2016): 68–83.

Gratarolo, Guglielmo. *A direction for the health of Magistrates and Studentes . . . Written in Latin by Guilielmus Gratarolus, and Englished, by T. N.* London, 1574.

Grazia, Margreta de. "Teleology, Delay, and the 'Old Mole.'" *Shakespeare Quarterly* 50 no. 3 (1999): 251–67.

Greene, Robert. *Orpharion*. London, 1589.

Greene, Roland. *Five Words: Critical Semantics in the Age of Shakespeare and Cervantes*. Chicago: University of Chicago Press, 2013.

Grislis, Egil. "Menno Simons on Conversion: Compared with Martin Luther and Jean Calvin." *Journal of Mennonite Studies* 11 (1993): 55–75.

Guenther, Genevieve. "Why Devils Came When Faustus Called Them." *Modern Philology* 109 no. 1 (2011): 46–70.

Gunton, Timothy. *An extemporary answer to a cluster of drunkards, met together at Schiedam: made by Timothy Gunton, who was compelled thereto, upon his refusall to drink the kings health*. London, 1648.

Gwalther, Rudolf. *An hundred, threescore and fiftene homelyes or sermons, vppon the Actes of the Apostles, written by Saint Luke. Made by Radulpe Gualthere Tigurine, and translated out of Latine into our tongue, for the commoditie of the Englishe reader.* London: Henrie Denham, 1572.

Hailwood, Mark. *Alehouses and Good Fellowship in Early Modern England*. London: Boydell and Brewer, 2014.

Hale, Sir Matthew. *Historia placitorum coronae: The History of the Pleas of the Crown*. Vol. 1. Philadelphia: R. H. Small, 1847.

Halpern, Richard. "Marlowe's Theater of Night: *Doctor Faustus* and Capital." *ELH* 71 no. 2 (2004): 455–95.

Hamlin, William M. "Casting Doubt in Marlowe's 'Doctor Faustus.'" *Studies in English Literature, 1500–1900* 41 no. 2 (Spring 2001): 257–75.

Hance, E. M. and T. N. Norton, eds. *Transactions of the Historical Society of Lancashire and Cheshire*. Vol. 36. Liverpool: Adam Holden, 1887.

[Hancock, John]. *The Touchstone, or Trial of Tobacco*. London, 1676.

Hardacre, Paul H. *The Royalists During the Puritan Revolution*. The Hague: Martinus Nijhoff, 1956.

Harris, Robert. *Drunkard's Cup*. London, 1619.

Harrison, Brian. *Drink and the Victorians*. London: Faber and Faber, 1971.

Hartley, T. E., ed. *Proceedings in the Parliaments of Elizabeth 1*. Vol. 3, *1593–1601*. London: Leicester University Press, 1995.

Hasler, P. W., ed. *History of Parliament: The House of Commons, 1558–1603*. London: Secker and Warburg, 1981.

Hattaway, Michael. "The Theology of Marlowe's *Doctor Faustus*." *Renaissance Drama*, new series, 3 (1970): 51–78.

Healy, Margaret. *Fictions of Disease in Early Modern England: Bodies, Plague, and Politics*. Basingstoke: Palgrave, 2001.

Heath, Robert. *Clarastella; together with Poems Occasional, Elegies, Epigrams, Satyrs*. London: Humphrey Moseley, 1650.

Helms, N. R. "Conceiving Ambiguity: Dynamic Mindreading in Shakespeare's *Twelfth Night*." *Philosophy and Literature* 36 no. 1 (2012): 122–35.

Helmholz, R. H., and J. H. Baker, eds. *The Oxford History of the Laws of England*. Oxford: Oxford University Press, 2004.

Henry VIII. *A glasse of the truthe*. London, 1532.

Herman, Peter C. "Leaky Ladies and Droopy Dames: The Grotesque Realism of Skelton's The Tunnynge of Elynour Rummynge." In *Rethinking the Henrician Era: Essays on Early Tudor Texts and Contexts*, ed. Peter C. Herman, 145–67. Chicago: University of Illinois Press, 1993.

Herring, Jonathan, Ciaran Regan, Darin Weinberg, and Phil Withington, eds. *Intoxication and Society: Problematic Pleasures of Drugs and Alcohol*. Basingstoke: Palgrave Macmillan, 2013.

———. "Starting the Conversation." In *Intoxication and Society: Problematic Pleasures of Drugs and Alcohol*, ed. Jonathan Herring, Ciaran Regan, Darin Weinberg, and Phil Withington, 1–32. Basingstoke: Palgrave Macmillan, 2013.

Heywood, Thomas. *Philocothonista or the Drunkard, Opened, Dissected, and Anatomized*. London: Robert Raworth, 1635.

Hickman, Timothy A. "Target America: Visual Culture, Neuroimaging, and the 'Hijacked Brain' Theory of Addiction." *Past and Present* 222 no. S9 (2014): 207–26.

Highley, Christopher, and John N. King, eds. *John Foxe and His World*. Aldershot: Ashgate, 2002.

Hoenselaars, A. J. *Images of Englishmen and Foreigners in the Drama of Shakespeare and His Contemporaries*. Rutherford, NJ: Fairleigh Dickinson University Press, 1992.

Holinshed, Raphael. *The firste volume of the chronicles of England, Scotlande, and Irelande*. London, 1577.

The holie Bible. Faithfully translated into English, out of the authentical Latin. By the English College of Doway. Douai, 1609–10.

Honderich, Pauline. "John Calvin and *Doctor Faustus*." *Modern Language Review* 68 no. 1 (1973): 1–13.

Hooker, Richard. *Laws of Ecclesiastical Polity: Preface, Book I, Book VIII*. Ed. Arthur Stephen McGrade. Cambridge: Cambridge University Press, 1989.

Horder, Jeremy. "Pleading Involuntary Lack of Capacity." *Cambridge Law Journal* 52 no. 2 (July 1993): 298–319.

Hornby, William. *The Scourge of Drunkennes*. London, 1618.

Hoskins, John. *Sermons preached at Pauls Crosse*. London, 1615.

Howard, Jean E., and Phyllis Rackin. *Engendering a Nation: A Feminist Account of Shakespeare's English Histories*. New York: Routledge, 1997.

Huarte, John. *The examination of mens wits*. London, 1594.

Hunt, Maurice. "Malvolio, Viola, and the Question of Instrumentality: Defining Providence in *Twelfth Night*." *Studies in Philology* 90 no. 3 (Summer 1993): 277–97.

Hunter, Judith. "English Inns, Taverns, Alehouses and Brandy Shops: The Legislative Framework, 1495–1797." In *The World of the Tavern: Public Houses in Early Modern Europe*, ed. Beat Kümin and B. Ann Tlusty, 65–82. Aldershot: Ashgate, 2002.

Hutson, Lorna. "On Not Being Deceived: Rhetoric and the Body in *Twelfth Night*." *Texas Studies in Literature and Language* 38 no. 2 (Summer 1996): 140–74.

Hutton, Ronald. *The Rise and Fall of Merry England: The Ritual Year 1400–1700*. Oxford: Oxford University Press, 2005.

Hyman, Steven E., Robert C. Malenka, and Eric J. Nestler, "Neural Mechanisms of Addiction:

The Role of Reward-Related Learning and Memory." *Annual Review of Neuroscience* 29 (2006): 565–98.

Ingram, Martin. "Reformation of Manners in Early Modern England." In *The Experience of Authority in Early Modern England*, ed. Paul Griffiths, Adam Fox, and Steve Hindle, 47–88. London: Macmillan Press, 1996.

Isbell, H., and W. M. White, "Clinical Characteristics of Addictions." *American Journal of Medicine* 14 (1953): 558–65.

Isler, Alan D. "Falstaff's Heroic Sherris." *Shakespeare Quarterly* 22 no. 2 (Spring 1971): 186–88.

I. T. *The haven of pleasure containing a freemans felicitie, and a true direction how to liue well.* London, 1597.

James, Heather. "Dido's Ear: Tragedy and the Politics of Response." *Shakespeare Quarterly* 52 (2001): 360–82.

Jameson, Storm. *The Decline of Merry England.* London: Cassell and Company, 1930.

Jeanneret, Michael. *A Feast of Words: Banquets and Table Talk in the Renaissance.* Trans. Jeremy Whiteley and Emma Hughes. Cambridge: Polity Press, 1991.

———. *Des Mets et des mots: banquets et propos de table à la Renaissance.* Paris: Librairie José Corti, 1987.

Jellinek, E. Morton. *The Disease Concept of Alcoholism.* New Haven, CT: Hillhouse Press, 1960.

Jessey, Henry. *The Lords loud call to England: being a true relation of some late, various, and wonderful judgments, or handy-works of God . . . As also of the odious sin of drinking healths.* London, 1660.

Jewel, John. *The second tome of homilees of such matters as were promised, and intituled in the former part of homilees. Set out by the aucthoritie of the Queenes Maiestie: and to be read in euery parishe church agreeably.* London, 1571.

Jones, Angela McShane. "Roaring Royalists and Ranting Brewers: The Politicization of Drink and Drunkenness in Political Broadside Ballads from 1640 to 1689." In *A Pleasing Sinne: Drink and Conviviality in Seventeenth-Century England*, ed. Adam Smyth, 69–88. London: D. S. Brewer, 2004.

Jones, John Henry, ed. *The English Faust Book.* Cambridge: Cambridge University Press, 1994.

Jonson, Ben. *Bartholomew Fair.* In *The Selected Plays of Ben Jonson*, vol. 2, ed. Martin Butler, 147–298. Cambridge: Cambridge University Press, 1989.

———. *Epicene, or The Silent Woman.* Ed. Richard Dutton. The Revels Plays. Manchester: Manchester University Press, 2003.

———. *Every Man Out of His Humor.* Ed. Helen Ostovich. The Revels Plays. Manchester: Manchester University Press, 2001.

The Journals of all the Parliaments of Queen Elizabeth, both of the House of Lords and House of Commons. Comp. by Sir Simonds D'Ewes of Stow-Hall in the County of Suffolk. Revised and published by Paul Bowes, of the Middle-Temple, London. London: John Starkey, 1682.

Journal of the House of Commons. Vol. 1, *1547–1629.* London, 1802.

Joye, George. *An apolgye.* London, 1535.

———. *The prophete Isaye, translated into englysshe.* London: 1531.

———. *The Psalter of Dauid in Englyshe.* London: Thomas Godfray, 1534.

———. *The Unitie and Scisme of the Olde Chirche.* Antwerp, 1543.

Kahn, Coppélia. *Man's Estate: Masculine Identity in Shakespeare.* Berkeley: University of California Press, 1981.

Kahn, Victoria. "'The Duty to Love': Passion and Obligation in Early Modern Political Theory." *Representations* 68 (Autumn 1999): 84–107.

Kalivas, Peter W., and Nora D. Voldow. "The Neural Basis of Addiction: A Pathology of Motivation and Choice." *American Journal of Psychiatry* 162 no. 8 (August 2005): 1403–13.

Kastan, David Scott. Introduction to *King Henry IV, Part 1*. Ed. David Scott Kastan, 1–131. Arden Shakespeare, 3rd series. London: Thomson Learning, 2002.

Keblusek, Marika. "Wine for Comfort: Drinking and the Royalist Exile Experience, 1642–1660." In *A Pleasing Sinne: Drink and Conviviality in Seventeenth-Century England*, ed. Adam Smyth, 55–68. London: D. S. Brewer, 2004.

Kendall, R. T. *Calvin and English Calvinism to 1649*. Oxford: Oxford University Press, 1979.

Kennedy, William J. *The Site of Petrarchism: Early Modern National Sentiment in Italy, France, and England.* Baltimore, MD: Johns Hopkins University Press, 2003.

Kezar, Dennis. "Shakespeare's Addictions." *Critical Inquiry* 30 no. 1 (2003): 31–62.

King, John N. *Foxe's "Book of Martyrs" and Early Modern Print Culture*. Cambridge: Cambridge University Press, 2006.

Kinney, Arthur. "Shakespeare's Falstaff as Parody." *Connotations* 12 no. 2–3 (2002–3): 105–25.

Kitzes, Adam. *The Politics of Melancholy from Spenser to Milton*. London: Routledge, 2006.

Klausner David N., ed. *The Castle of Perseverance*. Kalamazoo, MI: Medieval Institute Publications, 2010.

Knecht, Ross. "The Grammar of Passion: Language and Affect in Early Modern Literature." Ph.D. diss., New York University, 2011.

———. "'Invaded by the World': Passion, Passivity, and the Object of Desire in Petrarch's *Rime sparse*." *Comparative Literature* 63 no. 3 (2011): 235–52.

Knott, John R. *Discourses of Martyrdom in English Literature, 1563–1694*. Cambridge: Cambridge University Press, 1993.

Kümin, Beat. *Drinking Matters: Public Houses and Social Exchange in Early Modern Central Europe*. Basingstoke: Palgrave Macmillan, 2007.

Kümin, Beat, and B. Ann Tlusty, eds. *The World of the Tavern: Public Houses in Early Modern Europe*. Farnham, Surrey: Ashgate, 2002.

Kushner, Howard. "Taking Biology Seriously: The Next Task for Historians of Addiction?," *Bulletin of the History of Medicine* 80, no. 1 (Spring 2006): 115–43.

Lake, Peter. *Anglicans and Puritans? Presbyterianism and English Conformist Thought from Whitgift to Hooker*. London: Unwin Hyman, 1988.

———. "Calvinism and the English Church, 1570–1635." *Past and Present* 114 (1987): 32–76.

———. "Defining Puritanism—Again?" *Puritanism: Transatlantic Perspectives on a Seventeenth-Century Anglo-American Faith*, ed. Francis J. Bremer, 3–29. Boston: Massachusetts Historical Society, 1993.

Lake, Peter, with Michael C. Questier, *The Anti-Christ's Lewd Hat: Protestants, Papists and Players in Post-Reformation England.* New Haven, CT: Yale University Press, 2002.

Laroque, François. "Shakespeare's 'Battle of Carnival and Lent.'" In *Shakespeare and Carnival*, ed. Ronald Knowles, 83–96. New York: Macmillan, 1998.

———. *Shakespeare's Festive World: Elizabethan Seasonal Entertainment and the Professional Stage.* Trans. Janet Lloyd. Cambridge: Cambridge University Press, 1991.

Lemnius, Levinus. *The sanctuarie of saluation, helmet of health, and mirrour of modestie and good maners*. London, 1592.

Lemon, Rebecca. "Compulsory Conviviality in Early Modern England." *English Literary Renaissance* 43 no. 3 (September 2013): 381–414.

———. "Incapacitated Will." In *Staged Transgression: Performing Disorder in Early Modern England*, ed. Rory Loughnane and Edel Semple, 170–92. New York: Palgrave Press, 2013.

———. "Sacking Falstaff." In *Culinary Shakespeare*, ed. David B. Goldstein and Amy L. Tigner, 113–34. Pittsburgh: Duquesne University Press, 2016.

———. "Scholarly Addiction: *Doctor Faustus* and the Drama of Devotion." *Renaissance Quarterly* 69 no. 3 (2016): 865–98.

———. *Treason by Words: Literature, Law, and Rebellion in Shakespeare's England*. Ithaca, NY: Cornell University Press, 2006.

Lessius, Leonardus. *Hygiasticon: or, the right course of preserving Life and Health unto extreme old age: together with soundnesse and integritie of the senses, judgement, and memorie. Written in Latin by Leonard Lessius, and now done into English*. Cambridge: Roger Daniel, 1634.

Levenson, Jill L. "Shakespeare's Falstaff: 'The Cause that Wit Is in Other Men.'" *University of Toronto Quarterly* 74 no. 2 (2005): 722–28.

Levine, Harry G. "The Discovery of Addiction: Changing Concepts of Habitual Drunkenness in America." *Journal of Studies on Alcohol* 39 no. 1 (1978): 143–74.

Levy-Navarro, Elena. *The Culture of Obesity in Early and Late Modernity*. Basingstoke: Palgrave, 2008.

Little, Arthur L. "Re-Historicizing Race, White Melancholia, and the Shakespearean Property." *Shakespeare Quarterly* 67 no. 1 (2016): 84–103.

Lodge, Thomas. *Wits Misery, and the Worlds Madnesse*. London, 1596.

A looking glasse for drunkards: or, The hunting of drunkennesse Wherein drunkards are vnmasked to the view of the world. Very conuenient and vsefull for all people to ruminate on in this drunken age. London: M. Flesher for F. C[oules], 1627.

Loughnan, Arlie. "Mental Incapacity Doctrines in Criminal Law." *New Criminal Law Review* 15 no. 1 (Winter 2012): 1–31.

Lowes, John Livingston. "Chaucer and the Seven Deadly Sins." *PMLA* 30, no. 2 (1915): 237–371.

Loxley, James. *Royalism and Poetry in the English Civil Wars: The Drawn Sword*. Houndmills, Basingstoke: Macmillan Press, 1997.

Ludington, Charles. *The Politics of Wine in Britain: A New Cultural History*. Basingstoke: Palgrave, 2013.

———. "'Sometimes to your country true:' The Politics of Wine in England, 1660–1714." In *A Pleasing Sinne: Drink and Conviviality in Seventeenth-Century England*, ed. Adam Smyth, 89–108. London: D. S. Brewer, 2004.

Lupton, Donald. *The glory of their times: or, The liues of ye primitiue fathers*. London, 1640.

Lupton, Julia Reinhard. *Citizen Saints: Shakespeare and Political Theology*. Chicago: University of Chicago Press, 2005.

MacLeod, Roy M. "The Edge of Hope: Social Policy and Chronic Alcoholism, 1870–1900." *Journal of the History of Medicine and Allied Sciences* 22 (1967): 215–45.

Magennis, Hugh. *Anglo-Saxon Appetites: Food and Drink and Their Consumption in Old English and Related Literature*. Dublin: Four Courts Press, 1999.

Maguire, Laurie. "*Othello,* Theatre Boundaries, and Audience Cognition." In *Othello: The State of Play*, ed. Lena Cowen Orlin, 17–44. London: Bloomsbury Arden Shakespeare, 2014.

Mancall, Peter. *Deadly Medicine: Indians and Alcohol in Early America*. Ithaca, NY: Cornell University Press, 1997.

Marcus, Leah. *The Politics of Mirth: Jonson, Herrick, Milton, Marvell and the Defense of Old Holiday Pastimes*. Chicago: University of Chicago Press, 1986.

———. "Textual Indeterminacy and Ideological Difference: The Case of 'Doctor Faustus.'" *Renaissance Drama*, new series, 20 (1989): 1–29.

Markham, Gervase. *Cauelarice, or The English horseman contayning all the arte of horse-manship, as much as is necessary for any man to vnderstand*. London, 1607.

Marlowe, Christopher. *Doctor Faustus*. Ed. Michael Keefer. Toronto: Broadview Press, 2007.

Martin, A. Lynn. *Alcohol, Sex, and Gender in Late Medieval and Early Modern Europe*. Basingstoke: Palgrave Macmillian, 2001.

———. *Alcohol, Violence, and Disorder in Traditional Europe*. Kirksville, Missouri: Truman State University Press, 2009.

———. "Drinking and Alehouses in the Diary of an English Mercer's Apprentice, 1663–1674." In *Alcohol: A Social and Cultural History*, ed. Mack P. Holt, 93–106. New York: Berg, 2006.

Masten, Jeffrey. *Queer Philologies: Sex, Language, and Affect in Shakespeare's Time*. Philadelphia: University of Pennsylvania Press, 2016.

———. *Textual Intercourse: Collaboration, Authorship, and Sexualities in Renaissance Drama*. Cambridge: Cambridge University Press, 1997.

———. "Toward a Queer Address: The Taste of Letters and Early Modern Male Friendship." *GLQ: A Journal of Gay and Lesbian Studies* 10 no. 3 (2004): 367–84.

Maus, Katharine Eisaman. "Why Read Herrick?" In *'Lords of Wine and Oile': Community and Conviviality in the Poetry of Robert Herrick*, ed. Ruth Connolly and Tom Cain, 25–38. Oxford: Oxford University Press, 2011.

Maynwaringe, Everard. *Vita sana & longa the preservation of health and prolongation of life proposed and proved*. London, 1669.

Mazzotta, Guiseppe. *The Worlds of Petrarch*. Durham, NC: Duke University Press, 1993.

McCord, David. "The English and American history of voluntary intoxication to negate *Mens Rea*." *Journal of Legal History* 11 no. 3 (1990): 372–95.

McGuire, Laurie. "*Othello*, Theatre Boundaries, and Audience Cognition." In *Othello: The State of Play*, ed. Lena Cowen Orlin, 17–44. London: Bloomsbury Arden Shakespeare, 2014.

McHugh, James T. "What Is the Difference between a 'Person' and a 'Human Being' Within the Law." Special issue on public law, *Review of Politics* 54 no. 3, (Summer 1992): 445–61.

McIntosh, Marjorie Keniston. *Controlling Misbehavior in England, 1370–1600*. Cambridge: Cambridge University Press, 1998.

McNeill, John T. *The History and Character of Calvinism*. Oxford: Oxford University Press, 1954.

McShane, Angela. "Material Culture and 'Political Drinking' in Seventeenth-Century England." *Past and Present*, 222 no. S9 (2014): 247–76.

———. *See also* Jones, Angela McShane.

Meads, Chris. *Banquets Set Forth: Banqueting in English Renaissance Drama*. Manchester: Manchester University Press, 2002.

Melanchthon, Philipp. *A famous and godly history contaynyng the lyues a[nd] actes of three renowmed reformers of the Christia[n] Church, Martine Luther, Iohn Ecolampadius, and Huldericke Zuinglius. . . . Newly Englished by Henry Bennet Callesian*. London, 1561.

Merry Drollerie or, a collection of jovial poems, merry songs, witty drolleries. London: J. W., for P. H., 1661.

Miège, Guy. *A New Dictionary French and English, with another English and French*. London, 1677.

Milton, Anthony. *Catholic and Reformed: The Roman and Protestant Churches in English Protestant Thought, 1600–1640*. Cambridge: Cambridge University Press, 1995.

Miner, Earl. *The Cavalier Mode from Jonson to Cotton*. Princeton, NJ: Princeton University Press, 1971.

Mirabella, Bella. "'A Wording Poet': Othello among the Mountebanks." *Medieval and Renaissance Drama in England* 24 (2011): 150–75.

Montaigne, Michel de. *Essays*. Trans. John Florio. London: Printed by Melch. Bradwood for Edward Blount and William Barret, 1613.

———. *Les Essais, Livre III* (1595): *Traduction en français modern du texte d'édition de 1595*. Paris: Guy de Penon, 2009.

More, Thomas. *The Four Last Things* (1522). Ed. D. O'Connor. London: Burns Oates and Washbourne, 1935.

Morris, Rev. R., ed. *Old English Homilies of the Twelfth Century: From the Unique Ms. B.14.52 in the Library of Trinity College, Cambridge*. 2nd series. London: Early English Text Society, 1873.

Morse, Steven J. "Addiction and criminal responsibility." In *Addiction and Responsibility*, ed. Jeffrey Poland and George Graham, 159–200. Cambridge, MA: MIT Press, 2011.

[Morton, Charles or William Prynne]. *The Great Evil of Health-Drinking: or a discourse wherein the original, evil, and mischief of drinking of healths, are discovered and detected; and the practice opposed*. London: Jonathan Robinson, 1684.

Moulton, Ian Frederick. "Fat Knight, or What You Will: Inimitable Falstaff." In *A Companion to Shakespeare's Works*, ed. Richard Dutton and Jean Howard, 223–42. Oxford: Blackwell, 2003.

Muldrew, Craig. *The Economy of Obligation: The Culture of Credit and Social Relations in Early Modern England*. Basingstoke: Palgrave, 1998.

Naqvi, Nasir H., David Rudrauf, Hanna Damasio, and Antoine Bechara. "Damage to the Insula Disrupts Addiction to Cigarette Smoking." *Science*, new series, 315 no. 5811 (January 26, 2007): 531–34.

Nardizzi, Vin. "Grafted to Falstaff and Compounded with Catherine: Mingling Hal in the Second Tetralogy." In *Queer Renaissance Historiography: Backward Gaze*, ed. Vin Nardizzi, Stephen Guy-Bray, and Will Stockton, 149–69. Farnham, Surrey: Ashgate 2009.

Nashe, Thomas. *Pierce Penniless*. London, 1592.

———. *The Unfortunate Traveller and Other Works*. Ed. J. B. Steane. London: Penguin Classics, 1972.

Nedelsky, Jennifer. "Law, Boundaries, and the Bounded Self." *Representations* 30 (Spring 1990): 162–89.

Neil, Michael. *"Servile Ministers": Othello, King Lear and the Sacralization of Service*. Vancouver: Ronsdale Press, 2004.

Nelson, Maggie. *The Argonauts*. Minneapolis: Greywolf Press, 2015.

Nichols, Tom. "Double Vision: The Ambivalent Imagery of Drunkenness in Early Modern Europe." *Past and Present* 222 no. S9 (2014): 146–67.

Nicolls, Philip. *The copie of a letter sente to one maister Chrispyne chanon of Exceter for that he denied ye scripture to be the touche stone or trial of al other doctrines*. London, [1548?].

Nutall, A. D. *The Alternative Trinity: Gnostic Heresy in Marlowe, Milton, and Blake*. Oxford: Clarendon Press, 1998.

Nyquist, Mary. *Arbitrary Rule: Slavery, Tyranny, and the Power of Life and Death*. Chicago: University of Chicago Press, 2013.

O'Brien, Margaret Ann. "Christian Belief in *Doctor Faustus*." *ELH* 37 (1970): 1–11.

O'Callaghan, Michelle. *The English Wits: Literature and Sociability in Early Modern England*. Cambridge: Cambridge University Press, 2007.

———. "Tavern Societies, the Inns of Court, and the Culture of Conviviality in Early Seventeenth-Century London." In *A Pleasing Sinne: Drink and Conviviality in Seventeenth-Century England*, ed. Adam Smyth, 37–54. London: D. S. Brewer, 2004.

Okerlund, A. N. "The Intellectual Folly of Dr. Faustus." *Studies in Philology* 74 no. 3 (1977): 258–78.

Olson, Rebecca. "'Too Gentle': Jealousy and Class in *Othello*." *Journal for Early Modern Cultural Studies* 15 no. 1 (Winter 2015): 3–25.

Orgel, Stephen. "Tobacco and Boys: How Queer was Marlowe." In *The Authentic Shakespeare*, 211–30. New York: Routledge, 2002.

Orlin, Lena Cowen. "Desdemona's Disposition." In *Shakespearean Tragedy and Gender*, ed. Shirley Nelson Garner and Madelon Sprengnether, 171–92. Bloomington: Indiana University Press, 1996.

———, ed. *Othello: The State of Play*. London: Bloomsbury Arden Shakespeare, 2014.

Ornstein, Robert. "Marlowe and God: The Tragic Theology of *Dr. Faustus*." *PMLA* 83 (1968): 1378–85.

Park, Thomas, ed. *The Poetical Works of Edmund Waller*. London: J. Sharpe, 1806.

Parliament of Victoria Law Reform Committee. *Criminal Liability for Self-Induced Intoxication*. Melbourne: Government Printer, 1999.

Pask, Kevin. *The Emergence of the English Author: Scripting the Life of the Poet in Early Modern England*. Cambridge: Cambridge University Press, 1996.

Paster, Gail Kern. *The Body Embarrassed: Drama and the Disciplines of Shame in Early Modern Europe*. Ithaca, NY: Cornell University Press, 1993.

———. *Humoring the Body: Emotions and the Shakespearean Stage*. Chicago: University of Chicago Press, 2004.

Paster, Gail Kern, Katherine Rowe, and Mary Floyd-Wilson, eds. *Reading the Early Modern Passions: Essays in the Cultural History of Emotion*. Philadelphia: University of Pennsylvania Press, 2004.

Peat, Derek. "Falstaff Gets the Sack." *Shakespeare Quarterly* (Fall 2002): 379–85.

Percy, C. H. Herford and Evelyn Simpson, eds. *Ben Jonson*. Vol. VIII. Oxford: Clarendon Press, 1947.

Perkins, William. *A commentarie or exposition, upon the five first chapters of the Epistle to the Galatians*. London, 1604.

———. *A godly and learned exposition or commentarie upon the three first chapters of the Revelation*. London, 1595.

———. *A godlie and learned exposition upon the whole epistle of Iude, containing threescore and sixe sermons preached in Cambridge*. London, 1606.

———. *A treatise of man's imaginations*. London, 1607.

———. *A treatise tending vnto a declaration whether a man be in the estate of damnation or in the estate of grace and if he be in the first, how he may in time come out of it: if in the second, how he maie discerne it, and perseuere in the same to the end. The points that are handled are set downe in the page following*. London, 1590.

———. *The combat betweene Christ and the Diuell displayed: or A commentarie vpon the temptations of Christ*. London, 1606.

———. *The works of that famous and worthie Minister of Christ, in the University of Cambridge, M. W. Perkins*. Cambridge, 1605.

Phayre, Thomas. *The Regiment of Life*. London: Edward Allde, 1596.

Plowden, Edmund. *The commentaries, or Reports of Edmund Plowden: containing divers cases upon matters of law, argued and adjudged in the several reigns of King Edward VI., Queen Mary, King and Queen Philip and Mary, and Queen Elizabeth [1548–1579]*. 1551. Reprint, London: H. Watts and W. Jones, 1792.

Poland, Jeffrey, and George Graham, eds. *Addiction and Responsibility*. Cambridge, MA: MIT Press, 2011.

Pollard, Tanya. *Drugs and Theatre in Early Modern England*. Oxford: Oxford University Press, 2005.

Poole, Kristen. "Dr. Faustus and Reformation Theology." In *Early Modern English Drama*, ed. Garrett Sullivan, Patrick Cheney, and Andrew Hadfield, 96–107. Oxford: Oxford University Press, 2006.

———. *Radical Religion from Shakespeare to Milton: Figures of Nonconformity in Early Modern England*. Cambridge: Cambridge University Press, 2000.

———. *Supernatural Environments in Shakespeare's England: Spaces of Demonism, Divinity, and Drama*. Cambridge: Cambridge University Press, 2011.

Porter, Roy. *Disease, Medicine and Society in England, 1550–1860*. 2nd edition. Cambridge: Cambridge University Press, 1995.

———. "The Drinking Man's Disease: The 'Pre-History' of Alcoholism in Georgian Britain," *British Journal of Addiction* 80 no. 4 (1985): 385–96.

———. Introduction to Thomas Trotter, *An Essay, Medical, Philosophical, and Chemical, on Drunkenness, and Its Effects on the Human Body*, ed. Roy Porter. New York: Routledge, 1988.

Porter, Roy, and Mikulás Teich, eds. *Drugs and Narcotics in History*. Cambridge: Cambridge University Press, 1995.

Posner, Richard A. *Law and Literature*. 3rd edition. Cambridge, MA: Harvard University Press, 2009.

Potter, Lois. *Secret Rites and Secret Writing: Royalist Literature, 1641–1660*. Cambridge: Cambridge University Press, 1989.

Proceedings and Debates of the House of Commons in 1620 and 1621. Collected by a Member of that House. Vol 1. Oxford: Clarendon Press, 1766.

A Proclamation against Vicious, Debauch'd, and Prophane Persons, by the King. At our court at Whitehal, the 30th day of May, in the twelfth year of our reign. London: Christopher Barker, 1660.

Prynne, William. *Healthes Sicknesse. Or a compendious and brief discourse; proving the drinking and pledging of Healthes, to be Sinfull, and utterly Unlawfull unto Christians*. London, 1628.

———. *Histrio-mastix. The players scourge, or, actors tragædie, divided into two parts*. London, 1633.

Purchas, Samuel. *Purchas his pilgrimage*. London: Printed by William Stansby for Henrie Fetherstone, 1613.

Raber, Karen. "Shakespeare's Fluid Mechanics." In *Culinary Shakespeare*, ed. David B. Goldstein and Amy L. Tigner, 75–96. Pittsburgh: Duquesne University Press, 2016.

Radoilska, Lubomira. *Addiction and Weakness of Will*. Oxford: Oxford University Press, 2013.

Rasmussen, Eric. Introduction to *Doctor Faustus*, ed. David Bevington and Eric Rasmussen. The Revels Plays. Manchester: Manchester University Press, 1993.

Raylor, Timothy. *Cavaliers, Clubs, and Literary Culture: Sir John Mennes, James Smith, and the Order of the Fancy*. Newark: University of Delaware Press, 1994.

Reeser, Todd W. *Moderating Masculinity in Early Modern Culture*. Chapel Hill: University of North Carolina Press, 2006.

Reiss, Timothy. *Mirages of the Selfe: Patterns of Personhood in Ancient and Early Modern Europe*. Stanford, CA: Stanford University Press, 2003.

Revard, Stella P. "Thomas Stanley and 'A Register of Friends.'" In *Literary Circles and Cultural Communities in Renaissance England*, ed. Ted-Larry Pebworth, 148–72. London: University of Missouri Press, 2000.

Rhodes, Neil. *Elizabethan Grotesque*. New York: Routledge, 1980.

Richards, Jennifer. "Health, Intoxication, and Civil Conversation in Renaissance England." *Past and Present* 222 no. S9 (2014): 168–86.

Riche, Barnabe. *Rich His Farewell to the Military Profession*. London, 1583.

Riggs, David. "Marlowe's Quarrel with God." In *Critical Essays on Christopher Marlowe*, ed. Emily C. Bartels, 39–58. London: Prentice Hall, 1997.

Rivlin, Elizabeth. *The Aesthetics of Service in Early Modern England*. Evanston, IL: Northwestern University Press, 2012.

Roach, Joseph R. *The Player's Passion: Studies in the Science of Acting*. Newark: University of Delaware Press, 1985.

Roberts, Benjamin. "Drinking like a Man: The Paradox of Excessive Drinking for Seventeenth-Century Dutch Youths." *Journal of Family History* 29 no. 3 (July 2004): 237–52.

Rogers, Thomas. *Of the imitation of Christ, three, both for wisedome, and godlines, most excellent bookes; made 170 yeeres since by one Thomas of Kempis, and for the worthines thereof oft since translated out of Latine into sundrie languages*. London, 1580.

Roxburghe Ballads. British Library Roxburghe Collection, Vol. 1. London: British Library.

Rubenfeld, Jed. "The Riddle of Rape-by-Deception and the Myth of Sexual Autonomy." *Yale Law Journal* 122 no. 6 (April 2013): 1372–1443.

Ruiter, David. *Shakespeare's Festive History: Feasting, Festivity, Fasting and Lent in the Second Henriad*. Farnham, Surrey: Ashgate, 2003.

Rump: or, an Exact Collection of the Choycest Poems and Songs relating to the Late Times. London: Henry Brome, 1662.

Rush, Benjamin. *An Inquiry into the Effects of Ardent Spirits upon the Human Body and Mind* (1814). 8th edition. Boston: James Loring, 1823.

———. *Medical Inquiries and Observations upon the Diseases of the Mind* (1812). 5th edition. Philadelphia: Grigg and Eliot, 1835.

Sala, Angelus. *Opiologia: or, A treatise concerning the nature, properties, true preparation and safe vse and administration of opium*. Trans. Thomas Bretnor. London: Nicholas Okes, 1618.

Sanchez, Melissa E. *Erotic Subjects: The Sexuality of Politics in Early Modern English Literature*. Oxford: Oxford University Press, 2011.

———. "'In My Selfe the Smart I Try': Female Promiscuity in Astrophil and Stella." *ELH* 80, no. 1 (Spring 2013): 1–27.

Schaler, Jeffrey A. *Addiction Is a Choice*. Chicago: Open Court, 2000.

Schalkwyk, David. "The Discourses of Friendship and the Structural Imagination of Shakespeare's Theater: Montaigne, *Twelfth Night*, De Gournay." *Renaissance Drama* 38 (2010): 141–71.

———. "Is Love an Emotion: Shakespeare's *Twelfth Night* and *Antony and Cleopatra*." *symplokē* 18 no. 1–2 (2010): 99–130.

———. "Love and Service in Twelfth Night and the Sonnets." *Shakespeare Quarterly* 56 no. 1 (2005): 76–100.

———. *Shakespeare, Love and Service.* Cambridge: Cambridge University Press, 2008.

Schama, Simon. *The Embarrassment of Riches: An Interpretation of Dutch Culture in the Golden Age.* New York: Vintage Books, 1987.

Schoenfeldt, Michael C. *Bodies and Selves in Early Modern England: Physiology and Inwardness in Spenser, Shakespeare, Herbert, and Milton.* Cambridge: Cambridge University Press, 1999.

Schwarz, Kathryn. *What You Will: Gender, Contract, and Shakespearean Social Space.* Philadelphia: University of Pennsylvania Press, 2011.

Scodel, Joshua. *Excess and the Mean in Early Modern English Literature.* Princeton, NJ: Princeton University Press, 2002.

Scott-Warren, Jason. "When Theaters Were Bear-Gardens; or, What's at Stake in the Comedy of Humors." *Shakespeare Quarterly* 54 no. 1 (Spring 2003): 63–82.

Seaver, P. S. "Downame, John." In *Oxford Dictionary of National Biography.* Oxford: Oxford University Press, 2004.

Sebek, Barbara. "'Wine and sugar of the best and fairest': Canary, the Canaries and the Global in Windsor." In *Culinary Shakespeare,* ed. David B. Goldstein and Amy L. Tigner, 41–56. Pittsburgh: Duquesne University Press, 2016.

Sedgwick, Eve Kosofsky. "Epidemics of the Will." In *Tendencies,* 129–40. New York: Routledge, 1994.

Seneca, Lucius Annaeus. *Ad Lucilium Epistulae Morales.* Trans. Richard M. Gummere. Loeb Classical Library. London: Heinemann, 1920.

———. *Annaei Senecae Philosophi Opera, Quae Exstant Omnia, A Iusto Lipsio emendata, et Scholiis illustrate.* Antwerp: Plantijn-Moretus, 1605.

———. *Moral Essays.* Vol. 1. Trans. John W. Basore. Loeb Classical Library. London: Heinemann, 1928.

———. *The workes of Lucius Annaeus Seneca, both morrall and natural.* Trans. Thomas Lodge. London, 1614.

Serpieri, Alessandro. "*Othello* and Venice: Discrimination and Projection." In *Visions of Venice in Shakespeare,* ed. Laura Tosi and Shaul Bassi, 185–96. London: Routledge, 2011.

Shakespeare, William. *Hamlet.* Ed. Ann Thompson and Neil Taylor. Arden Shakespeare, 3rd series. London: Cengage Learning, 2006.

———. *King Henry IV, Part 1.* Ed. David Scott Kastan. Arden Shakespeare, 3rd series. London: Thomson Learning, 2002.

———. *King Henry IV, Part 2.* Ed. James C. Bulman. Arden Shakespeare, 3rd series. London: Bloomsbury Arden Shakespeare, 2016.

———. *King Henry V.* Ed. T. W. Craik. Arden Shakespeare, 3rd series. London: Bloomsbury Arden Shakespeare, 1995.

———. *Othello.* Ed. E. A. J. Honigmann, with a new introduction by Ayanna Thompson. Arden Shakespeare, revised 3rd series. London: Bloomsbury Arden Shakespeare, 2016.

———. *Richard II.* Ed. Charles R. Forker. Arden Shakespeare, 3rd series. London: Bloomsbury Arden Shakespeare, 2002.

———. *The Taming of the Shrew: Texts and Contexts.* Ed. Frances E. Dolan. Boston: Bedford/St. Martin's, 1996.

———. *Twelfth Night, Or What You Will.* Ed. Keir Elam. Arden Shakespeare. 3rd series. London: Bloomsbury Arden Shakespeare, 2008.

Shannon, Laurie. "Nature's Bias: Renaissance Homonormativity and Elizabethan Comic Likeness." *Modern Philology* 98 no. 2 (November 2000): 183–210.

———. *Sovereign Amity: Figures of Friendship in Shakespearean Contexts*. Chicago: University of Chicago, 2002.

Shepard, Alexandra. *Meanings of Manhood in Early Modern England*. Oxford: Oxford University Press, 2006.

———. "'Swil-bolls and tos-pots': Drink Culture and Male Bonding in England, c.1560–1640." In *Love, Friendship and Faith in Europe, 1300–1800*, ed. Laura Gowing, Michael Hunter, and Miri Rubin, 110–30. New York: Palgrave Macmillan, 2005.

Shrank, Cathy. "Beastly Metamorphoses: Losing Control in Early Modern Literary Culture." In *Intoxication and Society: Problematic Pleasures of Drugs and Alcohol*, ed. Jonathan Herring, Ciaran Regan, Darin Weinberg, and Phil Withington, 193–209. Basingstoke: Palgrave, 2013.

Sidney, Philip. *Sir Philip Sidney: The Major Works*. Ed. Katherine Duncan-Jones. Oxford: Oxford University Press, 2008.

Siemon, James. "Making Ambition Virtue? *Othello*, Small Wars, and Marital Profession." In *Othello: The State of Play*, ed. Lena Cowen Orlin, 177–202. London: Bloomsbury Arden Shakespeare, 2014.

Sinfield, Alan. *Faultlines: Cultural Materialism and the Politics of Dissident Reading*. Berkeley: University of California Press, 1992.

Singh, R. U. "History of the Defence of Drunkenness in English Criminal Law." *Law Quarterly Review* no. 196 (October 1933): 528–46.

Sisson, Andrew. "*Othello* and the Unweaponed City." *Shakespeare Quarterly* 66 no. 2 (Summer 2015): 137–66.

Skelton, John. "The Tunning of Elynour Rumming." In *John Skelton: The Complete English Poems*, ed. John Scattergood. New York: Penguin Classics, 1992.

Smith, Amy L., and Elizabeth Hodgson. "'A Cypress, not a bosom, hides my heart': Olivia's Veiled Conversions." *Early Modern Literary Studies* 15 no. 1 (2009–2010). http://purl.oclc.org/emls/15-1/olivveil.htm.

Smith, Bruce R. Introduction to *Twelfth Night: Texts and Contexts*. New York: Palgrave, 2001.

———. *Homosexual Desire in Shakespeare's England: A Cultural Poetics*. Chicago: University of Chicago Press, 1991.

———. *Shakespeare and Masculinity*. Oxford: Oxford University Press, 2000.

Smith, Ian. "Othello's Black Handkerchief." *Shakespeare Quarterly* 64 no. 1 (2013): 1–25.

———. "We Are Othello: Speaking of Race in Early Modern Studies." *Shakespeare Quarterly* 67 no. 1 (2016): 104–24.

Smyth, Adam, ed. *A Pleasing Sinne: Drink and Conviviality in Seventeenth-Century England*. London: D. S. Brewer, 2004.

———. "'It were far better to be a *Toad* or a *Serpant*, then a Drunkard': Writing About Drunkenness." In *A Pleasing Sinne: Drink and Conviviality in Seventeenth-Century England*, ed. Adam Smyth, 193–210. London: D. S. Brewer, 2004.

Snow, Edward. "Marlowe's *Doctor Faustus* and the Ends of Desire." In *Two Renaissance Mythmakers: Christopher Marlowe and Ben Jonson*, ed. Alvin Kernan, 70–110. Baltimore: Johns Hopkins University Press, 1977.

Sofer, Andrew. "Felt Absences: The Stage Properties of *Othello's* Handkerchief." *Comparative Drama* 31 no. 3 (Fall 1997): 367–93.

Sokolov, Danila. *Renaissance Texts, Medieval Subjectivities: Rethinking Petrarchan Desire from Wyatt to Shakespeare*. Pittsburgh, PA: Duquesne University Press, 2017.

Spiller, Elizabeth. "Marlowe's Libraries: A History of Reading." In *Christopher Marlowe in Context*, ed. Emily C. Bartels and Emma Smith, 101–9. Cambridge: Cambridge University Press, 2013.

Spufford, Margaret. "Puritanism and Social Control?" In *Order and Disorder in Early Modern England*, ed. Anthony Fletcher and John Stevenson, 41–57. Cambridge: Cambridge University Press, 1985.

Stachniewski, John. *The Persecutory Imagination: English Puritanism and the Literature of Religious Despair*. Oxford: Clarendon Press, 1991.

Stallybrass, Peter, and Allon White. *The Politics and Poetics of Transgression*. Ithaca, NY: Cornell University Press, 1986.

Stam, David Harry. "England's Calvin: A Study of the Publication of John Calvin's Works in Tudor England." Ph.D. diss., Northwestern University, 1978.

Stanislavsky, Konstantin. *An Actor Prepares*. Trans. Elizabeth Reynolds Hapgood. New York: Taylor and Francis, 1989.

———. *An Actor's Work*. Trans. Jean Benedetti. New York: Routledge, 2008.

A Statute against Drunkennesse. London, 1644.

The Statutes at Large, from the First Year of King James the First to the Tenth Year of the Reign of King William the Third. . . . Volume the Third. London, 1786.

Steggle, Matthew. "Charles Chester and Ben Jonson." *SEL* 39 no. 2 (Spring 1999): 313–26.

Stewart, Alan. *Close Readers: Humanism and Sodomy in Early Modern England*. Princeton, NJ: Princeton University Press, 1997.

Streete, Adrian. "Calvinist Conceptions of Hell in Marlowe's *Doctor Faustus*." *Notes and Queries* (December 2000): 430–32.

Strier, Richard. "Against the Rule of Reason: Praise of Passion from Petrarch to Luther to Shakespeare to Herbert." In *Reading the Early Modern Passions: Essays in the Cultural History of Emotion*, ed. Gail Kern Paster, Katherine Rowe, and Mary Floyd-Wilson, 23–42. Philadelphia: University of Pennsylvania Press, 2004.

———. "Shakespeare Against Morality." In *Reading Renaissance Ethics*, ed. Marshall Grossman, 206–25. New York: Routledge, 2007.

———. *Unrepentant Renaissance: From Petrarch to Shakespeare to Milton*. Chicago: University of Chicago Press, 2011.

Stubbes, Philip. *The Anatomie of Abuses: contayning a discoverie, or briefe summarie of such Notable Vices and Imperfections, as now raigne in many Christian Countreyes of the Worlde*. London: Richard Jones, 1583.

Szabados, Bela, and Kenneth G. Probert, eds. *Writing Addiction: Towards a Poetics of Desire and Its Others*. Regina, Canada: University of Regina Press, 2004.

Szalavitz, Maia. *Unbroken Brain: A Revolutionary New Way of Understanding Addiction*. New York: St. Martin's Press, 2016.

Szasz, Thomas. "The Discovery of Drug Addiction." In *Classic Contributions in the Addictions*, ed. Howard Shaffer and Milton Earl Burglass, 35–49. New York: Brunner/Mazel, 1981.

Tadmore, Naomi. *Family and Friends in Eighteenth-Century England: Household, Kinship and Patronage*. Cambridge: Cambridge University Press, 2001.

Taunton, Nina. "Food, Time and Age: Falstaff's Dietaries and Tropes of Nourishment in *The Comedy of Errors*." *Shakespeare Jahrbuch* 145 (2009): 91–105.

Taylor, John. *Drinke and Welcome*. London: Anne Griffin, 1637.

———. *The parable of the sower and of the seed*. London, 1621.

Thirsk, Joan. *Food in Early Modern England: Phases, Fads, Fashions 1500–1760*. London: Hambledon Continuum, 2007.

Thomas, Hugh M. *The English and the Normans: Ethnic Hostility, Assimilation, and Identity 1066–c.1220*. Oxford: Oxford University Press, 2005.

Thomas, Thomas. *Dictionarium Linguae Latinae et Anglicanae*. Cambridge, 1587.

Thompson, Ayanna. Introduction to *Othello*, ed. E. A. J. Honigmann, with a new introduction by Ayanna Thompson, 1–118. Arden Shakespeare, revised 3rd series. London: Bloomsbury Arden Shakespeare, 2016.

Thompson, Thomas. *A Diet for a Drunkard, delivered in two sermons at St. Nicholas Church in Bristoll, 1608*. London: Richard Backworth, 1612.

Thoms, William J., ed. *Anecdotes and Traditions*. London: Camden Society, 1839.

Tilmouth, Christopher. "Passion and Intersubjectivity in Early Modern Literature." In *Passions and Subjectivity in Early Modern Culture*, ed. Brian Cummings and Freya Sierhuis, 13–32. New York: Routledge, 2013.

Tlusty, B. Ann. *Bacchus and Civic Order: The Culture of Drink in Early Modern Germany*. Charlottesville: University of Virginia Press, 2001.

———. "The Public House and Military Culture in Early Modern Germany." In *The World of the Tavern*, ed. Beat Kümin and B. Ann Tlusty, 137–53. Aldershot: Ashgate, 2002.

Tracy, Sarah W. *Alcoholism in America: From Reconstruction to Prohibition*. Baltimore: Johns Hopkins University Press, 2005.

Traub, Valerie. *Desire and Anxiety: Circulations of Sexuality in Shakespearean Drama*. New York: Routledge, 1992.

———. *The Renaissance of Lesbianism in Early Modern England*. Philadelphia: University of Pennsylvania Press, 2002.

Treharne, Elaine. "Gluttons for Punishment: The Drunk and Disorderly in Early English Homilies." 24th Annual Brixworth Lecture. 2nd series, no. 6. Leicester: University of Leicester, 2007.

Trevor, Douglas. *The Poetics of Melancholy in Early Modern England*. Cambridge: Cambridge University Press, 2004.

Trotter, Thomas. *An Essay, Medical, Philosophical, and Chemical, on Drunkenness, and Its Effects on the Human Body*. 4th edition. London: Longman, Hurst, Kees, and Orme, 1810.

———. *A View of the Nervous Temperament; Being a Practical Enquiry into the Increasing Prevalence, Prevention, and Treatment of Those Diseases Commonly Called Nervous, Bilious, Stomach, and Liver Complaints; Indigestion; Low Spirits, Gout, etc.* 2nd edition. London, 1786.

T. W. [Thomas Weaver]. *Songs and Poems of Love and Drollery*. London, 1654.

Two broad-sides against tobacco. London, 1672.

Twyne, Thomas. *The schoolemaster, or teacher of table philosophie*. London: Richarde Iones, 1576.

Tyacke, Nicholas. *Anti-Calvinists: The Rise of English Arminianism c. 1590–1640*. Oxford: Oxford University Press, 1990.

Tyndale, William. *The Newe Testament dylygently corrected and compared with the Greke by Wil-lyam Tindale, and fynesshed in the yere of our Lorde God A.M.D. & xxxiiij. in the moneth of Nouember*. 1534.

Underdown, David E. *Fire from Heaven: The Life of an English Town in the Seventeenth Century*. London: Harper Collins, 1992.

Valverde, Mariana. *Diseases of the Will: Alcohol and the Dilemmas of Freedom*. Cambridge: Cambridge University Press, 1998.

Vaughan, William. *Natural and Artificial Directions for health, derived from the best Philosophers, as well moderne, as ancient*. London: Richard Bradocke, 1600.

Vaught, Jennifer C., ed. *Rhetorics of Bodily Disease and Health in Medieval and Early Modern England*. Farnham, Surrey: Ashgate Press, 2010.

Vergil, Polidore. *An abridgement of the notable woorke of Polidore Vergile. Gathered by Thomas Langley*. London, 1546.

Vicars, John. *The Looking-Glass for Malignants, enlarged: or, The second part of Gods hand against God-haters* (1643). London: I. Rothwell, 1645.

Vice, Sue, Matthew Campbell, and Tim Armstrong, eds. *Beyond the Pleasure Dome: Writing and Addiction from the Romantics*. Sheffield: Sheffield Academy, 1994.

Vieth, David M., ed. *The Complete Poems of John Wilmot, Earl of Rochester*. Yale Nota Bene. New Haven, CT: Yale University Press, 1962.

Voak, Nigel. *Richard Hooker and Reformed Theology*. Oxford: Oxford University Press, 2003.

Walkington, John. *The Optic Glass of Humours*. London, 1631.

Wallace, R. Jay. "Addiction as Defect of the Will: Some Philosophical Reflections." In *Normativity and the Will: Selected Papers on Moral Psychology and Practical Reason*, 165–89. Oxford: Oxford University Press, 2006.

Waller, Edmund. *Poems upon several occasions*. London: Thomas Walkley, 1645.

Walsham, Alexandra. "Skeletons in the Cupboard: Relics After the English Reformation." *Past and Present* 206 no. S5 (2010): 121–43.

Ward, Samuel. *Woe to Drunkards: A sermon by Samuel Ward, preacher of Ipswich*. London: A. Math for John Marriott, 1622.

Ward, Samuel, and Samuel Clark, *A Warning-piece to all Drunkards and Health-Drinkers*. London, 1682.

Warner, Jessica. "'Before There Was Alcoholism': Lessons from the Medieval Experience with Alcohol." *Contemporary Drug Problems* 19 (1992): 409–29.

———. *Craze: Gin and Debauchery in an Age of Reason*. London: Profile Books, 2003.

———. "'Resolv'd to Drink No More': Addiction as a Preindustrial Concept." *Journal of Studies on Alcohol* 55 (1994): 685–91.

Watkins, John. "The Allegorical Theatre: Moralities, Interludes, and Protestant Drama." In *The Cambridge History of Medieval English Literature*, ed. David Wallace, 767–93. Cambridge: Cambridge University Press, 1999.

Wear, Andrew. *Knowledge and Practice in English Medicine, 1550—1680*. Cambridge: Cambridge University Press, 2000.

Weimann, Robert. *Shakespeare and the Popular Tradition in the Theater*. Ed. Robert Schwartz. Baltimore: Johns Hopkins University Press, 1978.

Wendel, Françoise. *Calvin: The Origins and Development of His Religious Thought*. Trans. Philip Mairet. New York: Harper & Row, 1963.

Wenzel, Siegfried. "The Seven Deadly Sins: Some Problems of Research." *Speculum* 43, no. 1 (January 1968): 1–22.

West, Robert, and Jamie Brown. *The Theory of Addiction*. 2nd edition. Oxford: Wiley Blackwell, 2013.

Wheeler, Richard P. "Deaths in the Family: The Loss of a Son and the Rise of Shakespearean Comedy." *Shakespeare Quarterly* 51 no. 2 (Summer 2000): 127–53.

White, Jonathan. "'The Slow but Sure Poison': The Representation of Gin and Its Drinkers, 1736–1751." *Journal of British Studies* 42 (2003): 35–64.

Whitney, Charles. "Festivity and Topicality in the Coventry Scene of *1 Henry IV*." *English Literature Renaissance* 24 no. 2 (Spring 1994): 410–48.

Wilcher, Robert. *The Writing of Royalism, 1628–1660*. Cambridge: Cambridge University Press, 2001.

Wilcox, Jonathan, ed. *Aelfric's Prefaces*. Durham Medieval Texts, no.9. Durham: University of Durham, 1994.

Wilkinson, C. H., ed. *The Poems of Richard Lovelace*. Oxford: Oxford University Press, 1930.

Williams, Rebecca. "The Current Law of Intoxication: Rules and Problems." In *Intoxication and Society: Problematic Pleasures of Drugs and Alcohol*, ed. Jonathan Herring, Ciaran Regan, Darin Weinberg, and Phil Withington, 267–82. Basingstoke: Palgrave Macmillan, 2013.

Willis, Deborah. "*Doctor Faustus* and the Early Modern Language of Addiction." In *Placing the Plays of Christopher Marlowe*, ed. Sara Munson Deats and Robert A. Logan, 136–48. Farnham, Surrey: Ashgate, 2008.

Wilson, Dover. *The Fortunes of Falstaff*. Cambridge: Cambridge University Press, 1961.

Wilson, Luke. *Theaters of Intention: Drama and the Law in Early Modern England*. Stanford: Stanford University Press, 2000.

Withington, Phil. "Company and Sociability in Early Modern England." *Social History* 32 no. 3 (2007): 291–307.

———. "Introduction: Cultures of Intoxication." *Past and Present* 222 no. S9 (2014): 9–33.

———. "Renaissance Drinking Cultures and Popular Print." In *Intoxication and Society: Problematic Pleasures of Drugs and Alcohol*, ed. Jonathan Herring, Ciaran Regan, Darin Weinberg, and Phil Withington, 135–52. London: Palgrave, 2013.

———. *Society in Early Modern England: The Vernacular Origin of Some Powerful Ideas*. Cambridge: Polity Press, 2010.

Womersley, David. "Why Is Falstaff Fat?" *Review of English Studies: A Quarterly Journal of English Literature and the English Language* 47 no. 185 (1996): 1–22.

Wood, Andy. "The Place of Custom in Plebeian Political Culture: England, 1550–1800." *Social History* 22 (1997): 46–60.

Worcestershire County Records. Division 1. Documents Relating to Quarter Sessions. Calendar of the Quarter Sessions Papers. Vol. 1, *1591–1643*, ed. J. W. Willis Bund. Worcester: Ebenr Baylis and Son, 1900.

Worthen, William B. *The Idea of the Actor*. Princeton, NJ: Princeton University Press, 2014.

Wrightson, Keith Edwin. "Alehouses, Order, and Reformation in Rural England, 1590–1660." In *Popular Culture and Class Conflict, 1590–1914: Explorations in the History of Labour and Leisure*, ed. Eileen Yeo and Stephen Yeo, 1–27. Hassocks, Sussex: Harvester Press, 1981.

———. "Postscript: Terling Revisited." In *Poverty and Piety in an English Village: Terling, 1525–1700*, ed. Keith Wrightson and David Levine, 2nd edition, 186–220. Oxford: Oxford University Press, 1995.

———. "The Puritan Reformation of Manners, with Special Reference to the Counties of Lancashire and Essex, 1640–1660." Ph.D. diss., Cambridge University, 1973.

Wrightson, Keith, and David Levine, eds. *Poverty and Piety in an English Village: Terling, 1525–1700*. 2nd edition. Oxford: Oxford University Press, 1995.

Yachnin, Paul. "Wonder-Effects: Othello's Handkerchief." In *Staged Properties in Early Modern English Drama*, ed. Jonathan Gil Harris and Natasha Korda, 316–34. Cambridge: Cambridge University Press, 2002.

Yeager, Daniel B. "Marlowe's *Faustus:* Contract as Metaphor?" *University of Chicago Law School Roundtable* 2 no. 2 (1995): 599–617.

Young, Alexander. *History of the Netherlands*. Boston: Estes and Lauriate, 1884.

Younge, Richard [R. Junius, pseud.]. *The Drunkard's Character: or, A True Drunkard with such sinnes as raigne in him*. London, 1638.

Zamir, Tzachi. *Double Vision: Moral Philosophy and Shakespearean Drama*. Princeton, NJ: Princeton University Press, 2007.

Index

Page numbers in *italics* indicate an illustration.

Acknowledgments

I am grateful for a lively community of scholars who supported and interrogated this project from its inception to its end. William Fisher, Penelope Geng, and Bruce R. Smith read the manuscript from start to finish. They deserve special mention for their long-term engagements with the book. My colleague Peter Mancall offered early and then sustained interest in the project, sharing his own work on alcohol in early America, and co-organizing "Concepts of Addiction in the Early Modern World" with me, a conference we held at the Huntington Library at the inception of this project. Steve Pincus has been a bright light of support throughout, and his insights during two talks at Yale—and over congenial holiday dinners and Huntington lunches—helped shape the project's contours. Cynthia Herrup and Dympna Callaghan supported me as letter writers and friends, encouraging me forward. Judith Bennett's work on drinking helped my thinking as this project began to take shape, and I miss both her and Cynthia as colleagues. Heather James proved, as always, a generous interlocutor and offered helpful manuscript references along the way. She, with Heidi Brayman, invited me to present at the Renaissance Literature Seminar, where they provided crucial support and commentary.

A series of writing groups helped propel the project forward. My long-term writing group, and especially its organizing force, Will Fisher, deserves special recognition: Carla Mazzio, Will Fisher, Tiffany Werth, Penny Geng, Andy Fleck, Heidi Brayman, Marjorie Rubright, and Cynthia Nazarian all engaged with the project in ways they will recognize in my revised pages. My debt to them is ongoing. Many summers at the Huntington with Bob Darcy, Carla, Penny, and Will provided the friendship and support I needed during the long process of writing. Will and I found ourselves wrestling with similar questions as we shaped our projects, and our conversations have been transformative. A writing group with USC colleagues Hilary Schor and Emily Anderson provided the space needed for writing my *Doctor Faustus* chapter, and their enthusiasm for the unexpected turn in my project helped open up my thinking in transformative ways. With the Conclusion, Genevieve Love, Tiffany Werth, Joe Boone, Tania Modleski, Bruce Smith, and Lucia Martinez provided inspired guidance. Monday and Friday conversations with Genny at the Getty Research Institute, during a very busy teaching year, encouraged me to keep the project on the front burner. Further conversations with Tania Modleski reinforced my sense of the project's broadest possibilities; together Tania and Genny provided just the right amount of diversion to keep me sane during the revision process.

At the University of Southern California I am surrounded by brilliant early modernists—present and past—who inspire my own work: Emily Anderson, Bruce Smith, Heather James, Leo Braudy, Anna Rosensweig, Antonia Szabari, Natania Meeker, Deb Harkness, Lindsey O'Neill, Daniela Bleichmar, Peter Mancall, Jessica Rosenberg, Larry Green, Lisa Bitel, Nathan Perl-Rosenthal, Ed McGann, Meg Russett, David Rollo, Thea Cervone, Angus Fletcher, Carla Della Gatta, Hilary Schor, and Devin Griffiths. USC's Center for Law, History and Culture has been a productive home, led by Hilary Schor, Ariela Gross, Nomi Stolzenberg. USC also boasts a number of research centers on addiction. I was lucky to find Michael Quick and Anuj Aggerwal as helpful interlocutors on the neuroscience of addictions and to craft, with Giddeon Yaffe, an early grant proposal that helped to broaden my investigation of the law and science of addiction. Colleagues from across the university keep me alive to the broader stakes of the academic enterprise. I'm grateful to Joe Boone, Gordon Davis, David St. John, Dana Johnson, Brian Ingram, Karen Tongson, Dorothy Braudy, Aimee Bender, Viet Nguyen, Tim Gustafson, Bill Handley, Geoff Dyer, Phil Ethington, Robin Romans, and Colin Dickey. Joe deserves special thanks for his friendship, offering inspiring diversion during two summers in Istanbul, and ideal companionship during a December in India.

The Huntington Library supplied engaging interlocutors. I'm grateful to Steven Hindle for his support and for introducing me to Mark Hailwood during a very productive Huntington summer. That summer's highlight included a reading group organized by Steve Pincus, in which I received precisely the kind of feedback on the project that helped clarify what I was trying to achieve. At the Huntington Tanya Pollard generously agreed to attend the Early Modern Addiction conference I organized, and her encouragement, in person and through her research on drugs and theater, helped me see how a project on addition might be possible. I am also keen to acknowledge Urvashi Chakravorty, Deborah Willis, Stefanie Sobelle, Sharon Oster, Lindsay O'Neill,

Bill Sherman, Tom Cogswell, Chris Kyle, David Cressy, Molly Murray, Martine van Elk, Lloyd Kermode, Tobias Gregory, András Kiséry, Paul Hammer, Lara Lee Hullinghorst, Henry Turner, Matthew Growhoski, Stephanie Elsky, Ari Friedlander, Jennifer Anderson, Barbara Zimbalist, Pamela Allen Brown, Adam Zucker, Melissa Sanchez, Ambereen Dadabhoy, Rosemary Englander, Karen Cunningham, Arthur Little, Rebecca Laroche, and Mary Robertson. J. K. Barret and Tiffany Werth both offered insight on the project's chapters during their residencies at the Huntington, and I am in their debt. I'm also grateful to the Huntington Library staff, past and present, including Steve Tabor, Juan Gomez, Kadin Henningson, Catherine Wehrey-Miller, Meredith Jones, Sara Ash Georgi, and Claire Kennedy.

London has been a second home. I have enjoyed the friendship of Phiroze Vasunia and Miriam Leonard over many years, and particularly during three very happy spring semesters in London: our tea breaks at the British Library helped propel me forward during long days of reading and note taking; and their intelligent political commentary always reminds me of better possibilities. Nic Bilham has been a best friend and one-man support team for decades, offering scientific and psychological engagements with the project. I can't imagine life without him. I am especially grateful for his move to Brockley, where he offers me, truly, a room of my own. Mona Benjamin and Sanja Perovic offered further friendship. From our early bonding in college over my love of her grandfather's work, Mona has been a kindred spirit. So, too, with Sanja, where our time at the Stanford Humanities Center led to serendipitous intersections and overlaps. The magical Lena Orlin provided the ideal writing space in a quiet corner of very central London. Peter Lake helped sharpen my thinking over a series of conversations on the Reformation of Manners. Angela McShane and Phil Withington invited me to join one session of their stunning Intoxicants and Intoxication working group, where Jenny Richards proved an especially welcoming interlocutor, in addition to both Angela and Phil themselves. I'm especially grateful to all three of them for their research into drinking and consumption, which has influenced and shaped my own approaches. Other supportive friends in London include Lucy Munro, Claire McManus, Matthew Growhoski, David Schalkwyk, Doug Pfiffer, Michelle O'Callaghan, Russ McDonald, Kevin Sharpe, Roger and Penny Bilham, Philippa Bilham, Becky Beasley, and Markman Ellis. Russ, Kevin, and Philippa continue to be missed.

A group of friends at the annual meeting of the Shakespeare Association of America (SAA)—and beyond—offers the extended network of intelligent conversation needed to generate public work. Garrett Sullivan, Bob Darcy, Alan Stewart, and Stephen Guy-Bray helped provide the annual levity and astute commentary needed to keep me going. So has Susanne Wofford, over our annual dinners. I'm also grateful to Elizabeth Rivlin, Dennis Britton, Diane Purkiss, Melissa Walter, Cora Fox, Paul Yachnin, Adam Kitzes, Jeff Theis, Richelle Munkhoff, Garrett Sullivan, Simone Chess, Ayanna Thompson, Doug Treavor, Liz Pentland, Jim Marino, Richard Preiss, Deanne Williams, Adam Smyth, Liza Blake, Dympna Callaghan, Henry Turner, Mary Thomas Crane, Carolyn Sale, David Goldstein, Amy Tigner, Katherine Eggert, Jessica Rosenberg, Amanda Bailey, Natasha Korda, and Lena Orlin. Alan Stewart has been an especially long-term supporter, engaging with the project from a very early stage at a Renaissance Society of America conference. David Goldstein—as a fellow traveler in the world of food and drink studies—has been an invaluable friend and interlocutor, organizing a Berlin panel, and a Prague seminar. I continue to be grateful for the insights that he and Amy Tigner offered on my Falstaff materials in their earlier form for their collection *Culinary Shakespeare*. At the World Shakespeare Congress in Stratford, Suzanne Wofford, Will West, Gina Bloom, and Natasha Korda helped me fine-tune my thinking in the final stages; their interventions were more crucial than they know. Conversations with Gina about drinking culture were especially helpful, as was the panel she organized at the Modern Language Association of America (MLA) conference.

I'm grateful to my graduate students at USC for their patience. My preoccupation with this project required their understanding, and they generously offered it. Barbara Mello, Megan Herrold, Devin Toohey, Katy Karlin, Lauren Weindling, Megan Mercer, Brooke Carlson, Rich Edinger, and Penelope Geng all lived with the project as long as I have, and I am forever grateful to them for writing in tandem with me. Barbara Mello deserves special mention as the world's best research assistant. Lauren's insights on *Romeo and Juliet*—along with Heather James's—led to an unfinished chapter on that play, a prospect that continue to tantalize me even as the book heads to press. Steven Minas, Betsy Sullivan, Amanda Ruud, and Michael Benitez are newer to the project but have been equally patient and engaged. I'm appreciative, especially from my current vantage point, of the graduate students who took—some years ago—my two seminars on drinking culture, one on early modern taverns, inns, and alehouses (with class visits to the Red Lion, if memory serves); and another—arguably more successfully designed—on early modern spirits: in retrospect I realize that the first seminar's focus on the material culture of early modern drinking might not have compelled everyone else as much as me, but the participants forged ahead and found ways of illuminating everything we studied. Teaching assistants Alex Young, Brian Ingram, Betsy Sullivan, Lisa Lee, and Chinmayi Sirsi were invaluable. I'm indebted to three USC undergraduates who assisted me in the project's final stages. Morgan Millender read, with insight, precision, and speed, the entire manuscript before I sent it to press—I can't imagine asking for more. Emma Dyson read through the manuscript with its copyedited corrections, and somehow managed to see her way through the mire of notations to offer further improvements. I'm in her debt. Morgan, Emma, and David Norton were all able to assist me thanks to the USC SHURE program, which offers funding for undergraduates to work with professors. To the Dornsife dean responsible for this program, thank you.

I received fellowship support for this project, and it is pleasure to record that debt to the Huntington Library, the USC-Huntington Early Modern Studies Institute, the McElderry Fund, and to an USC Advancing Scholarship in the Humanities and Social Sciences grant for an early sabbatical fellowship. The USC-Huntington Early Modern Studies Institute deserves special mention for providing the fellowship support necessary to write the book. The Huntington Library supplied a gorgeous—and gigantic—office on Mahogany Row, as well as a Francis Bacon Foundation fellowship. I'm also grateful to the librarians at those institutions where I spent my fellowship and sabbatical time: the British Library, Wellcome Collection, Folger Shakespeare Library, and most especially the Huntington Library and the Getty Research Institute.

Academic audiences know just how necessary the public airing of one's project can be. I thank the audiences and my hosts at the University of Warwick, the University of Cambridge, Sheffield Hallam University, Yale University's Macmillan Center and CHESS seminar, Boise State University, and the University of Colorado, Boulder, as well as the Huntington Library Renaissance Literature Seminar; and the Huntington Library British History Seminar, where this book in part began. I also thank audiences and participants at the Shakespeare and science seminar I ran at the SAA, as well as audiences at numerous SAAs, MLAs, North American Conference on British Studies and Pacific Coast Conference on British Studies meetings; and at the RSA in Berlin, and the World Shakespeare Congress in Prague. For their role in organizing these visits, and for their insights, I thank Jonathan Bate, Subha Mukherji, John Kerrigan, Lisa Hopkins, Steve Pincus, Julia Adams, Kelsey Champagne,

David Froomkin, Flynn Cratty, Edward "Mac" Test, Paul Hammer, Katherine Eggert, Richelle Munkhuff, Nan Goodman, William Kuskin, Genny Love, Heidi Brayman, Heather James, Cynthia Herrup, Judith Bennett, David Cressy, Lindsay O'Neill, Lori Anne Ferrell, Tom Cogswell, Henry Turner, Mary Thomas Crane, Carla Mazzio, Katherine Eggert, Liza Blake, David Goldstein, Amy Tigner, Ayesha Ramachandran, Garrett Sullivan, and Meredith Evans. I'm especially grateful to Adam Smyth, for a conference on conviviality that first gave me the idea for this project.

Earlier versions of sections of this book were published previously and have been revised and expanded. I thank the journal and collection editors for their insightful feedback, including Arthur Kinney, Nicholas Terpstra, Michelle Legro, Amy Tigner, David Goldstein, Edel Semple, and Rory Loughnane. I'm grateful for permission to reprint portions of this material: "Compulsory Conviviality in Early Modern England," *English Literary Renaissance* 43 no. 3 (September 2013): 381–414, an earlier version of Chapter 6; "Incapacitated Will," in *Staged Transgression: Performing Disorder in Early Modern England*, ed. Rory Loughnane and Edel Semple, Palgrave Shakespeare Studies (Basingstoke: Palgrave Press, 2013), 170–92, for materials in Chapter 5; "Sacking Falstaff," in *Culinary Shakespeare*, ed. Amy Tigner and David Goldstein (Pittsburgh: Duquesne University Press, 2016), 113–34, for an earlier version of Chapter 4; and "Scholarly Addiction: *Doctor Faustus* and the Drama of Devotion," *Renaissance Quarterly* 69 no. 3 (2016): 865–98, for a version of Chapter 2. I am also grateful to the Huntington Library and the British Library for permission to print images from their collections.

To Penn Press, and most especially to Jerry Singerman and the two no-longer-anonymous reviewers, Amanda Bailey and Adam Smyth, I am not sure my thanks will ever be sufficient. Jerry was supportive and patient through every stage of the process, approaching the project with interest and just the right amount of skepticism. As a sign of his deep insight into the project's possibilities, he found two readers who approached my work from quite different angles and in the process strengthened the book beyond what I might have imagined. Adam's capacious knowledge and astute feedback on the project helped expand my sense of the project's potential. And Amanda will see the fruits of her brilliant and extensive commentary on this project on nearly every page. I'm not quite sure what to say by way of thanks, except to record my lasting gratitude.

Every book relies on a support staff. This one is no exception. Larry Green, our Director of Undergraduate Studies in the Department of English at USC, helped to grant my teaching wishes, to the best of his ability; the front office at USC helped provide good working conditions, including Nellie Ayala-Reyes, Flora Ruiz, Janalynn Bliss, Laura Hough, Tim Gotimer, and Kaye Watson. My support team also includes Rachel Putter, Joshua Kartsch, Marcela Widrig, Jasper's caregivers, including Happyland Preschool and Step by Step, as well as Jaquie Fuente, Gino Conti, Gabriel Raspigi, Cornelia Hanson, David Schoenberger, and Dick Lemon and Karen McCauley, both of whom watched Jasper for extended periods of time so I could travel for work—or conduct it at home without interruption. Gino deserves a special shout-out as an academic colleague as well as a friend; I take inspiration from him, since he traveled this path of parenthood before me with grace. I'm grateful to my family for enduring the project over holidays for so many years: Dick Lemon, Heidi Lemon Justice, Karen McCauley, Arthur Okner, Kristin Lemon, Evie Lemon, Aaron Lemon-Strauss, Samantha Trepel, Jacob Lemon-Strauss, Christine Elliot, Graham Justice, Betsy Strauss, Linda and Mark Champagne, Karen Koch, Gretchen and Mike Schnitzer, Ashley Champagne, and Wesley Jackson. I'm delighted to welcome my cousin Ashley into the profession, and grateful to my sister-in-law and friend Sam for leaving her work at the Justice Department long enough to hear me give a paper at her alma mater. My family extends to other loved ones too far away for my liking, including Emma Mason, Jared Farmer and Magda Maczynska with their daughter Zosia, Bonnie-Jeanne Casey and Maggie O'Grady with their son Thomas, Bob Darcy, and Nic Bilham. Bonnie-Jeanne deserves special mention for offering—for the last thirty years—the kind of friendship that helps keep me grounded and hopeful at the same time. Karen McCauley read through the entire manuscript in preparation for the first round of press readers and offered her characteristically insightful response; my father, Dick Lemon, has been unwavering in his engagement and enthusiasm for the project, and I am forever grateful for that. From its inception through its conclusion, he has encouraged me forward and waited patiently for the book to appear.

My final thanks goes to Marc and Jasper. To Jasper, the apple of my eye, I will say simply, thank you for appearing and for being yourself. To Marc, I owe a different kind of thanks, for his numerous mornings and afternoons, especially during Jasper's first years before preschool, spent entertaining the little guy so that I could get writing time. Marc's willingness to do precisely those things that fray me—navigating traffic for pickup and drop off, grocery shopping (or indeed any shopping), night driving, and the list goes on—makes him the most sympathetic partner I can imagine. His ability to connect with me—from finding a concert, record, or a TV a show we might share, to making me a lovely cocktail, to mimicking Carolyn's high voice from *Little House on the Prairie*—always keeps me smiling. I could, by way of thanks, offer to avoid pursuing another big, preoccupying book project. But I'll be more realistic, and simply dedicate this book to him—and to our son.